SWITCHING THEORY FOR LOGIC SYNTHESIS

SWITCHING THEORY FOR LOGIC SYNTHESIS

Tsutomu SASAO

Kyushu Institute of Technology

Iizuka, Japan

KLUWER ACADEMIC PUBLISHERS

Boston/London/Dordrecht

Distributors for North, Central and South America:
Kluwer Academic Publishers
101 Philip Drive, Assinippi Park
Norwell, Massachusetts 02061 USA
Telephone (781) 871-6600
Fax (781) 871-6528
E-Mail <kluwer@wkap.com>

Distributors for all other countries:
Kluwer Academic Publishers Group
Distribution Centre
Post Office Box 322
3300 AH Dordrecht, THE NETHERLANDS
Telephone 31 78 6392 392
Fax 31 78 6546 474
E-Mail <orderdept@wkap.nl>

Electronic Services <http://www.wkap.nl>

Library of Congress Cataloging-in-Publication Data

Sasao, Tsutomu, 1950-
Switching theory for logic synthesis / Tsutomu Sasao.
p. cm
Includes bibliographical references and index.
ISBN 0-7923-8456-3 (alk. paper)
1. Logic design--Data processing. 2. Sequential machine theory.
3. Switching theory. I. Title.
TK78683L6S26 1999
621.39'5--dc21 99-11572
CIP

Printed on acid-free paper.

Printed in the United States of America

CONTENTS

PREFACE

This book presents theories for design and analysis of logic networks. It assumes that the reader has an introductory background in digital networks.

This book consists of three parts: **The first** part (Chapters 1–5) shows mathematical foundations, such as sets and relations, Boolean algebra, representations of logic functions, optimization of two-level logic networks, and various classes of logic functions.

The second part (Chapters 6–8) is an introduction to sequential circuits, optimization of sequential machines, and asynchronous sequential circuits. The first two parts deal with only basic subjects, and contain many examples.

The third part (Chapters 9–14) treats advanced subjects such as multi-valued input two-valued output functions, multi-level logic synthesis, logic design using modules, logic design using EXOR gates, and complexity of logic networks.

Many researchers studied switching theory for use in the design of logic networks, and many textbooks have been published. For most classes, the textbooks without "special features" are easy to use. So, in the first two parts, I tried to remove such "features" to make the textbook standard. These parts are indispensable and will remain unchanged for many years. On the other hand, in some areas, the design methods have changed drastically with the advancement of computers and semi-conductor technology. Now, logic networks are designed by computers.

The third part, which is the main feature of the book, introduces theories on logic synthesis: multi-valued input two-valued output functions, logic design for PLDs/ FPGAs, EXOR-based design, and complexity theories of logic networks.

The first and second parts are suitable for undergraduate courses, while the third part is suitable for graduate or advanced courses. Thus, graduate students or engineers who need to study modern logic design can intensively use the third part.

Acknowledgments

This book is the outcome of the lectures at the Kyushu Institute of Technology, Japan as well as seminars for CAD engineers of Japanese industries. In writing this book, I consulted many books and papers. Most of them are shown in the references at the end of the book. The third part of the book is based on the research, which was supported by the Ministry of Education, Science, Culture and Sports of Japan, as well as many Japanese companies.

Prof. Hiroshi Ozaki and Prof. Kozo Kinoshita introduced me to switching theory and logic design. Profs. Hideo Fujiwara, Tatsuo Higuchi, Masao Mukaidono, Saburo Muroga, Hiroaki Terada, Yoshihiro Thoma, and Shuzo Yajima encouraged me in various occasions.

Prof. Takashi Nanya recommended me to write the Japanese version of this book. He read the draft of the Japanese version, and provided useful comments. Dr. Tomoyuki Fujita, Prof. Norio Koda, Dr. Yusuke Matsunaga, Dr. Shinichi Minato, and Prof. Hiroyuki Ochi also reviewed the draft of the Japanese version and provided various comments.

To the Japanese first edition, many students offered me valuable suggestions. Comments of Profs. Kiyoshi Furuya, Teruhiko Yamada, Etsuro Moriya, and Terumine Hayashi are also appreciated.

The drafts of the English version were reviewed by Prof. Jon T. Butler, Prof. Radomir Stanković, and Dr. Debatosh Debnath. Especially, Prof. Butler read through the entire manuscript repeatedly and made important corrections and improvements.

Finally, Mr. Munehiro Matsuura patiently composed LaTeX files of both the Japanese and the English version.

The author thanks the President of Kindai-Kagaku-Sha book company for her willingness to publish the English version.

Tsutomu Sasao

About Exercises

Most problems should be solved in a short time. Problems with (M) signs are more difficult. Problems with (D) signs are quite difficult, and appropriate only for challenging students.

1

MATHEMATICAL FOUNDATION

This chapter introduces mathematical concepts necessary to understand this book. Because the index lists important key words, readers can refer to the words whenever necessary.

1.1 SET

A **set** is a collection of objects. An object in the set is an **element** of the set. Usually, sets are denoted by upper case letters A, B, etc., and elements are denoted by the lower case letters a, b, etc. If a is an element of A, then "a belongs to the set A," and denoted by $a \in A$. If a is not an element of A, then $a \notin A$. A set is a **finite set** if the number of elements is finite, else it is an **infinite set**. The set without any element is an **empty set**, denoted by ϕ. ϕ is a finite set. $|A|$ denotes the number of elements in a finite set A. Two methods exist to describing sets: 1) The **enumeration method** explicitly lists all the elements enclosed by {and}. 2) The **general expression and its condition method** uses {general expression | condition}.

Example 1.1

1. The prime numbers between 1 and 12 constitute a set with five elements. By the enumeration method, it is represented as {2,3,5,7,11}.
2. A set of natural numbers is represented as $\{x \mid x$ is natural number$\}$ or {1,2,3,...}. ∎

Subset

If every element in a set A is also an element of a set B, then A is a **subset** of B, and denoted by $A \subseteq B$. In this case, B **contains** A. If $A \subseteq B$ does not hold, then $A \not\subseteq B$. An empty set is a subset of an arbitrary set. The set that consists of all the elements in the discussion is the **universal set**, which denoted by the symbol U. All the sets in the discussion are subsets of U. Let A and B be sets. If $A \subseteq B$ and $B \subseteq A$ hold, then A and B are the **equal sets**, and we denote it by $A = B$. If $A \subseteq B$ and $A \neq B$, then A is a **proper subset** of B, and we denote it by $A \subset B$. For any set A, $A \subseteq A$ holds.

Example 1.2

1. Let A be a set of men, and B be a set of human, then $A \subset B$.
2. Let $A = \{0, 1\}$ and $B = \{0, 1, 0, 1\}$, then $A = B$. In other words, sets are the same even if the elements appears more than once. Also, sets are the same even if the order of the elements are different. ∎

Power Set

The set of subsets of A is the **power set** of A, denoted by $P(A)$. Especially, $\phi \in P(A)$ and $A \in P(A)$.

Example 1.3 Let $A = \{0, 1\}$. Then, $P(A) = \{\phi, \{0\}, \{1\}, \{0, 1\}\}$.

∎

Union, Intersection, Complement

Let A and B be sets. The set of elements that belong to A or B or both is the **union** of A and B, denoted by $A \cup B$. The set of elements that belong to both A and B is the **intersection** of A and B, and denoted by $A \cap B$. The set of elements in the universal set U, but not the elements in A, is the **complement** of A, and it is denoted by $\overline{A}$ or A^c.

By using symbols, these sets are represented as follows:

$$
\begin{aligned}
A \cup B &= \{x \mid x \in A \text{ or } x \in B\}; \\
A \cap B &= \{x \mid x \in A \text{ and } x \in B\}; \\
\overline{A} &= \{x \mid x \notin A \text{ and } x \in U\}.
\end{aligned}
$$

Especially, $\overline{U} = \phi$ and $\overline{\phi} = U$. A and B are **disjoint** if they have no common element, i.e., if $A \cap B = \phi$.

Example 1.4 Let U={0,1,2,3}, A={1,2}, and B={2,3}. Then, $A \cup B$={1,2,3}, $A \cap B$={2}, $\overline{A}$={0,3}. ∎

Properties of Sets

Let A, B, and C be sets. Then, the following equations hold:

(1) **Idempotent laws:** $A \cup A = A$, $A \cap A = A$;
(2) **Commutative laws:** $A \cup B = B \cup A$, $A \cap B = B \cap A$;
(3) **Associative laws:** $A \cup (B \cup C) = (A \cup B) \cup C$, $A \cap (B \cap C) = (A \cap B) \cap C$;
(4) **Absorption laws:** $A \cup (A \cap B) = A$, $A \cap (A \cup B) = A$;
(5) **Distributive laws:** $A \cup (B \cap C) = (A \cup B) \cap (A \cup C)$,
$A \cap (B \cup C) = (A \cap B) \cup (A \cap C)$;
(6) **Involution law:** $\overline{(\overline{A})} = A$;
(7) $A \cup \overline{A} = U$, $A \cap \overline{A} = \phi$;
(8) $A \cup \phi = A$, $A \cap U = A$;
(9) $A \cup U = U$, $A \cap \phi = \phi$;
(10) **De Morgan's laws:** $\overline{(A \cup B)} = \overline{A} \cap \overline{B}$, $\overline{(A \cap B)} = \overline{A} \cup \overline{B}$.

Because of the associative law, the representations $A \cup B \cup C$ or $A \cap B \cap C$, etc., have no ambiguity. However, a representation such as $A \cup B \cap C$ is not permitted, since $(A \cup B) \cap C \neq A \cup (B \cap C)$, and the results depend on the placement of parentheses. Let A and B be finite sets, then $|A \cup B| = |A| + |B| - |A \cap B|$.

Venn's Diagram

Venn's diagram is a graphical representation of sets. The universal set is represented by the rectangle. Sets A, B, etc., are represented by interiors of the closed domains, such as circles. In a Venn's diagram, the sets $A \cup B$, $A \cap B$, and $\overline{A}$ are represented as the shaded parts of Fig. 1.1.

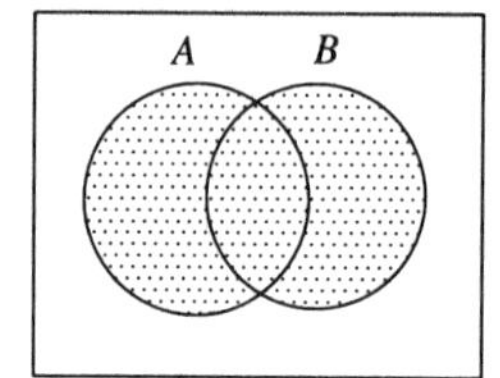

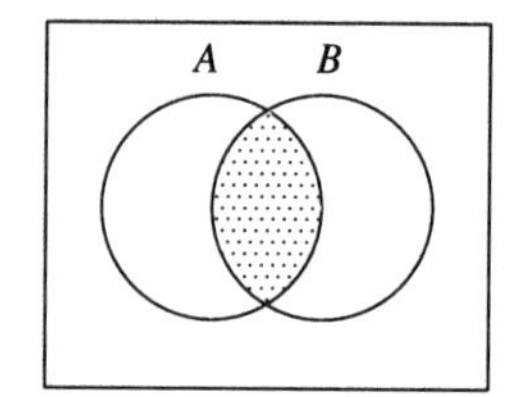

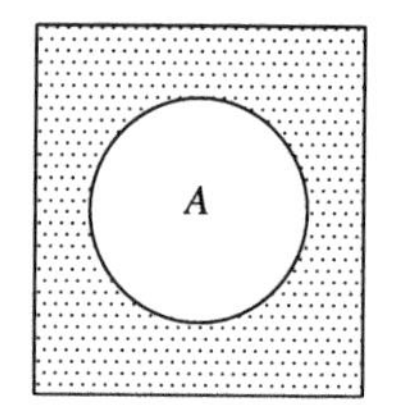

Figure 1.1 Venn's diagram for $A \cup B$, $A \cap B$, and $\overline{A}$.

1.2 RELATION

Pair and n-tuple

A tuple (a, b) of two elements arranged in a fixed order is a **pair** or an **ordered pair**. In general, a tuple of n elements $a_1, a_2 \ldots, a_n$, that considers the order $(a_1, a_2, \ldots, a_n)$ is an **n-tuple**. Two n-tuples $(a_1, a_2, \ldots, a_n)$ and $(b_1, b_2, \ldots, b_n)$ are equal if and only if $a_i = b_i$ for all i.

Direct Product

Let A and B be sets, and a and b be elements of A and B, respectively. The set that consists of all the ordered pair of (a, b) is the **direct product** or the **Cartesian product** of A and B, and is denoted by $A \times B$, i.e.,

$$A \times B = \{(a, b) \mid a \in A, b \in B\}.$$

If A or B is a null set, then $A \times B$ is also a null set. When $A = B$, $A \times A$ is abbreviated by A^2. For n sets $\{A_1, A_2, \ldots, A_n\}$, the direct product $A_1 \times A_2 \times \cdots \times A_n$ is defined as follows:

$$A_1 \times A_2 \times \cdots \times A_n = \{(a_1, a_2, \ldots, a_n) \mid a_i \in A_i, i = 1, 2, \ldots, n\}.$$

Example 1.5

1. Let A=$\{0,1\}$ and B=$\{0,1,2\}$. Then,
   $A \times B = \{(0,0), (0,1), (0,2), (1,0), (1,1), (1,2)\}$.
2. Let R be a set of points in a straight line. Then, R^2 denotes a set of points in a (two-dimensional) plane, and R^3 denotes a set of points in a (three-dimensional) cube. ∎

Relation

Let A and B be sets. Subset R of direct product $A \times B$ is a **binary relation** from A to B. That is, if $R \subseteq A \times B$, $a_i \in A$, $b_j \in B$, and $(a_i, b_j) \in R$, then "a_i and b_j are in the relation R," or "relation R holds." If $(a_i, b_j) \notin R$, then "relation R does not hold." If $(a_i, b_j) \in R$, then we write $a_i R b_j$. A binary relation from A to A is a **binary relation on the set** $\boldsymbol{A}$. A subset of a direct product of n sets $A_1 \times A_2 \times \cdots \times A_n$ is called an $\boldsymbol{n}$**-ary relation**. Let R be a relation from A to B. The set where the order of the elements are interchanged in the ordered pair R is an **inverse relation** of R, and is denoted by R^{-1}. In other words, $R^{-1} = \{(b_j, a_i) \mid (a_i, b_j) \in R\}$.

Example 1.6

1. Let A={stone, scissors, paper }. In a toss, let the relation "α wins over β" be denoted by $R = \{(\alpha, \beta) \mid \alpha, \beta \in A\}$. Then, we have R={(stone, scissors),(scissors, paper),(paper, stone)}. The inverse relation is R^{-1}= {(scissors, stone),(paper, scissors),(stone, paper)}. It is "α lose to β."
2. Let $B = \{0, 1, 2, 3\}$. If the relation "α is equal to or less than β" is denoted by "$\leq$," then "$\leq$"={(0,0),(0,1),(0,2),(0,3),(1,1),(1,2),(1,3),(2,2),(2,3), (3,3)}. If the inverse relation is denoted by the symbol "$\geq$," then "$\geq$"= {(0,0),(1,0),(2,0),(3,0),(1,1),(2,1),(3,1),(2,2),(3,2),(3,3)}. It is "α is equal to or greater than β." ∎

1.3 EQUIVALENCE CLASS

Equivalence Relation

Let R be a binary relation on a set A. For all elements a, b, and c in A, if the following three conditions hold, then R is an **equivalence relation** on A.

(1) **Reflective law:** aRa.
(2) **Symmetric law:** If aRb, then bRa.
(3) **Transitive law:** If aRb and bRc, then aRc.

When R is an equivalence relation and aRb, we say "a and b are equivalent in the relation R."

Example 1.7

1. The relation "equal"$(=)$ in the mathematics is an equivalence relation.
2. Relations such as "have the same family name," "are the same sex," "are the same age," and "graduated from the same school" are equivalence relations. ∎

Equivalence Class

Let R be an equivalence relation on the set A. Then, we can partition the set A into some blocks, such that the equivalent elements in R belong to the same block: $[a] = \{x \mid aRx, x \in A\}$. The set $[a]$ is the **equivalence class** of the set A that contains a in relation R. In this case, a is a **representative** of the equivalence class $[a]$. A set A can be partitioned into equivalence classes by the equivalence relation R. Also, each equivalence class does not have common part with another class. Moreover, an arbitrary element of A is an element of exactly one equivalence class. The set of all the equivalence classes of the equivalence relation R on A is a **quotient set** of A with respect to R, which is denoted by A/R. The number of equivalence classes is the **rank** of the relation R.

Logic Notation

The following logic symbols are used to represent conditions concisely. In a sentence where the meaning is clear, if we can decide whether the sentence is true or false objectively, then the sentence is a **proposition**. Let P and Q be propositions. "If P holds, then Q holds" is denoted by $P \Rightarrow Q$. Also, "P is true if and only if Q is true" (P iff Q) is denoted by $P \Leftrightarrow Q$. When $P \Rightarrow Q$, the proposition P is called a **sufficient condition** of the proposition Q. Moreover, the proposition Q is a **necessary condition** of the proposition P. When $P \Leftrightarrow Q$, the proposition P is a **necessary and sufficient condition** of the proposition Q. Even if $P \Rightarrow Q$ holds, $Q \Rightarrow P$ does not necessarily hold. $Q \Rightarrow P$ is called a **converse** of $P \Rightarrow Q$. To prove $P \Rightarrow Q$ is equivalent to prove the **contraposition** $\overline{Q} \Rightarrow \overline{P}$.

Example 1.8 Let Z be the set of integers. The relation $(\equiv)$ "For $n, m \in Z$, $n \equiv m \Leftrightarrow (m - n)$ is a multiple of k" is an equivalence relation, where the multiple contains 0. When $n \equiv m$, "n and m are equivalent modulo k."

Especially when $k = 2$,

$$\begin{aligned}[0] &= \{2n \mid n \in Z\} = \text{Set of the even numbers;} \\ [1] &= \{2n+1 \mid n \in Z\} = \text{Set of the odd numbers.}\end{aligned}$$

And, the rank of $Z/\equiv$ is two. ∎

Example 1.9 Let Z be the set of integers. If the binary relation $\sim$ on $Z \times Z$ is defined by

$$(a, b) \sim (c, d) \Leftrightarrow a + b = c + d$$

then, $\sim$ is an equivalence relation. ∎

Partition

Let A be a set, and let A_1, A_2,..., and A_n be subsets of A. If $A_1 \cup A_2 \cup \cdots \cup A_n = A$, and $A_i \cap A_j = \phi$, for all i, j $(i \neq j)$, then A is said to be **partitioned** into A_1, A_2,..., and A_n. When an equivalence relation R is defined on A, A is partitioned into the equivalence classes by the relation R. Conversely, for an arbitrary partition of A, we can define an equivalence relation R of A as follows: $a_i \in [a]$ and $a_j \in [a]$ iff $a_i R a_j$. When A is partitioned into A_1, A_2,..., and A_n, the elements of A can be represented as shown in Fig. 1.2.

Example 1.10 Let A be the set of people who are less than or equal to 100 years old. If the relation "same age" is denoted by R, then R is an equivalence relation. In this case, A is partitioned into $A_0, A_1, \ldots, A_{100}$, where A_i denotes the set of people who are i years old. ∎

Refinement

Let R_1 and R_2 be two equivalence relations on the set A. For arbitrary elements x and y in A, if

$$xR_1y \Rightarrow xR_2y$$

holds, then R_1 is a **refinement** of R_2, and denoted by $R_1 \subseteq R_2$.

Example 1.11 Let $A = \{011, 100, 110, 111\}$. Let R_0 be the relation that "all the corresponding bits are the same," R_1 be the relation that "right two bits are the same," and R_2 be the relation that "the rightmost bits are the same." In this case, we have

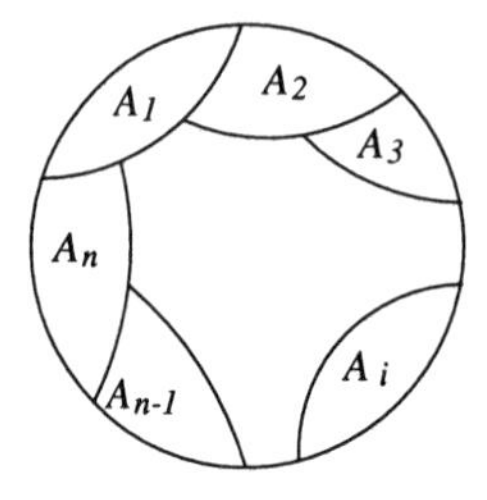

Figure 1.2 Partition.

011	100
111	110

(a) R_0

011	100
111	110

(b) R_1

011	100
111	110

(c) R_2

Figure 1.3 An example of refinement.

$$
\begin{aligned}
R_0 &= \{(011, 011), (100, 100), (110, 110), (111, 111)\}, \\
R_1 &= \{(011, 011), (011, 111), (100, 100), (110, 110), (111, 011), (111, 111)\}, \\
R_2 &= \{(011, 011), (011, 111), (100, 100), (110, 110), (100, 110), (110, 100), \\
&\qquad (111, 011), (111, 111)\}.
\end{aligned}
$$

R_0 is a refinement of R_1, and R_1 is a refinement of R_2 (See Fig. 1.3). ∎

1.4 FUNCTION

Function

Let A and B be sets, and f be a binary relation from A to B. For each element a in A, if there exists unique element b in B such that afb, then f is a **function** from A to B, or f is a **mapping** from A to B, and denoted by $f : A \rightarrow B$. A is a **domain** of f. If an element a_i of A corresponds to an element b_j of B, then we denote it by $f(a_i) = b_j$. In this case, b_j is the **value of function** f with respect to a_i. $b = f(a) \in B$ is an **image** of $a \in A$. The whole image of the domain is a **range**, and is denoted by $f(A)$. In this case, $f(A) \subseteq B$.
A function is a special case of a relation. We can define a relation R_f from the function f as follows:

For the function $f : A \rightarrow B$, $f(a_i) = b_j$ iff $(a_i, b_j) \in R_f$.

Let f^{-1} denote the **inverse relation** of the function $f : A \rightarrow B$. Then, f^{-1} is, in general, not a function. Let $b = f(c)$. Then, $f^{-1}(b)$ is, in general, a subset of A. $f^{-1}(b)$ is an **inverse image** of b.

Example 1.12 Let Z be the set of integers. $f(x) = x^2$ is a function from Z to Z. In this case, $f = \{\ldots, (-2,4), (-1,1), (0,0), (1,1), (2,4), \ldots\}$. Although f is a function, the inverse relation is not a function. Specifically, for some y, there exist two x such that $y = f(x)$. Moreover, there is no x such that $f(x) = 3$. ∎

A rule f that assigns an element of B to each element of the direct product $A_1 \times A_2 \times \cdots \times A_n$ is an **n-variable function.** We denote f by $f : A_1 \times A_2 \times \cdots \times A_n \to B$. If an n-tuple $(a_1, a_2, \ldots, a_n)$ $(a_i \in A_i, i = 1, 2, \ldots, n)$ corresponds to $b \in B$, then we denote it by $f(a_1, a_2, \ldots, a_n) = b$. Let f be a function from X to Y. If $A \subseteq X$, then $f(| A)$ represents the function where the domain of f is A. It is called a **restriction** of f to A.

Example 1.13 Let $A_1 = \{0,1\}$, $A_2 = \{0,1,2\}$, $A_3 = \{0,1,2,3\}$, and $B = \{0,1,2,3,4\}$. In this case, the number of functions $f : A_1 \times A_2 \times A_3 \to B$ is 5^{24}. That is, for each element in the set of $2 \times 3 \times 4 = 24$ elements, there are five ways to choose an element in B. ∎

Operation

A function from the set A to A is often called as a **unary operation** of A. An example of a unary operation is denoted by the symbol $\bar{\ }$. In this case, $\bar{a}_i = a_j$ denotes that the element a_i correspond the element a_j. A two-variable function from $A \times A$ to A is often called as a **binary operation** of A. In a binary operation, the correspondence of a_k to (a_i, a_j) is denoted by $a_i * a_j = a_k$.

Example 1.14 Let B={0,1}. Let $\bar{\ }$ be a unary operation, and $\wedge$, $\vee$, and $\oplus$ be binary operations on B as follows: For arbitrary elements a and b in B,

$$\begin{aligned} \bar{a} &= 1 - a \\ a \wedge b &= a \cdot b \\ a \vee b &= a + b - a \cdot b \\ a \oplus b &= a + b \pmod 2 \\ &= a + b - 2ab. \end{aligned}$$

In this case, the symbols $+$, $-$, $\cdot$ denote ordinary addition, subtraction, and multiplication, respectively. Table 1.1 shows the functions represented by $\bar{\ }$, $\wedge$, $\vee$, and $\oplus$. ∎

In Chapter 14, when the complexity of logic networks is analyzed, the following notation is used. Let the function $g(x)$ be defined for all positive numbers x,

Table 1.1 Various operations.

a	b	$\bar{a}$	$a \wedge b$	$a \vee b$	$a \oplus b$
0	0	1	0	0	0
0	1	1	0	1	1
1	0	0	0	1	1
1	1	0	1	1	0

and let $g(x) \geq 0$. Suppose that a function $f(x)$ is defined for a set S of positive numbers.

$$f(x) = O(g(x)) \Leftrightarrow \text{There exist a constant } M \text{ such that } \frac{|f(x)|}{g(x)} < M.$$

$$f(x) = o(g(x)) \Leftrightarrow \lim_{\substack{x \to \infty \\ x \in S}} \frac{f(x)}{g(x)} = 0$$

$$f(x) \sim g(x) \Leftrightarrow \lim_{\substack{x \to \infty \\ x \in S}} \frac{f(x)}{g(x)} = 1$$

Usually, S is a set of natural numbers.

Example 1.15

1. Let $f(x) = a_0 + a_1 x^1 + a_2 x^2 + \cdots + a_k x^k$. Then, $f(x) = O(x^k)$.
2. Let $f(x) = x^k$, (k is an arbitrary number). Then, $f(x) = o(2^x)$.
3. Let $f(x) = x^3 - x$, and $g(x) = x^3 + x$. Then $f(x) \sim g(x)$. ∎

1.5 ORDERED SET

Ordered Relation

Let R be a binary relation on a set A. If any elements a, b, c in A satisfy the following three conditions:

(1) **Reflective law**: aRa
(2) **Anti-symmetric law**: If aRb and bRa, then $a = b$
(3) **Transitive law**: If aRb and bRc, then aRc

then R is an **ordered relation** or is a **partial order relation**. Especially, if R is a partial order relation, and if

(4) For all $a, b \in A$, aRb or bRa

then, R is a **total order relation**. An ordered relation R is usually represented by $a \leq_R b$ instead of aRb.

Example 1.16

1. Let P={{0},{1},{0,1}}, and A and B be elements of P. Let $A \subseteq B$ denote the relation "A is contained by B," then $\subseteq$ is a partial order relation. There is no relation $\subseteq$ between $\{0\}$ and $\{1\}$.
2. Let Z be the set of integers. If $a \leq b$ denotes the relation "a is less than or equal to b," then $\leq$ is a total order relation. ∎

Ordered Set

Let $\leq_R$ be an ordered relation defined on a set A. A pair of A and $\leq_R$, $\langle A, \leq_R \rangle$, is an **ordered set**. Especially if $\leq_R$ is a partial order relation, then $\langle A, \leq_R \rangle$ is a **partially ordered set**. If $\leq_R$ is a total order relation, then $\langle A, \leq_R \rangle$ is a **totally ordered set**.

Example 1.17

1. Let $P(A)$ be the power set on a set A. Then, $\langle P(A), \subseteq \rangle$ is a partially ordered set.
2. Let Z be a set of the integers. Then, $\langle Z, \leq \rangle$ is a totally ordered set. ∎

Hasse Diagram

Let A be a finite set, and $\leq_R$ be an ordered relation on A. Let a, b be two elements in A such that $a \leq_R b$ and $a \neq b$. If there is no element c such that $a \leq_R c$, $c \leq_R b$, where c is different from a and b, then b **covers** a. When b covers a, the diagram which is obtained by writing b above a, and by connecting b and a by a straight line is the **Hasse diagram**.

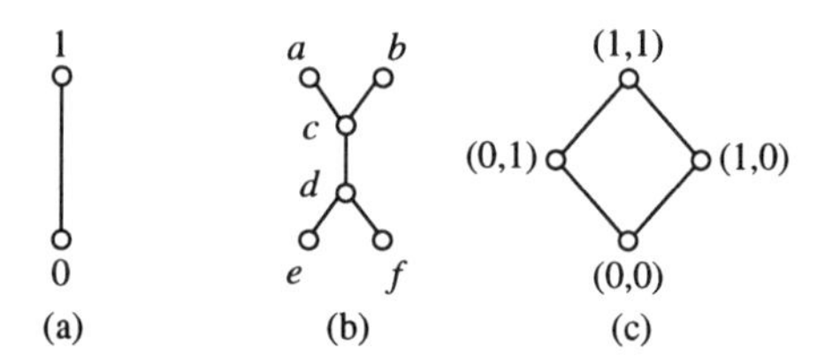

Figure 1.4 Examples of Hasse diagrams.

Example 1.18 Fig. 1.4 shows examples of a Hasse diagram. In a Hasse diagram, we can reach from b to a downward by tracing the connected lines, iff $a \leq_R b$. ∎

Example 1.19 In Fig. 1.4(a), 1 is the maximum element, and 0 is the minimum element. In Fig. 1.4(c), (1,1) is the maximum element, and (0,0) is the minimum element. In Fig. 1.4(b), there is no maximum element nor minimum element. ∎

Maximal Element, Minimal Element

Let $\langle A, \leq_R \rangle$ be an ordered set, and let a_0 be an element of A. If there is no element a in A such that $a_0 \leq_R a$, and $a_0 \neq a$, then a_0 is a **maximal element** of A. If there is no element a in A such that $a \leq_R a_0$, and $a_0 \neq a$, then a_0 is a **minimal element** of A. The maximal element and the minimal element may not exist. However, in a finite set, they always exist. Sometimes, there are more than one maximal and/or minimal elements.

Example 1.20 In Fig. 1.4(a), 1 is the maximal element, and 0 is the minimal element. In Fig. 1.4(b), a and b are maximal elements, while e and f are minimal elements. In Fig. 1.4(c), (1,1) is the maximal element, and (0,0) is the minimal element. ∎

Maximum Element, Minimum Element

Let $\langle A, \leq_R \rangle$ be an ordered set, and let a_0 be an element of A. For each element a in A, if $a \leq_R a_0$, then a_0 is the **maximum element** of A. For each element a in A, if $a_0 \leq_R a$, then a_0 is the **minimum element** of A. The maximum element or the minimum element may not exist.

Least Upper Bound, Greatest Lower Bound

Let $\langle A, \leq_R \rangle$ be an ordered set, and let $B \subseteq A$. The element a in A is an **upper bound** of B, if $b \leq_R a$ holds for each element b in B. The element a in A is a **lower bound** of B, if $a \leq_R b$ holds for each element b in B. If there is the minimum element in the set of the upper bounds of B, then it is the **least upper bound** of B. If there is the maximum element in the set of the lower bounds of B, then it is the **greatest lower bound** of B. Especially, for two elements a and b, if the least upper bound and the greatest lower bound of $\{a, b\}$ exist, then they are denoted by $a \vee b$ and $a \cdot b$, respectively.

Example 1.21 In Fig. 1.4(b), there is no least upper bound of $\{a, b\}$. However, the greatest lower bound of $\{a, b\}$ is $a \cdot b = c$. Also, the least upper bound of $\{e, f\}$ is $e \vee f = d$. However, there is no greatest lower bound of $\{e, f\}$. ∎

Bibliographical Notes

For more detailed discussion, see textbooks on discrete mathematics [228, 229, 319], textbooks on computational complexity [135, 376], and textbooks on graph theory [156].

Exercises

1.1 Let N be a set of natural numbers, and let $D(n) = \{m \mid m \in N,\ m \text{ is a divisor of } n\}$. Then what is the set $A = \{n \mid |D(n)| = 2,\ n \in N\}$?

1.2 Let A, B, and C be sets. Prove that $A \cup (B \cap C) = (A \cup B) \cap (A \cup C)$ holds.

1.3 Let A, B, and C be sets. Does each of the following holds? If it holds, then prove it, otherwise show a counterexample.
(1) $A \cup B = A \cup C \;\Rightarrow\; B = C$.
(2) $A \cap B = A \cap C \;\Rightarrow\; B = C$.
(3) $A \oplus B = A \oplus C \;\Rightarrow\; B = C$, where $A \oplus B = (A \cap \overline{B}) \cup (\overline{A} \cap B)$.
(4) $(A \cap \overline{B}) \cup (A \cap \overline{C}) = A \cap (\overline{B \cap C})$.

1.4 Let A, B, and C be finite sets. Prove the following:
(1) $|A \cup B| = |A| + |B| - |A \cap B|$.
(2) $|A \cup B \cup C| = |A| + |B| + |C| - |A \cap B| - |A \cap C| - |B \cap C| + |A \cap B \cap C|$.

1.5 Let $A, B, C \subseteq S$. Let $|S| = 60$, $|A| = 30$, $|B| = 28$, $|C| = 14$, $|A \cap B| = 11$, $|A \cap C| = 4$, $|B \cap C| = 3$, and $|A \cap B \cap C| = 2$. Then obtain the value of $|A \cup B \cup C|$.

1.6 A survey of the hobbies for n people shows the following: a people like music; b people like paintings; c people like sports; d people like both music and paintings; e people like both paintings and sports; and f people like both sports and music. Then, how many people like music, painting and sports? Assume that each person has at least one hobby.

1.7 Let A and B be subsets of U. Show that

$$|\overline{A} \cap \overline{B}| = |U| - |A| - |B| + |A \cap B|.$$

1.8 Let A and B be sets, and let the numbers of elements in A and B be N_A and N_B, respectively. How many binary relations are there from A to B?

1.9 Let L be the set of all lines in a plane, and let l_1, $l_2 \in L$. Is each of the following relation $\sim$, equivalence relation?
(a) $l_1 \sim l_2 \Leftrightarrow l_1$ is parallel to l_2.
(b) $l_1 \sim l_2 \Leftrightarrow l_1$ is perpendicular to l_2.

1.10 Show that $\sim$ in Example 1.9 is an equivalence relation.

1.11 (M) Consider the equivalence relations on $A = \{a_1, a_2, a_3, a_4\}$. How many equivalence relations are there?

1.12 (M) Let R and S be equivalence relations on set A. Is each of the following an equivalence relation? If it is an equivalence relation, then prove it, otherwise show a counterexample.
(1) $R \cup S$, (2) $R \cap S$.

1.13 Let X be a set consisting of d elements, and let Y be a set consisting of r elements. How many functions are there from X to Y?

1.14 Let $P_i = \{0, 1, \ldots, p_i - 1\}(i = 1, \ldots, n)$, $R = \{0, 1, \ldots, r - 1\}$, $p_i \geq 1$, and $r \geq 1$. Enumerate the number of mappings: $P_1 \times P_2 \times \cdots \times P_n \to R$.

1.15 How many unary operations on $L = \{0, 1, 2\}$? How many unary operations on $L = \{0, 1, 2, 3\}$?

1.16 Let i be a positive integer. Let $A(i)$ be the set of positive integers which are divisor of i. When a is a divisor of b, we define the binary relation $\leq_R$ by $a \leq_R b$. Then, show that $\langle A(120), \leq_R \rangle$ is an ordered set. Also draw the Hasse diagram.

1.17 Let $B = \{0, 1\}$, and let $\boldsymbol{a} = (a_1, a_2, a_3)$ and $\boldsymbol{b} = (b_1, b_2, b_3)$ be two elements in $B \times B \times B = B^3$. Let $\boldsymbol{a} \leq \boldsymbol{b}$ iff $a_1 \leq b_1$, $a_2 \leq b_2$, and $a_3 \leq b_3$. Show that $\langle B^3, \leq \rangle$ is an ordered set, and draw the Hasse diagram.

1.18 Draw the Hasse diagram of B^5.

1.19 Let $N = \{1, 2, \ldots, 10\}$. The binary relation $<\cdot$ on N is defined as $n <\cdot\, m$, where "$n < m$, and n and m have common divisor other than 1." Show the Hasse diagram of $\langle N, <\cdot \rangle$.

1.20 Obtain all possible partitions of $T = \{0, 1, 2\}$.

1.21 Let $R = \{(a, a), (b, b), (a, c), (c, a), (c, b)\}$ be a binary relation on $S = \{a, b, c\}$. Does R satisfy each of the following? If not, show a counterexample.

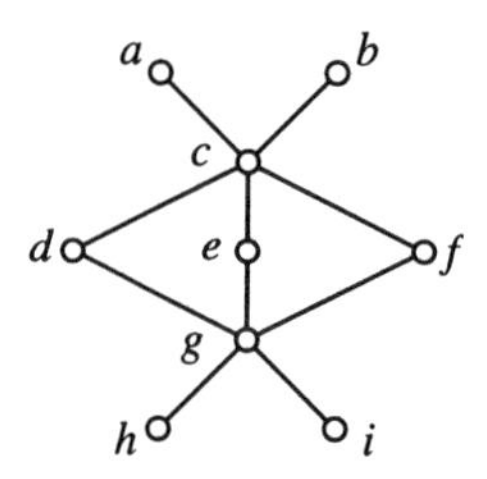

Figure 1.5

(a) reflective, (b) symmetric, (c) antisymmetric, (d) transitive, (e) partial order relation, (f) equivalence relation, (g) function.

1.22 Prove or disprove the following:
(a) $x \leq y \vee z \Leftrightarrow (x \leq y \text{ or } x \leq z)$.
(b) $x \leq y \cdot z \Leftrightarrow (x \leq y \text{ and } x \leq z)$.

1.23 Let n be an arbitrary non-negative integer. Prove that $n^3 + 2n$ is a multiple of 3.

1.24 In Fig. 1.5, obtain the upper bound, lower bound, least upper bound, and greatest lower bound of $\{d, e, f\}$.

1.25 Let N be a set of natural numbers containing 0. Let F be the set of all the functions from N to N. For functions $f,\ g \in F$, if $f(n) = g(n)$ holds excepts for the finite number of points in N, then we denote it by $f \sim g$. Prove that $\sim$ is an equivalence relation.

2

LATTICE AND BOOLEAN ALGEBRA

This chapter presents, lattice and Boolean algebra, which are basis of switching theory. Also presented are some algebraic systems such as groups, rings, and fields.

2.1 ALGEBRA

This book considers various **algebraic systems**. In this section, we present a general form of them. An algebraic system is defined by the tuple $\langle A, o_1, \ldots, o_k; R_1, \ldots, R_m; c_1, \ldots, c_k \rangle$, where, A is a non-empty set, o_i is a function $A^{p_i} \to A$, p_i is a positive integer, R_j is a relation on A, and c_i is an element of A.

Example 2.1 $\langle Z, + \rangle$ is an algebraic system consisting of a set of integers Z and addition $(+)$. $\langle Z, +, \leq \rangle$ is an algebraic system consisting of a set of integers Z, addition, and the relation "equal to or less than". ∎

2.2 LATTICE

The **lattice** is an algebraic system $\langle A, \vee, \cdot \rangle$ with two binary operations $\vee$ and $\cdot$, and arbitrary elements a, b, c in A satisfy the following four **axioms** (1)–(4):

(1) Idempotent laws: $a \vee a = a$, $a \cdot a = a$;
(2) Commutative laws: $a \vee b = b \vee a$, $a \cdot b = b \cdot a$;
(3) Associative laws: $a \vee (b \vee c) = (a \vee b) \vee c$, $a \cdot (b \cdot c) = (a \cdot b) \cdot c$;

(4) Absorption laws: $a \vee (a \cdot b) = a,\ a \cdot (a \vee b) = a.$

Example 2.2

1. Let $A = \{0, 1\}$. Let $a \vee b = \max\{a, b\}$ and $a \cdot b = \min\{a, b\}$ be binary operations on A. Then, the algebraic system $\langle A, \vee, \cdot \rangle$ satisfies the axioms of the lattice.
2. Let Z be the set the integers. Let $a \vee b = \max\{a, b\}$ and $a \cdot b = \min\{a, b\}$ be binary operations on Z. Then, the algebraic system $\langle Z, \vee, \cdot \rangle$ satisfies the axioms of the lattice.
3. Let $S = \{a, b\}$. Let $\cup$ and $\cap$ be binary operations on $P(S) = \{\phi, \{a\}, \{b\}, \{a, b\}\}$. Then, the algebraic system $\langle P(S), \cup, \cap \rangle$ satisfies the axioms of the lattice.
4. Let $A = \{1, 2, 3, 6\}$. Let $a \vee b =$(least common multiple of a and b), and $a \wedge b =$(greatest common divisor of a and b) be binary operations on A. Then, the algebraic system $\langle A, \vee, \wedge \rangle$ satisfies the axioms of the lattice. ∎

As shown in the above example, various algebraic systems satisfy the axioms of the lattice. Note that each example is a special case of the abstract algebraic system defined by the axioms. Such an example is called a **model** of the algebraic system. In general, many models satisfy the algebraic system defined by the axioms.

2.3 DISTRIBUTIVE LATTICE AND COMPLEMENTED LATTICE

The lattice $\langle A, \vee, \cdot \rangle$ satisfying the following axiom is a **distributive lattice.**

(5) Distributive laws: $a \vee (b \cdot c) = (a \vee b) \cdot (a \vee c),\ a \cdot (b \vee c) = (a \cdot b) \vee (a \cdot c).$

Example 2.3 The ordered set represented by the Hasse diagram in Fig. 2.1 is a distributive lattice. The ordered set represented by the Hasse diagram in Fig. 2.2 is a lattice, but not a distributive lattice. In Fig. 2.2(a), the distributive law is not satisfied since $a \cdot (b \vee c) = a \cdot 1 = a$, and $a \cdot b \vee a \cdot c = b \vee 0 = b$ ∎

Let a lattice $\langle A, \vee, \cdot \rangle$ have a maximum element 1 and a minimum element 0. For any element a in A, if there exists an element x_a such that $a \vee x_a = 1$ and $a \cdot x_a = 0$, then the lattice is a **complemented lattice**. In this case, x_a

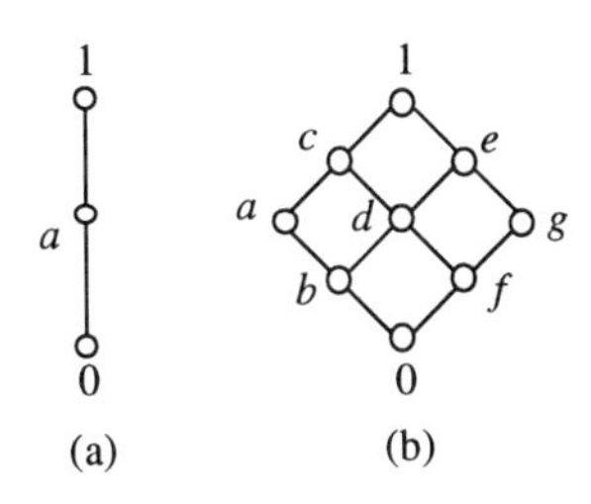

Figure 2.1 Examples of distributive lattice.

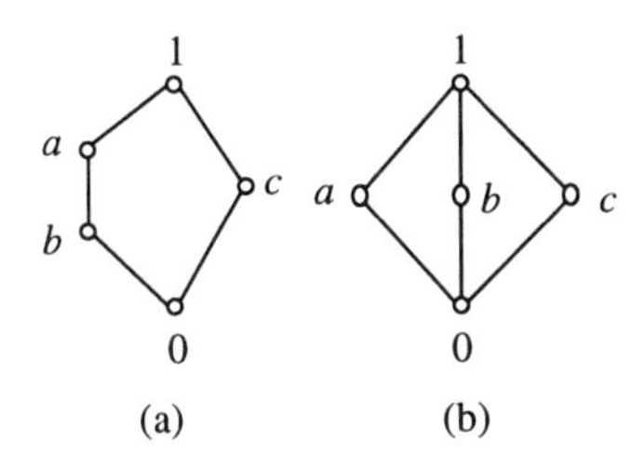

Figure 2.2 Examples of non-distributive lattice.

is a **complement** of a. A **complemented distributed lattice** is a Boolean algebra, which is introduced in the next part. In general, a complement is not unique as shown in the next.

Example 2.4

1. In Fig. 2.1(a), there is no complement for a.
2. In Fig. 2.2(a), the complement of a is c, and the complement of b is also c.
3. In Fig. 2.2(b), the complements of c are a and b. ∎

2.4 BOOLEAN ALGEBRA

2.4.1 Boolean Algebra

Let B be a set with at least two elements 0 and 1. Let two binary operations $\vee$ and $\cdot$, and a unary operation $^{-}$ are defined on B. The algebraic system $\langle B, \vee, \cdot, ^{-}, 0, 1\rangle$ is a **Boolean algebra**, if for arbitrary elements a, b and c in B the following postulates are satisfied:

(1) Idempotent laws: $a \vee a = a,\ a \cdot a = a;$

(2) Commutative laws: $a \vee b = b \vee a,\ a \cdot b = b \cdot a;$

(3) Associative laws: $a \vee (b \vee c) = (a \vee b) \vee c,$
$a \cdot (b \cdot c) = (a \cdot b) \cdot c;$

(4) Absorption laws: $a \vee (a \cdot b) = a,\ a \cdot (a \vee b) = a;$

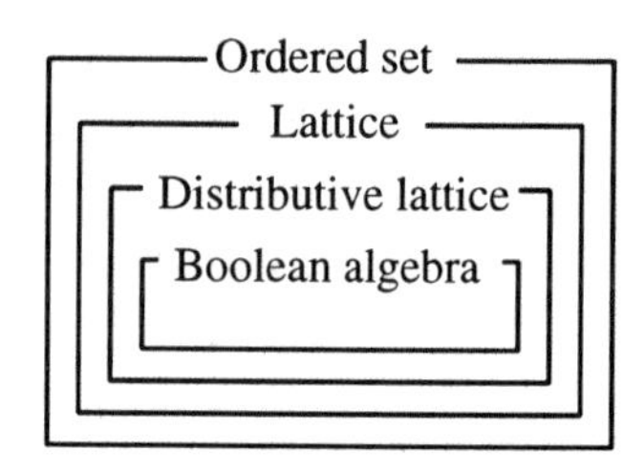

Figure 2.3 Relation of algebraic systems.

(5) Distributive laws: $a \vee (b \cdot c) = (a \vee b) \cdot (a \vee c)$,
$a \cdot (b \vee c) = (a \cdot b) \vee (a \cdot c)$;
(6) Involution: $\bar{\bar{a}} = a$;
(7) Complements: $a \vee \bar{a} = 1, \ a \cdot \bar{a} = 0$;
(8) Identities: $a \vee 0 = a, \ a \cdot 1 = a$;
(9) $a \vee 1 = 1, \ a \cdot 0 = 0$;
(10) De Morgan's laws: $\overline{a \vee b} = \bar{a} \cdot \bar{b}, \ \overline{a \cdot b} = \bar{a} \vee \bar{b}$.

In the above axioms, $\vee$, $\cdot$, *and* $\bar{}$ are called **Boolean sum**, **Boolean product**, and **complement**, respectively. A Boolean algebra is a distributive lattice satisfying the conditions (6)–(10) (Fig. 2.3).

Huntington's Postulates

Boolean algebra is the algebra satisfying the ten axioms in Section 2.4.1. However, to verify whether the given algebra is Boolean algebra or not, we need only to check the following four axioms, the **Huntington's postulates**.

Identities: $a \vee 0 = a, \ a \cdot 1 = a$;
Commutative laws: $a \vee b = b \vee a, \ a \cdot b = b \cdot a$;
Distributive laws: $a \vee (b \cdot c) = (a \vee b) \cdot (a \vee c), \ a \cdot (b \vee c) = (a \cdot b) \vee (a \cdot c)$;
Complements: $a \vee \bar{a} = 1, \ a \cdot \bar{a} = 0$.

From the above four axioms, we can derive the other axioms of the Boolean algebra.

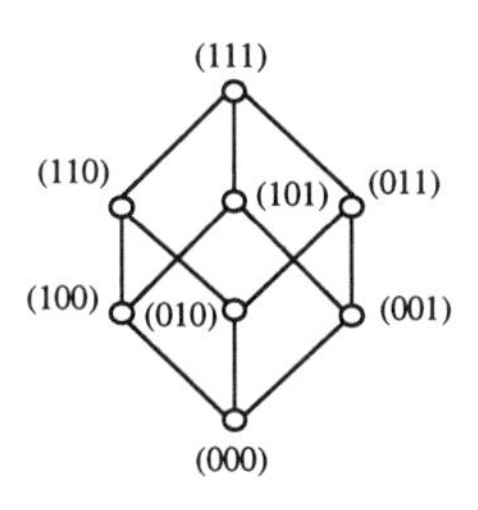

Figure 2.4
Hasse diagram of B^3.

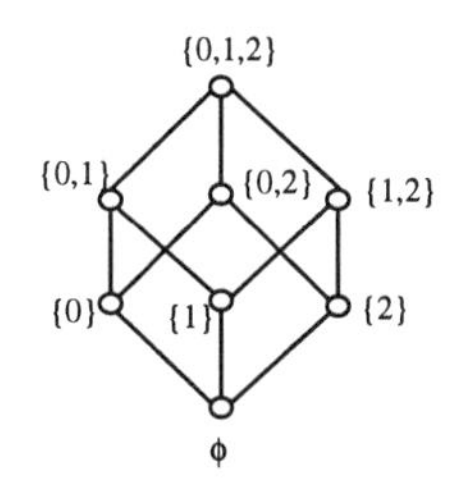

Figure 2.5
Hasse diagram of $P(\{0,1,2\})$.

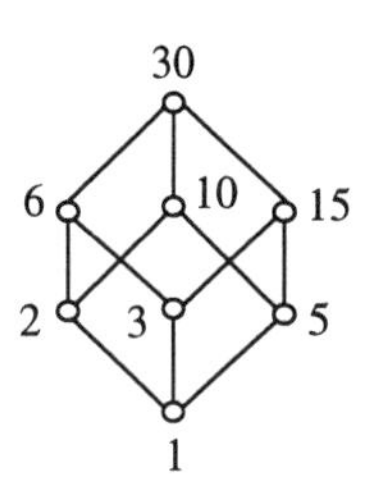

Figure 2.6
Hasse diagram of $A(30)$.

2.4.2 Models of Boolean Algebra

Boolean Algebra Over {0, 1}

Let B={0, 1}. $\langle B, \vee, \cdot, \ ^{-}\ , 0, 1\rangle$ is the simplest (model of the) Boolean algebra.

Boolean Algebra Over Boolean Vectors

In an n-dimensional vector $\boldsymbol{a} = (a_1, a_2, \ldots, a_n)$, if each element is 0 or 1, then $\boldsymbol{a}$ is an **n-dimensional Boolean vector**. Let $\boldsymbol{B}^n = \{(a_1, a_2, \ldots, a_n) \mid a_i \in \{0,1\}\}$ be the set of n-dimensional Boolean vectors. Let two elements in $\boldsymbol{B}^n$ be $\boldsymbol{a} = (a_1, a_2, \ldots, a_n)$ and $\boldsymbol{b} = (b_1, b_2, \ldots, b_n)$. Define the operations $\vee, \cdot,$ *and* $^{-}$ as follows: $\boldsymbol{a} \vee \boldsymbol{b} = (a_1 \vee b_1, a_2 \vee b_2, \ldots, a_n \vee b_n)$, $\boldsymbol{a} \cdot \boldsymbol{b} = (a_1 \cdot b_1, a_2 \cdot b_2, \ldots, a_n \cdot b_n)$, and $\bar{\boldsymbol{a}} = (\bar{a}_1, \bar{a}_2, \ldots, \bar{a}_n)$.
Then, $\langle B^n, \vee, \cdot, \ ^{-}\ , \boldsymbol{0}, \boldsymbol{1}\rangle$ is a (model of the) Boolean algebra, where, $\boldsymbol{0} = (0, 0, \ldots, 0)$, and $\boldsymbol{1} = (1, 1, \ldots, 1)$. Fig. 2.4 is the Hasse diagram of B^3.

Boolean Algebra Over Power Set

Let A be a non-empty set, and let $P(A)$ be a power set of A. For each element of $P(A)$, if the operations $\cup, \cap,$ and $^{-}$ correspond to the union, the intersection, and the complement operation, respectively, then the algebraic system $\langle P(A), \cup, \cap, \ ^{-}\ , \phi, A\rangle$ is a (model of the) Boolean algebra. Fig. 2.5 shows the Hasse diagram of $P(\{0,1,2\})$.

Example 2.5 Let $A(30)$ be the set of positive integers that are divisor of 30. Let $a \vee b$ =(the least common multiple of a and b), $a \wedge b$ =(the greatest common divisor of a and b), and $\bar{a} = 30/a$ (the quotient obtained by dividing 30 by a). Then, the algebraic system $\langle A(30), \vee, \wedge, ^{-}, 1, 30\rangle$ is a (model of the) Boolean algebra. Fig. 2.6 shows the Hasse diagram of $A(30)$. ∎

Isomorphic Boolean Algebra

Note that Figs. 2.4, 2.5, and 2.6 have the same structures. In this case, these Boolean algebra are **isomorphic** to each other. Two Boolean algebras $\langle A, \vee, \cdot, ^{-}, 0_A, 1_A\rangle$ and $\langle B, \vee, \cdot, ^{-}, 0_B, 1_B\rangle$ are isomorphic iff there is the mapping $f: A \rightarrow B$, such that

1) for arbitrary a, $b \in A$, $f(a \vee b) = f(a) \vee f(b)$, $f(a \cdot b) = f(a) \cdot f(b)$, and $f(\bar{a}) = \bar{f}(a)$, and
2) $f(0_A) = 0_B$, $f(1_A) = 1_B$ hold.

An arbitrary finite Boolean algebra is isomorphic to the Boolean algebra $\langle B^n, \vee, \cdot, ^{-}, 0, 1\rangle$, which consists of n-dimensional binary vectors, for some integer n. Therefore, if the number of the elements in the algebra is not the power of two, then it is not a Boolean algebra.

Example 2.6 Fig. 2.7 shows the Hasse diagram of B^4. ∎

2.4.3 De Morgan's Theorem

In the Boolean algebra, the De Morgan's laws or the **De Morgan's theorem** holds:
$\overline{a \cdot b} = \bar{a} \vee \bar{b}$, $\overline{a \vee b} = \bar{a} \cdot \bar{b}$.
These equations can be generalized for the n-variable case:
$\overline{x_1 \cdot x_2 \cdot \cdots \cdot x_n} = \bar{x}_1 \vee \bar{x}_2 \vee \cdots \vee \bar{x}_n$,
$\overline{x_1 \vee x_2 \vee \cdots \vee x_n} = \bar{x}_1 \cdot \bar{x}_2 \cdot \cdots \cdot \bar{x}_n$.

Definition 2.1 *Let $\langle B, \vee, \cdot, ^{-}, 0, 1\rangle$ be a Boolean algebra. The variable that takes arbitrary values in the set B is a* **Boolean variable**. *The expression that is obtained from the Boolean variables and constants by combining with the operators $\vee$, $\cdot$, $^{-}$ and parenthesis is a* **Boolean expression**. *If a map-*

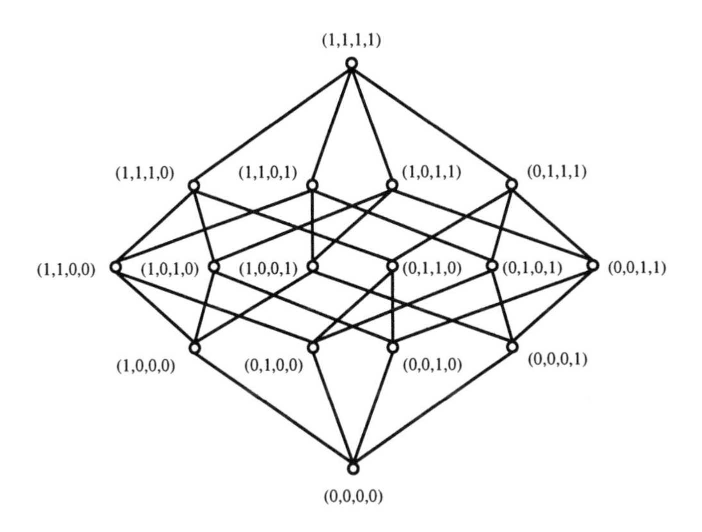

Figure 2.7 Hasse diagram of B^4.

ping f:$B^n \to B$ is represented by a Boolean expression, then f is a **Boolean function**. *However, not all the mappings $f : B^n \to B$ are Boolean functions.*

In Boolean expressions, the following **generalized De Morgan's theorem** holds.

Theorem 2.1 *Let $F(x_1, x_2, \ldots, x_n)$ be a Boolean expression. Then, the complement of the Boolean expression $\overline{F}(x_1, x_2, \ldots, x_n)$ is obtained from F as follows:*

1) *Add the parenthesis according to the order of operations.*
2) *Interchange $\vee$ with $\cdot$.*
3) *Interchange x_i with $\bar{x}_i$.*
4) *Interchange 0 with 1.*

Example 2.7 Let $F = x \vee \bar{y}z$. By applying the De Morgan's theorem, we have
$\overline{x \vee (\bar{y} \cdot z)} = \bar{x} \cdot (y \vee \bar{z})$. ∎

2.4.4 Principle of Duality

In the axioms of Boolean algebra (1)–(10), in an equation that contains $\vee$, $\cdot$, 0, or 1, if we interchange $\vee$ with $\cdot$, and/or 0 with 1, then the another equation holds. In general, this is true. In other words, given an equation in a Boolean algebra, the equation that is obtained from the equation by interchanging $\vee$ with $\cdot$, and/or 0 with 1 also holds. This property is the **principle of duality**.

Dual Boolean Expressions

Let A be a Boolean expression. The **dual** A^D is defined recursively as follows:

(1) $0^D = 1$.
(2) $1^D = 0$.
(3) If x_i is a variable, then ${x_i}^D = x_i (i = 1, \ldots, n)$.
(4) If A, B, and C are Boolean expressions, and $A = B \vee C$, then $A^D = B^D \cdot C^D$.
(5) If A, B, and C are Boolean expressions, and $A = B \cdot C$, then $A^D = B^D \vee C^D$.
(6) If A and B are Boolean expressions, and $A = \overline{B}$, then $A^D = \overline{(B^D)}$.

Applications of the Principle of Duality

Let A and B be Boolean expressions. The symbol $\equiv$ denotes that two expressions represent the same function. If $A \equiv B$, then $A^D \equiv B^D$. In this book, we distinguish a Boolean expression and the function represented by the expression. However, we often do not distinguish them, as in $F(x, y) = x \vee \bar{y} = \overline{(\bar{x} \cdot y)}$.

Example 2.8 Consider the identity: $xy \vee \bar{y}z = xy \vee \bar{y}z \vee xz$. If we apply the principle of duality to this equation, we have another identity: $(x \vee y)(\bar{y} \vee z) = (x \vee y)(\bar{y} \vee z)(x \vee z)$. ∎

As shown in the above example, the symbols for multiplication $\cdot$ are often omitted.

Example 2.9 Consider the Boolean algebra $B = \{0, 1, a, \bar{a}\}$. In this case, check whether the one-variable function $B \to B$ shown in Table 2.1 is a Boolean function or not. In a Boolean function, the following relation holds

Table 2.1

x	$f(x)$
0	a
1	1
a	$\bar{a}$
$\bar{a}$	1

Table 2.2

x_1	x_2	f	g	$f \vee g$	$f \cdot g$	$\bar{f}$	$\bar{g}$
0	0	0	0	0	0	1	1
0	1	1	0	1	0	0	1
1	0	1	0	1	0	0	1
1	1	0	1	1	0	1	0

(Problem 2.18): $f(x) = \bar{x}f(0) \vee xf(1)$. By assigning the values from Table 2.1, we have $f(x) = \bar{x} \cdot a \vee x \cdot 1$. Next, by assigning $x = a$, we have $f(a) = \bar{a} \cdot a \vee a \cdot 1 = 0 \vee a = a$. However, as shown in Table 2.1, $f(a) = \bar{a}$. Therefore, f is not a Boolean function. ∎

2.5 LOGIC FUNCTION

2.5.1 Two-valued Logic Function

Let $B = \{0, 1\}$. A mapping $B^n \rightarrow B$ is always represented by a Boolean expression. This mapping is a **two-valued logic function** or, a **switching function**. There are 2^n elements in B^n, and 2 elements in B. So, the total number of n-variable function is 2^{2^n}. Let us define the operations $\vee$, $\cdot$, and $\bar{}$ among logic functions as follows:

$$f \vee g = h \Leftrightarrow f(x_1, x_2, \ldots, x_n) \vee g(x_1, x_2, \ldots, x_n) = h(x_1, x_2, \ldots, x_n),$$

where x_i may take either 0 or 1, and the value of the function is computed by using the rule of the Boolean algebra {0, 1}. Similarly, $f \cdot g = h$ is also defined. And the complement of the function is defined as follows:

$$\bar{f} = g \Leftrightarrow \bar{f}(x_1, x_2, \ldots, x_n) = g(x_1, x_2, \ldots, x_n).$$

Example 2.10 Consider the case of n=2. The function f maps (0,1) and (1,0) to 1, and other combinations to 0 as shown in Table 2.2. On the other hand, g maps only (1,1) to 1, and other combinations to 0. In this case, $f \vee g$ maps (0,0) to 0, and other combinations to 1. $f \cdot g$ maps all the combinations to 0. $\bar{f}$ maps (0,0) and (1,1) to 1, and other combinations to 0. $\bar{g}$ maps (1,1) to 0, and other combinations to 1. ∎

2.5.2 Boolean Algebra Composed of Logic Functions

Let $\mathcal{F}_n$ be a set of n-variable logic functions. Then, $\langle \mathcal{F}_n, \vee, \cdot, ^{-}, 0, 1 \rangle$ is a Boolean algebra with 2^{2^n} elements. The constant 0 function maps all the n-tuples to 0. Similarly, the constant 1 function maps all the n-tuples to 1.

2.5.3 Recursive Definition of Logical Expressions

A **logical expression** is obtained from logic variables and constants 0 and 1, combined with operations $\vee$, $\cdot$, and $^{-}$. In other words, a logical expression is a Boolean expression where $B = \{0, 1\}$. For a human, to check whether the given expression is a logical expression or not is easy when the expression is simple. However, when the computer program manipulates the logical expressions or when the proof of theorem is necessary, the following **recursive definition** is more convenient.

Definition 2.2

1. *Constants 0 and 1 are logical expressions.*
2. *Variables $x_1, x_2, \ldots$, and x_n are logical expressions.*
3. *If E is a logical expression, then $(\overline{E})$ is also a logical expression.*
4. *If E_1 and E_2 are logical expressions, then $(E_1 \vee E_2)$ and $(E_1 \cdot E_2)$ are also logical expressions.*
5. *The logical expressions are obtained by finite applications of 1–4. In this case, parentheses may be deleted if it does not introduce ambiguity.*

2.5.4 Evaluation of Two-valued Logical Expressions

Given a logical expression and the values of the variables, we can evaluate the value of the expression. Formally, we have the following: An **assignment mapping** $\alpha : \{x_i\} \to \{0, 1\}$ $(i = 1, \ldots, n)$ is an assignment of logic values to all logical variables. For an assignment mapping α, the **valuation mapping** $| F |_\alpha$ of a logical expression F is defined recursively to obtain the value of the logic function.

(1) $| 0 |_\alpha = 0$ and $| 1 |_\alpha = 1$.
(2) If x_i is a variable, then $| x_i |_\alpha = \alpha(x_i)$ $(i = 1, 2, \ldots, n)$.
(3) If F is a logical expression, then $| \overline{F} |_\alpha = 1 \Leftrightarrow | F |_\alpha = 0$.
(4) If F and G are logical expressions, then $| F \vee G |_\alpha = 1 \Leftrightarrow (| F |_\alpha = 1$ or $| G |_\alpha = 1)$.
(5) If F and G are logical expressions, then $| F \cdot G |_\alpha = 1 \Leftrightarrow (| F |_\alpha = 1$ and $| G |_\alpha = 1)$.

Example 2.11 Let us evaluate the value of the logical expression $F : x \vee \bar{y} \cdot z$. Let the assignment α be $\alpha(x) = 0$, $\alpha(y) = 0$, and $\alpha(z) = 1$. Then, we have $| x \vee \bar{y} \cdot z |_\alpha = 1 \Leftrightarrow (| x |_\alpha = 1$ or $| \bar{y} \cdot z |_\alpha = 1)$. Next, since $| x |_\alpha = \alpha(x) = 0$, we have $| F |_\alpha = 1 \Leftrightarrow | \bar{y} \cdot z |_\alpha = 1$. Next, note that $| \bar{y} \cdot z |_\alpha = 1 \Leftrightarrow (| \bar{y} |_\alpha = 1$ and $| z |_\alpha = 1)$. Since, $\alpha(z) = 1$, we have $| F |_\alpha = 1 \Leftrightarrow | \bar{y} |_\alpha = 1$. Finally, since $| \bar{y} |_\alpha = 1 \Leftrightarrow \alpha(y) = 0$, and $| y |_\alpha = \alpha(y) = 0$, we have $| F |_\alpha = 1$. ∎

As shown in the above example, given a logical expression F and an assignment α, we can obtain the value of $| F |_\alpha$. The computations of $| F |_\alpha$ for the given logical expression and assignment often appear in logic design.

2.5.5 Equivalence of Logical Expressions

Let F and G be logical expressions. If $| F |_\alpha = | G |_\alpha$ holds for any assignment α, then F and G are **equivalent**, denoted by the symbol $F \equiv G$. The decision of equivalence for two logical expressions is very important in logic design and verification. Numerous logical expressions of n variables exist, and they can be classified into 2^{2^n} equivalence classes by the equivalence relation $(\equiv)$.

2.6 GROUP, RING, AND FIELD

In this section, we will study group, ring, and field. In a semigroup, only addition (or a multiplication) is defined. In a group, addition and subtraction (or multiplication and division) are defined. In a ring, addition, subtraction, and multiplication are defined. In a field, addition, subtraction, multiplication, and division are defined.

In switching theory, Boolean algebra is mainly used. However, in this section, we briefly introduce other algebraic systems. We assume that in these algebraic systems, "the algebra is **closed** under the operations". In other words, let S

be a set, and let $\oplus$ and $\cdot$ be the operations. If $x, y \in S$, then $x \oplus y \in S$ and $x \cdot y \in S$. The following axioms are basic in these algebraic systems.

A1. Associative law for addition: $(x \oplus y) \oplus z = x \oplus (y \oplus z)$.
A2. Zero element for addition: For all x, a unique 0 element exist such that $x \oplus 0 = 0 \oplus x = x$.
A3. Inverse element for addition: For any x, there exists an element y such that $x \oplus y = y \oplus x = 0$.
A4. Commutative law for addition: $x \oplus y = y \oplus x$.
M1. Associative law for multiplication: $(x \cdot y) \cdot z = x \cdot (y \cdot z)$.
M2. Unit element for multiplication: For any x, a unique 1 element exists such that $x \cdot 1 = 1 \cdot x = x$.
M3. Inverse element for multiplication: For any x, an element y exist such that $x \cdot y = y \cdot x = 1$.
M4. Commutative law for multiplication: $x \cdot y = y \cdot x$.
D1. Distributive law: Multiplication over addition. $x \cdot (y \oplus z) = x \cdot y \oplus x \cdot z$.
D2. Distributive law: Addition over multiplication. $(y \oplus z) \cdot x = y \cdot x \oplus z \cdot x$.

2.6.1 Semigroup

When an algebraic system $\langle S, \cdot, 1 \rangle$ satisfies the axiom M1, then it is a semigroup. If $\langle S, \oplus, 0 \rangle$ satisfies the axiom A1, then it is also a semigroup. When an algebraic system $\langle S, \cdot, 1 \rangle$ satisfies axioms M1 and M2, then it is a **semigroup with identity** or is a **monoid**. If $\langle S, \oplus, 0 \rangle$ satisfies the axioms A1 and A2, then it is also a monoid.

Example 2.12 Let the set of non-negative integers be $N = \{0, 1, \ldots\}$. Then, $\langle N, \cdot, 1 \rangle$ and $\langle N, +, 0 \rangle$ are monoids. ∎

2.6.2 Group

When an algebraic system $\langle G, \cdot, 1 \rangle$ satisfies the axioms M1, M2, and M3, then it is a **group**. When the group also satisfies M4, it is a **commutative group**, or an **Abelian group**. In this definition, addition instead of multiplication may be used. In other words, if an algebraic system satisfies axiom A1, A2, and A3, then it is also a group. In this case, if it satisfies A4, then it is a commutative group.

Table 2.3 Addition and multiplication in Z_3.

$\oplus$	0	1	2
0	0	1	2
1	1	2	0
2	2	0	1

$\cdot$	0	1	2
0	0	0	0
1	0	1	2
2	0	2	1

Example 2.13

1. Let Z be the set of integers. Then, $\langle Z, +, 0\rangle$ is a commutative group.
2. Let T be the set of multiples of 3. Then, $\langle T, +, 0\rangle$ is a commutative group.
3. Let $S = \{1, -1\}$. Then, $\langle S, \cdot, 1\rangle$ is a commutative group. ∎

2.6.3 Ring

When an algebraic system $\langle R, \oplus, \cdot, 0\rangle$ satisfies the axioms A1–A4, M1, D1 and D2, then it is a **ring**. If the ring satisfies M4, then it is a **commutative ring**. If a ring satisfies M2, then it is a **ring with identity**.

Example 2.14

1. Let Z be the set of integers. Then, $\langle Z, +, \cdot, 0\rangle$ is a commutative ring with a unit element.
2. Let $R[X]$ be the set of polynomials of X whose coefficients are real numbers. Then, $\langle R[X], +, \cdot, 0\rangle$ is a commutative ring. In this case the zero element is a polynomial where all the coefficients are 0s.
3. Let M be the set of square matrices whose elements are integers. Then, $\langle M, +, \cdot, 0\rangle$ is a ring. However, it is not commutative with respect to the multiplication. For example, let

$$M_1 = \begin{bmatrix} 1 & 1 \\ 0 & 0 \end{bmatrix}, \text{ and } M_2 = \begin{bmatrix} 1 & 0 \\ 1 & 0 \end{bmatrix}.$$

Then, $M_1 \cdot M_2 \neq M_2 \cdot M_1$. ∎

2.6.4 Field

When an algebraic system $\langle F, +, \cdot, 0, 1\rangle$ satisfies the axioms A1–A4, M1, M2, M4, D1, D2 and the following axiom M3*, then it is a **field**.

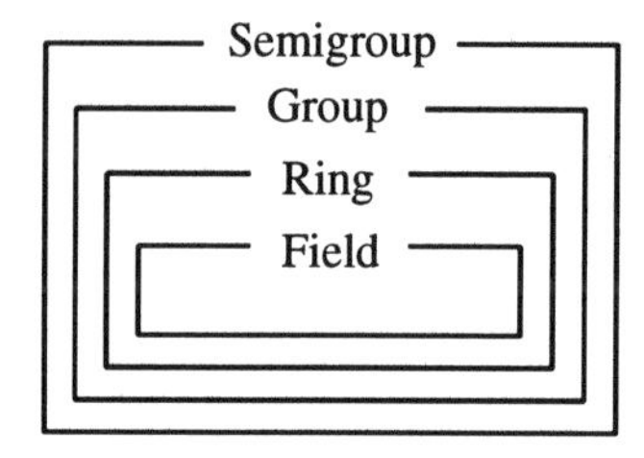

Figure 2.8 Relation of algebraic systems.

M3*. For an arbitrary non-zero element x, there exist y such that $x \cdot y = y \cdot x = 1$.

Example 2.15

1. Let Q be a set of the rational numbers. Then, $\langle Q, +, \cdot, 0, 1\rangle$ is a field.
2. Let R be a set of the real numbers. Then, $\langle R, +, \cdot, 0, 1\rangle$ is a field.
3. Let C be a set of the complex numbers. Then, $\langle C, +, \cdot, 0, 1\rangle$ is a field.
4. Let $Z_k = \{0, 1, \ldots, k-1\}$ $(k \geq 2)$. Then, $\langle Z_k, \oplus, \cdot, 0, 1\rangle$ is a field if k is a prime number. Where, addition $\oplus$ and multiplication $\cdot$ are modulo k operations. For example, when k=3, $Z_k = \{0, 1, 2\}$, and addition $\oplus$ and multiplication $\cdot$ can be defined as shown in Table 2.3.
5. Let R be the set of real numbers. Consider $R^2 = R \times R$. Let the addition $+$ and multiplication $\cdot$ in the set R^2 be $(x_1, x_2)+(y_1, y_2) = (x_1+y_1, x_2+y_2)$ and $(x_1, x_2) \cdot (y_1, y_2) = (x_1 \cdot y_1, x_2 \cdot y_2)$, respectively. Then, the algebraic system $\langle R^2, +, \cdot, 0, 1\rangle$ is a commutative ring with the unit element (1, 1) for the multiplication. Also, (0, 0) is the zero element for addition. This algebraic system is not a field, since the condition M3* does not hold.
6. In the above example, if we replace the multiplication operation with $(x_1, x_2) \cdot (y_1, y_2) = (x_1 \cdot y_1 - x_2 \cdot y_2, x_1 \cdot y_2 + x_2 \cdot y_1)$, then the algebraic system $\langle R^2, +, \cdot, 0, 1\rangle$ is a field. ∎

Fig. 2.8 shows the relations among semigroup, group, ring, and field.

Bibliographical Notes

Lattice theory is extensively described in the textbook [28]. A good, but formal reference book on Boolean algebra is [46].

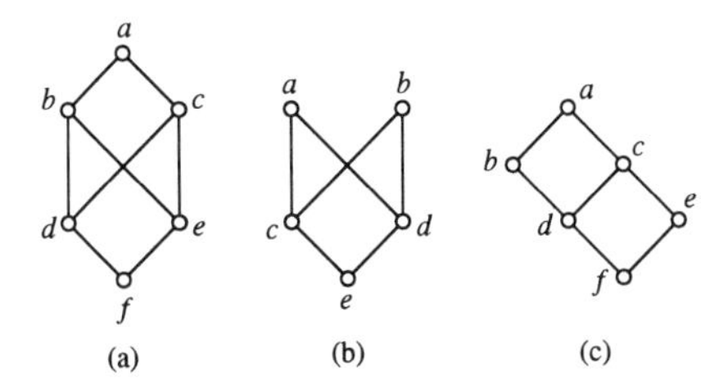

Figure 2.9

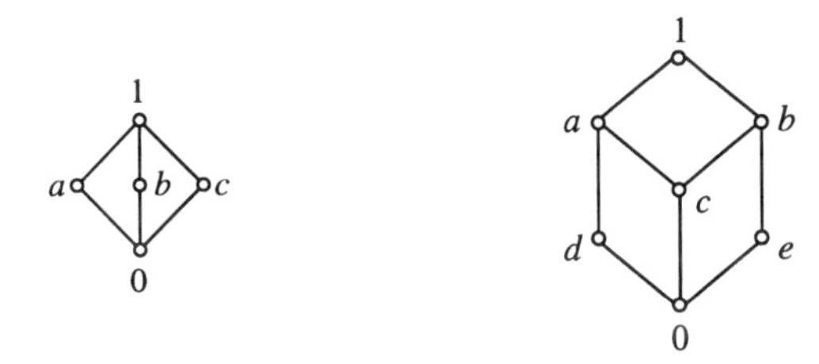

Figure 2.10

Figure 2.11

Exercises

2.1 Among the Hasse diagrams in Fig. 2.9, find a lattice. For each lattice show the operation tables.

2.2 A **modular lattice** is a lattice that satisfies the following conditions: If $x \geq z$, then $x \cdot (y \vee z) = (x \cdot y) \vee z$. Prove that the Hasse diagram in Fig. 2.10 represents a modular lattice. Is it a distributive lattice?

2.3 Does the Hasse diagram in Fig. 2.11 represent a modular lattice? Does it represent a distributive lattice?

2.4 Show that a distributive lattice is a modular lattice (See Exercise 2.2).

2.5 Let A be the set of positive integers that are divisors of 6. For $x, y \in A$, define the operations as follows:

$x \cdot y = GCD(x, y)$: The greatest common divisor of x and y.
$x \vee y = LCM(x, y)$: The least common multiple of x and y.

$\bar{x} = 6/x$: The quotient of 6 divided by x.
$I = 6$.
$O = 1$.
Show that $\langle A, \vee, \cdot, ^{-}, O, I \rangle$ is (a model of) a Boolean algebra.

2.6 Show the difference of a partially ordered set, a lattice, a distributive lattice, and a Boolean algebra. Show the relation between them by a Venn's diagram.

2.7 Consider the algebra on the set $A = \{0, 1, a\}$ that are defined in the following tables.

$\vee$	0	1	a
0	0	1	a
1	1	1	1
a	a	1	a

$\cdot$	0	1	a
0	0	0	0
1	0	1	a
a	0	a	a

x	$\bar{x}$
0	1
1	0
a	a

Check whether each of the axioms in the Huntington's postulates holds.

2.8 Define the operations $\cdot$, $\vee$, and $^{-}$ (complement) so that the Boolean algebra holds on a set $A = \{0, 1, a, b\}$.

2.9 Show that the following two algebras are isomorphic each other.
Algebra 1: $\langle A, \vee, \cdot, ^{-}, O_A, I_A \rangle$. Let A be the set of positive integers that are divisors of 120. For $x, y \in A$, define as follows:
$x \cdot y = GCD(x, y)$: the greatest common divisor of x and y.
$x \vee y = LCM(x, y)$: the least common multiple of x and y.
$\bar{x} = 120/x$: the quotient of 120 divided by x.
$I_A = 120$, $O_A = 1$.
Algebra 2: $\langle B, \vee, \cdot, ^{-}, O_B, I_B \rangle$. Let $B = \{0, 1, 2, 3\} \times \{0, 1\} \times \{0, 1\}$. Then, the vector $\boldsymbol{x} = (x_1, x_2, x_3)$ that satisfies $x_1 \in \{0, 1, 2, 3\}, x_2 \in \{0, 1\}, x_3 \in \{0, 1\}$ is an element of B. Let $\boldsymbol{x} = (x_1, x_2, x_3)$ and $\boldsymbol{y} = (y_1, y_2, y_3)$ be an element of B, define the algebra as follows:
$\boldsymbol{x} \cdot \boldsymbol{y} = (\min(x_1, y_1), \min(x_2, y_2), \min(x_3, y_3))$.
$\boldsymbol{x} \vee \boldsymbol{y} = (\max(x_1, y_1), \max(x_2, y_2), \max(x_3, y_3))$.
$\bar{\boldsymbol{x}} = (3 - x_1, 1 - x_2, 1 - x_3)$.
$I_B = (3, 1, 1)$, $O_B = (0, 0, 0)$.

2.10 Verify that the $GF(4)$ shown in Table 2.4 is a field.

2.11 Let $*$ be a binary operation on the set R of real numbers. For $a, b \in R$, define $a * b = a + b - a \cdot b$ (where, $+$ and $\cdot$ denote ordinary addition and

Table 2.4 Operation table of $GF(4)$.

$+$	0	1	a	b
0	0	1	a	b
1	1	0	b	a
a	a	b	0	1
b	b	a	1	0

$\cdot$	0	1	a	b
0	0	0	0	0
1	0	1	a	b
a	0	a	b	1
b	0	b	1	a

multiplication, respectively). Show that the operation $*$ satisfies the associative law and the commutative law.

2.12 (M) Prove the De Morgan's laws by using the Huntington's postulates.

2.13 Modify the function in Table 2.1 to make it a Boolean function.

2.14 Show the following examples by using Hasse diagrams:

a. A partially ordered set, but not a lattice.
b. A lattice, but not a distributive lattice.
c. A distributive lattice, but not a Boolean algebra.

2.15 Let $T = \{0, 1, 2\}$. Define the binary relation $\leq$ on T^2 be as follows:

$$(x_1, y_1) \leq (x_2, y_2) \Leftrightarrow (x_1 \leq x_2) \text{ and } (y_1 \leq y_2).$$

Then, $\langle T^2, \leq \rangle$ is a partially ordered set.

a. Draw the Hasse diagram of partially ordered set.
b. Is it a lattice?
c. Is it a Boolean algebra?

2.16 In a Boolean algebra, prove the following without using truth tables: If $a \vee b = a \vee c$ and $ab = ac$, then $b = c$.

2.17 Let $Z_4 = \{0, 1, 2, 3\}$. Let addition $\oplus$ and multiplication $\cdot$ be modulo 4 operations. Is $\langle Z_4, \oplus, \cdot, 0, 1 \rangle$ a field?

2.18 (M) Let B be a Boolean algebra, and let $f : B \to B$ be a Boolean function. Then, show that

$$f(x) = \bar{x} f(0) \vee x f(1).$$

Note that the proof of Theorem 3.1 in Section 3.4 is for logic function. Show that it is true for a general Boolean function.

3
LOGIC FUNCTIONS AND THEIR REPRESENTATIONS

This chapter first briefly describes logic elements. Then, as representation methods for logic functions, we introduce truth tables, sum-of-products expressions, Reed-Muller canonical expressions, multi-level logic networks, and binary decision diagrams.

3.1 LOGIC ELEMENTS AND LOGIC NETWORKS

A **two-valued logic element** performs logic operation, where two physical states correspond to logic values 0 and 1. This section introduces contact networks, where the presence or absence of the connection in the electric circuit correspond to logic values 1 and 0, respectively. Also presented are MOS gate circuits where the high and low of the voltage correspond to 1 and 0, respectively. Other two-valued logic elements utilize the presence or absence of current, direction of magnetization, presence or absence of electric charge, presence or absence of light, high or low of oil pressure or gas pressure, etc. In addition, some elements have more than two physical states to perform logic operations. Such elements are **multi-valued logic elements**. By using multi-valued logic elements, we can realize multi-valued logic networks.

Switch

Fig. 3.1 shows a **switch** with a single contact. When the switch is in the upper position, the contact network is open, and no current flows. When the switch

Figure 3.1 Switch (Make-contact).

Figure 3.2 Switch (Break-contact).

Figure 3.3 Transfer-contact.

is in the lower position, the contact network is closed, and current flows. Such a contact is a **make-contact**. Fig. 3.2 shows another switch: when it is in the upper position, it is closed, and when it is in the lower position, it is open. Such a contact is a **break-contact**. Fig. 3.3 shows the switch that has both make and break contacts. Such contact is a **transfer-contact**. Let x denote the make-contact. Then, when $x = 0$, the contact is open. And when $x = 1$, the contact is closed. Let $\bar{x}$ denote the break-contact. Then, when $\bar{x} = 0$, the contact is open, and when $\bar{x} = 1$, the contact is closed.

Relay

A **relay** consists of an **electromagnet** and contacts as shown in Fig. 3.4. In the make-contact relay shown in Fig. 3.4(a), the contact is open when no current flows in the coil of the electromagnet. When current flows in the coil, **armature** (the moving part activated by a magnetic field) is attracted to the electromagnet, and the contact closes. In the break-contact relay shown in Fig. 3.4(b), the contact closes when no current flows in coil, and opens when current flows.

Some relays have more than one contact in an armature. In the relay in Fig. 3.5 with a transfer-contact, when no current flows in the coil, terminals a and b are connected. When current flows in the coil, terminals a and c are connected. That is, when $x = 0$, a and b are connected, and when $x = 1$, a and c are connected.

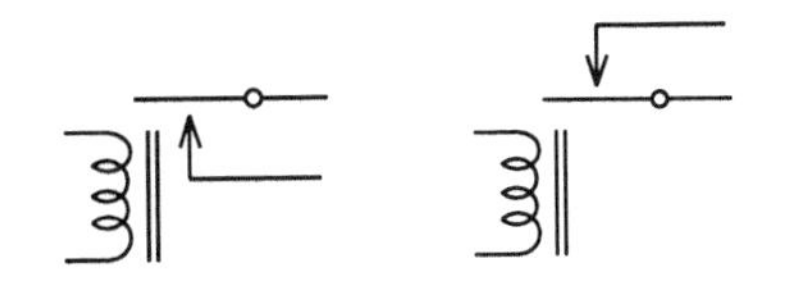

(a) Make-contact. (b) Break-contact.

Figure 3.4 Relay.

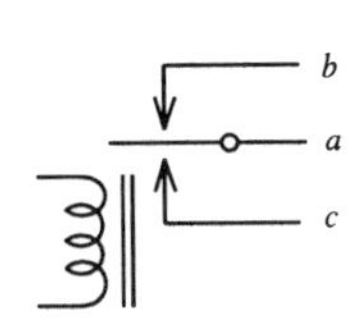

Figure 3.5 Relay (Transfer-contact).

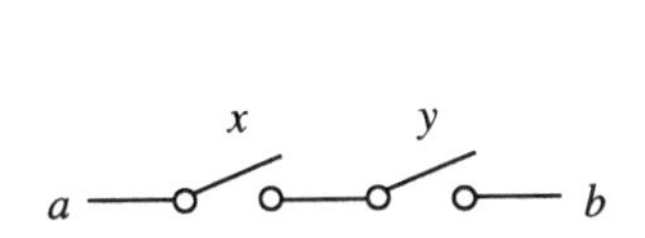

Figure 3.6 Series connection of contacts.

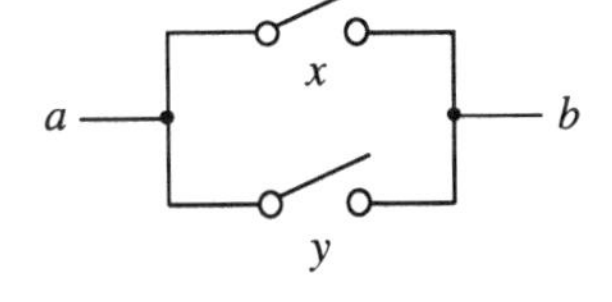

Figure 3.7 Parallel connection of contacts.

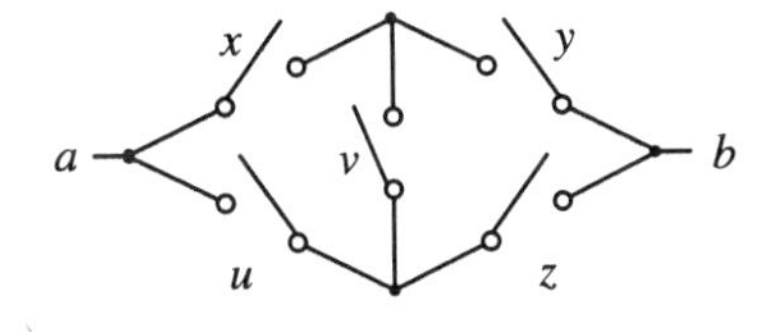

Figure 3.8 Example of non-series-parallel network.

Analysis and Synthesis of Contact Networks

Consider the network shown in Fig. 3.6, where two contacts are connected in series. The **transmission function** f shows whether there is a connection between terminals a and b. In Fig. 3.6, there is a connection between a and b only if $x = y = 1$, so the transmission function is $f = x \cdot y$. Next, consider the network shown in Fig. 3.7, where two contacts are connected in parallel. In Fig. 3.7, there is a connection between a and b when $x = 1$ or $y = 1$, so the transmission function is $f = x \vee y$.

Consider m networks whose transmission are T_1, $T_2, \ldots$, and T_m. The transmission function of the series connected m networks is $T_1 \cdot T_2 \cdot \cdots \cdot T_m$. And, the transmission function of the parallel connected m networks is $T_1 \vee T_2 \vee \cdots \vee T_m$.

A **series-parallel network** is a **contact network** which is obtained by iteratively connecting contacts in series or parallel. A **non-series-parallel network** is a contact network that cannot be obtained by iteratively connecting contacts in series or parallel. Fig. 3.8 shows an example of a non-series-parallel network. The transmission function of the contact network is the logical sum of all the logical products that represent the paths between a and b.

For example, the transmission function in Fig. 3.8 is $f = xy \vee xvz \vee uvy \vee uz$. Since the logical AND operation is equal to an ordinary multiplication operation, the symbol $\cdot$ is often omitted.

Given a transmission function f, to realize a series-parallel contact network for f is easy. However, to realize a series-parallel network with the minimum number of contacts is not so easy. Furthermore, to realize a general (non-series-parallel) network with the minimum number of contacts is very difficult.

MOS Gate Network

Fig. 3.9 shows a **MOSFET** (Metal Oxide Semiconductor Field Effect Transistor) gate. The upper transistor is the **load FET**, and acts as a load resistance. The lower transistor is the **driver FET**. When the power supply is 5V (Volts), the input voltage is either 0V or 5V. When the input voltage is 0V, the driver is open, and virtually no current flows, and the output voltage is 5V. When the input voltage is 5V, the driver FET conducts, and the output voltage is about 0V. Let the low voltage (0V) and the high voltage (5V) denote 0 and 1, respectively. Then, the network in Fig. 3.9 realizes the NOT operation.

Consider the series connection of two driver FETs and a load FET as shown in Fig. 3.10. In this case, when both x and y are high voltage, two driver FETs conduct and the output f is the low voltage. When at least one of x or y are low voltage, no current flows in the network, and the output f is the high voltage. Thus, the output function is represented as $f = \overline{x \cdot y}$. This is a NAND gate. Consider the parallel connection of two driver FETs as shown in Fig. 3.11. In this case, when either one of x or y is high, the output f is low. Thus, the output function is represented as $f = \overline{x \vee y}$. This is a NOR gate.

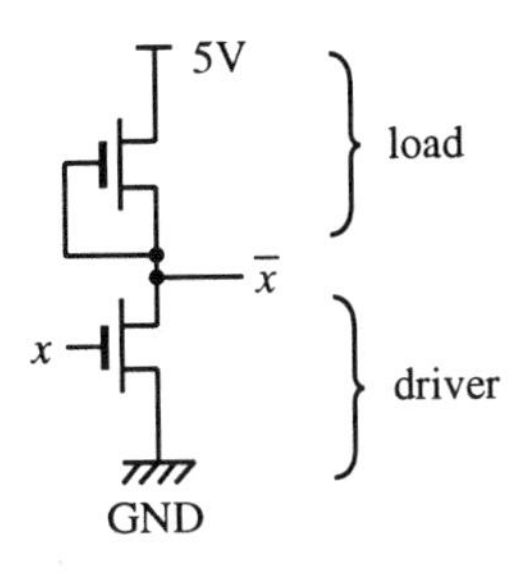

Figure 3.9 MOS gate network.

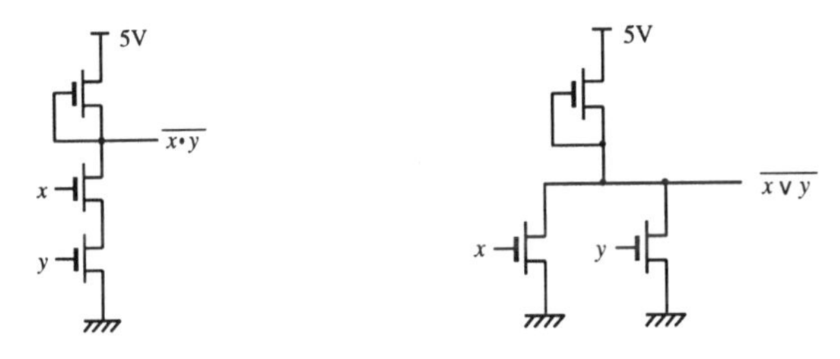

Figure 3.10 NAND gate. **Figure 3.11** NOR gate.

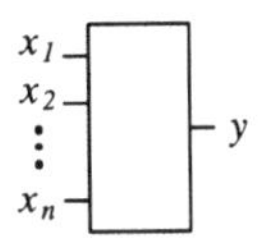

Figure 3.12 Combinational network.

3.2 LOGIC FUNCTIONS AND COMBINATIONAL NETWORKS

Fig. 3.12 shows the general form of a **combinational network**, where inputs $x_1, x_2, \ldots, x_n$ and output y take one of two physical values.

For example, if the voltage of a line takes high or low value, then these states are represented by 1 and 0, respectively. In this case, inputs $x_1, x_2, \ldots, x_n$ are represented by variables that take either 1 or 0. Such a variable is a **two-valued logic variable**. In this book, unless otherwise noted, we assume that variables

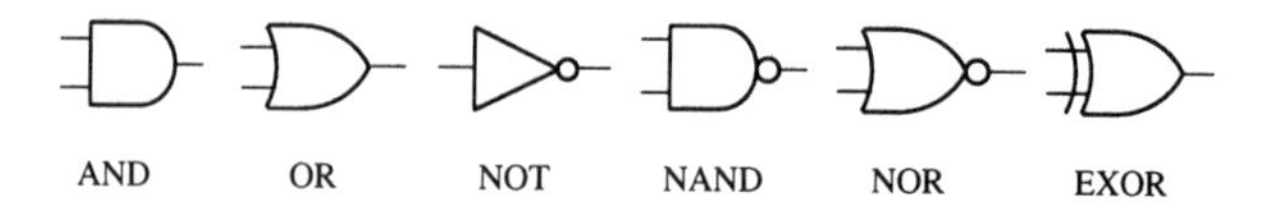

Figure 3.13 Logic symbols.

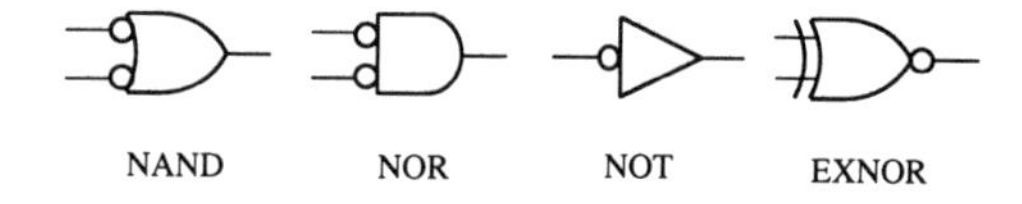

Figure 3.14 Alternative logic symbols.

are two-valued. (In Chapter 9, we will consider functions whose variables are multi-valued.) As the result of processing in the network, the value of the output y is determined. The output y is a function of inputs x_i. So, when all the values of x_i are set to either 0 or 1, then the value of y will be either 0 or 1. Such a function is a logic function: $\{0,1\}^n \rightarrow \{0,1\}$. There are 2^n different input combinations, and a **truth table** shows all input values and the corresponding outputs.

Logic Symbols and Schematic Diagrams

Fig. 3.13 shows the symbols for AND, OR, NOT (inverter), NAND, NOR, and EXOR gates, where small circles ∘ denote the complement operation. Table 3.1 shows truth tables of these operations. In the **AND operation**, if both inputs are 1, then the output is 1; otherwise the output is 0. In the **OR operation**, if both inputs are 0, then the output is 0; otherwise the output is 1. In the **NOT operation**, if the input is 1, then the output is 0; and if the input is 0, the output is 1. In the **NAND operation**, if both inputs are 1, then the output is 0; otherwise the output is 1. In the **NOR operation**, if both inputs are 0, the output is 1; otherwise the output is 0. In the **EXOR operation** (XOR operation or EOR operation), if the number of 1's in the inputs is an odd number, then the output is 1; otherwise the output is 0.

NAND, NOR, and NOT gates are also denoted by the symbols in Fig. 3.14. The validity of the alternate symbols are proved by the De Morgan's theorem: $\overline{x \cdot y} = \bar{x} \vee \bar{y}$, $\overline{x \vee y} = \bar{x} \cdot \bar{y}$. We will use these alternative symbols later in this chapter.

Table 3.1 Truth table for basic logic operations.

	AND	OR	NOT	NAND	NOR	EXOR
x y	$x \cdot y$	$x \vee y$	$\bar{x}$	$\overline{x \cdot y}$	$\overline{x \vee y}$	$x \oplus y$
0 0	0	0	1	1	1	0
0 1	0	1	1	1	0	1
1 0	0	1	0	1	0	1
1 1	1	1	0	0	0	0

The symbol of EXOR gate with a small o at the output denotes **EXNOR** (or, XNOR, ENOR). This gate, realizes $\overline{x \oplus y} = \bar{x} \oplus y = x \oplus \bar{y}$. Since the output is 1 iff the values of x and y are equal, it is called a **coincidence**.

The number of inputs for a gate is **fan-in**. The maximum number of inputs for a gate is the **maximum fan-in**. The number of connections from an output of a gate is **fan-out**. The maximum number of connections from an output of a gate is the **maximum fan-out**. The maximum fan-ins and the maximum fan-outs are often 3–5. When the maximum fan-in and the maximum fan-out are large, we can realize the network using fewer gates.

3.3 SOP AND POS

Many expressions exist for a logic function. In this part, we consider the canonical expressions.

Definition 3.1 *A (two-valued) variable x_i has two* **literals** *x_i and $\bar{x}_i$. A logical product where each variable is represented by at most one literal is a* **product** *or a* **product term** *or a* **term**. *A term can be a single literal. The number of literals in a product term is the* **degree**. *A logical sum of product terms forms a* **sum-of-products expression** *(***SOP***). A logical sum where each variable is represented by at most one literal is a* **sum term**. *A sum term can be a single literal. A logical product of sum terms forms a* **product-of-sums expression** *(***POS***).*

Example 3.1 $x_1 x_2$ is a product term. $x_1 \vee x_2$ is a sum term. $x_1 \bar{x}_2 \vee \bar{x}_1 x_2$ is an SOP. $(x_1 \vee \bar{x}_2) \cdot (\bar{x}_1 \vee x_2)$ is a POS. ∎

Table 3.2 Truth table for 3-variable function.

x_1	x_2	x_3	f
0	0	0	1
0	0	1	1
0	1	0	1
0	1	1	0
1	0	0	1
1	0	1	0
1	1	0	1
1	1	1	0

Definition 3.2

(1) A **minterm** *is a logical product of n literals where each variable occurs as exactly one literal. A* **canonical sum-of-products expression** *(***canonical SOP***) is a logical sum of minterms, where all the minterms are different. It is also called as a* **canonical disjunctive form** *or a* **minterm expansion***.*

(2) A **maxterm** *is a logical sum of n literals where each variable occurs as exactly one literal. A* **canonical product-of-sums expression** *(***canonical POS***) is a logical sum of maxterms, where all the maxterms are different. It is also called as a* **canonical conjunctive form** *or a* **maxterm expansion***.*

An arbitrary logic function is represented by a canonical SOP or a canonical POS, and the representation is unique for a given set of variables. The word **canonical** means that the representation is unique for a logic function. When two logic functions are represented by canonical forms, the decision of equivalence of two logic functions is easy. For logic function f, the canonical SOP is a logical sum of minterms such that $f = 1$. The canonical POS is a logical product of maxterms which are obtained by complementing the minterms such that $f = 0$.

Example 3.2 The canonical SOP for the truth table in Table 3.2 is

$$F = \bar{x}_1\bar{x}_2\bar{x}_3 \vee \bar{x}_1\bar{x}_2x_3 \vee \bar{x}_1x_2\bar{x}_3 \vee x_1\bar{x}_2\bar{x}_3 \vee x_1x_2\bar{x}_3,$$

and the canonical POS is

$$F = (x_1 \vee \bar{x}_2 \vee \bar{x}_3)(\bar{x}_1 \vee x_2 \vee \bar{x}_3)(\bar{x}_1 \vee \bar{x}_2 \vee \bar{x}_3).$$

∎

As illustrated in Example 3.2, the canonical POS is obtained by applying De Morgan's theorem to the canonical SOP for $\bar{f}$, the complement of f.

3.4 SHANNON EXPANSION

The **Shannon's expansion theorem** is useful in the analysis and the synthesis of logic networks. This theorem was presented by Boole in 1854, much earlier than Shannon, it should be called as the **Boole's expansion theorem**. However, since most people call it "the Shannon's expansion theorem," we adopt it.

Theorem 3.1 *An arbitrary logic function $f(x_1, x_2, \ldots, x_n)$ is expanded as follows:*

$$f(x_1, x_2, \ldots, x_n) = \bar{x}_1 f(0, x_2, x_3, \ldots, x_n) \vee x_1 f(1, x_2, x_3, \ldots, x_n).$$

(Proof)
When $x_1 = 1$,

$$\begin{aligned} &\text{(Right-hand side of the equation)} \\ &= 0 \cdot f(0, x_2, x_3, \ldots, x_n) \vee 1 \cdot f(1, x_2, x_3, \ldots, x_n) \\ &= f(1, x_2, x_3, \ldots, x_n) \\ &= \text{(Left-hand side of the equation).} \end{aligned}$$

When $x_1 = 0$,

$$\begin{aligned} &\text{(Right-hand side of the equation)} \\ &= 1 \cdot f(0, x_2, x_3, \ldots, x_n) \vee 0 \cdot f(1, x_2, x_3, \ldots, x_n) \\ &= f(0, x_2, x_3, \ldots, x_n) \\ &= \text{(Left-hand side of the equation).} \end{aligned}$$

Thus, we have the theorem. □

An n-variable function f is expanded into two $(n-1)$-variable functions: $f(0, x_2, x_3, \ldots, x_n)$ and $f(1, x_2, x_3, \ldots, x_n)$. Applying the similar expansions to the $(n-1)$-variable functions recursively, results in a sum of minterms:

Theorem 3.2 $x_i^{c_i}$ *denotes* x_i *when* $c_i = 1$, *and* $\bar{x}_i$ *when* $c_i = 0$. *An arbitrary logic function* $f(x_1, x_2, \ldots, x_n)$ *is uniquely expanded as follows:*

$$f(x_1, x_2, \ldots, x_n) = \bigvee_{(c_1,c_2,\ldots,c_n)} f(c_1, c_2, \ldots, c_n) \cdot {x_1}^{c_1} {x_2}^{c_2} \cdots {x_n}^{c_n},$$

where the symbol $\bigvee_{(c_1,c_2,\ldots,c_n)}$ *denotes the logical sum of* 2^n *elements such that* $(c_1, c_2, \ldots, c_n) \in B^n$.

Example 3.3 Let $n = 3$. An arbitrary three-variable function is represented as

$$\begin{aligned} f(x_1, x_2, x_3) &= f(0,0,0)\bar{x}_1\bar{x}_2\bar{x}_3 \vee f(0,0,1)\bar{x}_1\bar{x}_2 x_3 \\ &\vee f(0,1,0)\bar{x}_1 x_2\bar{x}_3 \vee f(0,1,1)\bar{x}_1 x_2 x_3 \\ &\vee f(1,0,0)x_1\bar{x}_2\bar{x}_3 \vee f(1,0,1)x_1\bar{x}_2 x_3 \\ &\vee f(1,1,0)x_1 x_2\bar{x}_3 \vee f(1,1,1)x_1 x_2 x_3. \end{aligned}$$

This is the **minterm expansion** of three variables. ∎

By considering the dual of Theorem 3.1, we have the following:

Theorem 3.3 *An arbitrary logic function* $f(x_1, x_2, \ldots, x_n)$ *is expanded as follows:*

$$f(x_1, x_2, \ldots, x_n) = (\bar{x}_1 \vee f(1, x_2, x_3, \ldots, x_n)) \cdot (x_1 \vee f(0, x_2, x_3, \ldots, x_n)).$$

3.5 REED-MULLER EXPRESSION

This section shows the representation of logic functions by using AND $(\cdot)$, EXOR $(\oplus)$ and constant 1. The EXOR operation has the following properties:

Lemma 3.1

$$\begin{aligned} (x \oplus y) \oplus z &= x \oplus (y \oplus z) && (\textit{Associative law}), \\ x(y \oplus z) &= xy \oplus xz && (\textit{Distributive law}), \\ x \oplus y &= y \oplus x && (\textit{Commutative law}), \\ x \oplus x &= 0, \\ x \oplus 1 &= \bar{x}. \end{aligned}$$

As the lemma shows, the algebraic system $\langle B, \oplus, \cdot, 0, 1\rangle$ satisfies the conditions of the ring. Thus, it is called the **Boolean ring**. When $x \neq 0$, there exist an element y such that $xy = yx = 1$. Thus, this algebraic system satisfies the condition of the field. This field is called the **Galois field**, and denoted by $GF(2)$.

Lemma 3.2 $xy = 0 \Leftrightarrow x \vee y = x \oplus y$.

(Proof)
($\Rightarrow$) Let $xy = 0$. Then, $x \oplus y = \bar{x}y \vee x\bar{y} = (\bar{x}y \vee xy) \vee (x\bar{y} \vee xy) = x \vee y$.
($\Leftarrow$) Let $xy \neq 0$, then $x = y = 1$. Thus, $x \oplus y = 0$, $x \vee y = 1$.
Therefore, $x \vee y \neq x \oplus y$. □

Starting with a canonical SOP, by applying the relation $\bar{x} = x \oplus 1$ to the complemented literals, and then by applying the distributive laws to expand the expressions, and finally by using relation $x \oplus x = 0$ to simplify the expression, we have the logical expression that consists of ANDs, EXORs, and constant 1.

An arbitrary two-variable function is represented by the canonical SOP:

$$f(x_1, x_2) = f(0,0)\bar{x}_1\bar{x}_2 \vee f(0,1)\bar{x}_1 x_2 \vee f(1,0)x_1\bar{x}_2 \vee f(1,1)x_1x_2.$$

Since each product term has no common minterms, the $\vee$ operators are replaced by $\oplus$ operators. And, the above expression is modified to

$$f(x_1, x_2) = f(0,0)\bar{x}_1\bar{x}_2 \oplus f(0,1)\bar{x}_1 x_2 \oplus f(1,0)x_1\bar{x}_2 \oplus f(1,1)x_1x_2.$$

Next, by applying $\bar{x}_1 = 1 \oplus x_1$ and $\bar{x}_2 = 1 \oplus x_2$, we have

$$\begin{aligned} \bar{x}_1\bar{x}_2 &= (x_1 \oplus 1)(x_2 \oplus 1) = x_1x_2 \oplus x_1 \oplus x_2 \oplus 1, \\ \bar{x}_1 x_2 &= (x_1 \oplus 1)x_2 = x_1x_2 \oplus x_2, \text{ and} \\ x_1\bar{x}_2 &= x_1(x_2 \oplus 1) = x_1x_2 \oplus x_1. \end{aligned}$$

By assigning these equations to the above expression, we have

$$\begin{aligned} f(x_1, x_2) = {} & f(0,0)(x_1x_2 \oplus x_1 \oplus x_2 \oplus 1) \oplus f(0,1)(x_1x_2 \oplus x_2) \\ & \oplus f(1,0)(x_1x_2 \oplus x_1) \oplus f(1,1)x_1x_2. \end{aligned}$$

By rearranging this, we have

$$\begin{aligned} f(x_1, x_2) = {} & f(0,0) \oplus x_1\{f(0,0) \oplus f(1,0)\} \oplus x_2\{f(0,0) \oplus f(0,1)\} \\ & \oplus x_1x_2\{f(0,0) \oplus f(0,1) \oplus f(1,0) \oplus f(1,1)\}. \end{aligned}$$

The above representation is the **Reed-Muller canonical expression** or the **positive polarity Reed-Muller expression** (PPRM) of the function. This canonical representation was published in Russia by Zhegalkin in 1927. However, it was not known in other countries. In 1954, Reed and Muller independently published papers in the U.S.A., subsequently this representation is called the Reed-Muller canonical expression in the western countries.

For n-variable function, we have the following:

Theorem 3.4 *An arbitrary n-variable function is uniquely represented as*

$$\begin{aligned} f(x_1, x_2, \ldots, x_n) &= a_0 \oplus (a_1 x_1 \oplus a_2 x_2 \oplus \cdots \oplus a_n x_n) \\ &\quad \oplus (a_{12} x_1 x_2 \oplus a_{13} x_1 x_3 \oplus \cdots \oplus a_{n-1,n} x_{n-1} x_n) \\ &\quad \oplus \cdots \oplus a_{12\cdots n} x_1 x_2 \cdots x_n, \end{aligned}$$

where $a_i = 0$ or 1.

This theorem shows that arbitrary logic function is realized by an AND-EXOR two-level logic network. In this case, the complement literals are not necessary, but constant 1 is necessary (depends on the function). For a given function, the Reed-Muller canonical expression is unique if we ignore the order of products. Thus, we cannot simplify the expression. AND-EXOR two-level logic networks based on Reed-Muller canonical expansions often requires more gates than AND-OR two-level logic networks. In Chapter 13, we will generalize the expansion method to reduce the number of gates.

3.6 LOGICAL EXPRESSIONS AND MULTI-LEVEL LOGIC NETWORKS

The logical expression $F_1 : x_1(x_2 \vee x_3 \vee x_4 \vee x_5) \vee x_2 x_3 x_4 x_5$ corresponds to the OR-AND-OR three-level network shown in Fig. 3.15. By expanding this logical expression using the distributive law, we have $F_2 : x_1 x_2 \vee x_1 x_3 \vee x_1 x_4 \vee x_1 x_5 \vee x_2 x_3 x_4 x_5$. This expression corresponds to the two-level network in Fig. 3.16. This network requires 6 gates, two more gates than the three-level network. Also, the number of connections in the two-level network is 17, which is five more connections than in the three-level network. As shown in this example, multi-level networks often requires fewer gates and fewer connections than two-level networks. The multi-level network which corresponds to a factored expression has special properties. As shown in Fig. 3.15, each output of a gate

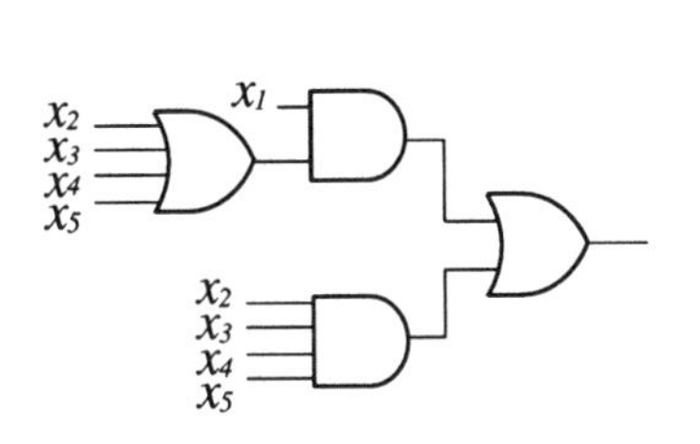

Figure 3.15 Three-level logic network.

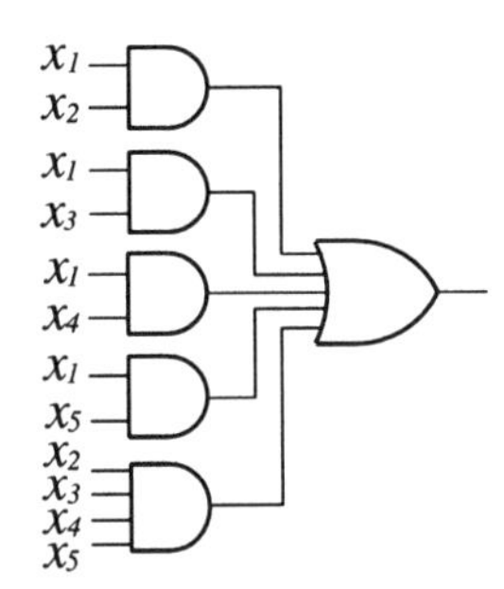

Figure 3.16 Two-level logic network.

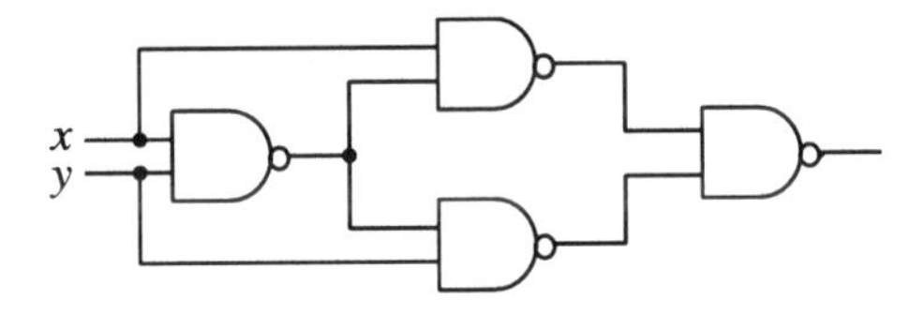

Figure 3.17 Network with fan-out.

is connected to at most one input of the other gate. That is, the fan-out of each gate is 1. Such a network is a **fan-out free network.** However, as shown in Fig. 3.17, a network may contain a gate whose fan-out is more than one. We will study design of multi-level logic networks in Chapter 11.

Example 3.4 Let us obtain the logic function realized by the NAND gate network in Fig. 3.18. The common method to analyze the network is to compute the logical expression for each gate from the inputs to the output. However, this method requires complicated calculations of logical expressions, and manual calculations often produce errors. Here, we will analyze the network by using the Shannon expansions.
As stated before, for a NAND gate, if one of the inputs is 0, then the output of the gate is 1, independent of value of other inputs. For example, let $x_1 = 0$. Then, the outputs for gates 1, 2, 4, 5, and 6 are all 1. Thus, these gates can be replaced by constants 1. Therefore, we have the simplified network shown in Fig. 3.19. In a NAND gate, a constant 1 input can be deleted without changing output function. So, Fig. 3.19 is simplified to Fig. 3.20. Next, convert NAND gates that are in the odd number of levels from the output, to OR gates with

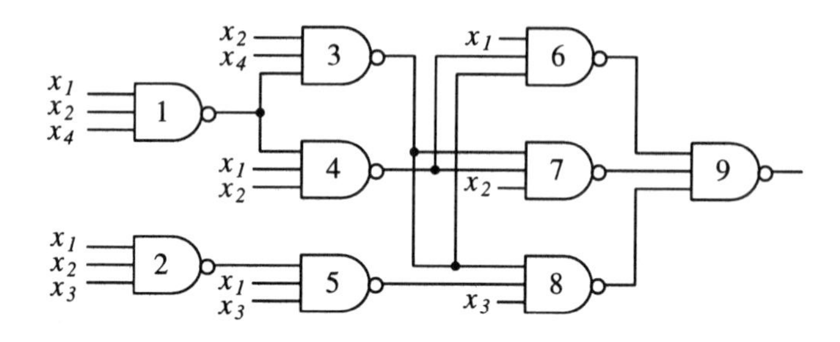

Figure 3.18 Analysis of NAND network.

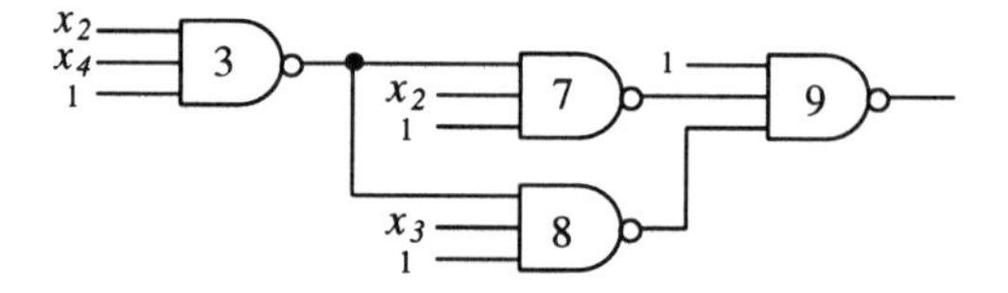

Figure 3.19 When $x_1 = 0$.

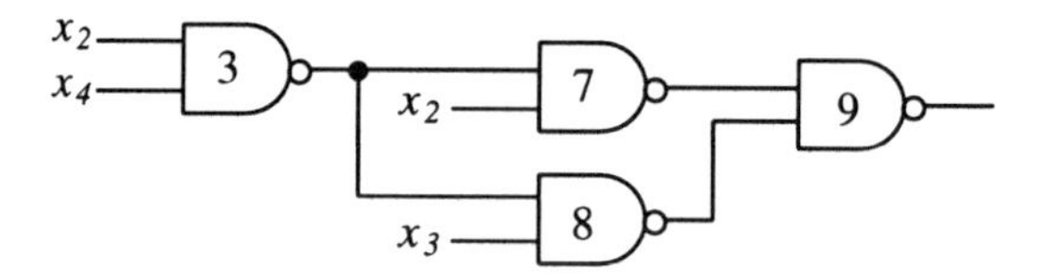

Figure 3.20 Network after deleting constant 1.

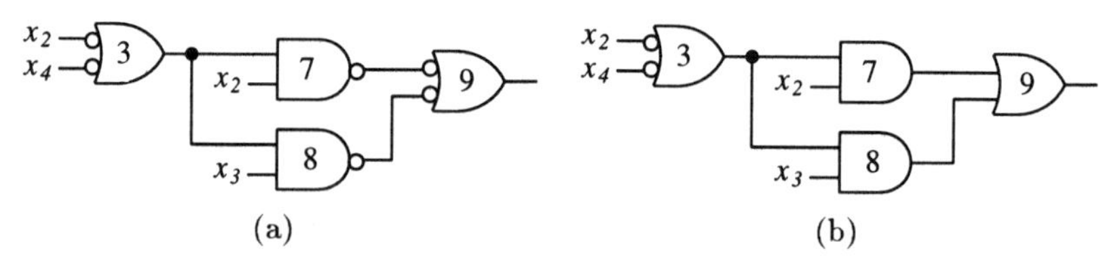

Figure 3.21 Replacement of NAND symbols.

inverted inputs. Then, we have the network in Fig. 3.21(a). In this network, by canceling the cascaded complements, we have the network in Fig. 3.21(b). Thus, we have

$$\begin{aligned} f(0, x_2, x_3, x_4) &= (\bar{x}_2 \vee \bar{x}_4)x_2 \vee x_3(\bar{x}_2 \vee \bar{x}_4) \\ &= x_2\bar{x}_4 \vee \bar{x}_2 x_3 \vee x_3\bar{x}_4 \end{aligned}$$

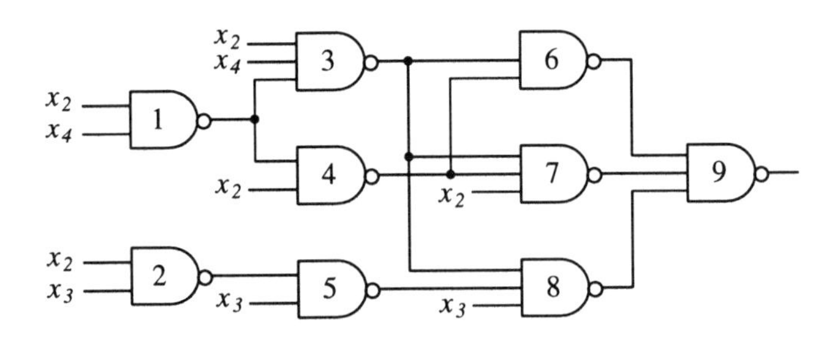

Figure 3.22 When $x_1 = 1$.

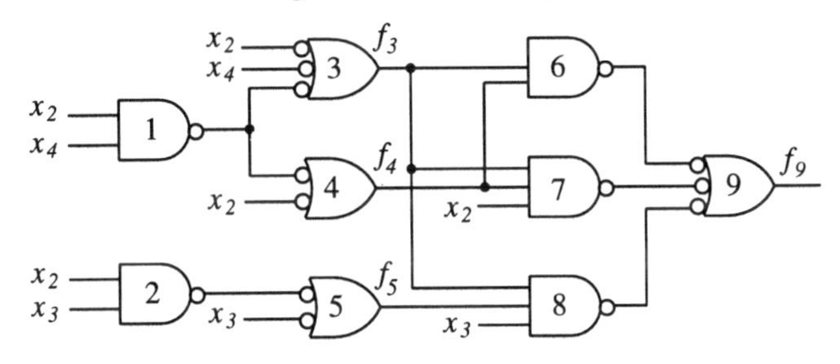

Figure 3.23 Replacement of NAND symbols.

$$= x_2\bar{x}_4 \vee \bar{x}_2 x_3.$$

Next, let $x_1 = 1$ in Fig. 3.18. Since constant 1 inputs in NAND gates can be deleted without changing the function, Fig. 3.18 is simplified to Fig. 3.22. Next, convert the symbols of the NAND gates that are in the odd number of levels from the output, with symbols in Fig. 3.14. Then, we have the network in Fig. 3.23. Thus, we have the following equations:

$$\begin{aligned} f_3 &= x_2x_4 \vee \bar{x}_2 \vee \bar{x}_4 = 1, \\ f_4 &= x_2x_4 \vee \bar{x}_2 = \bar{x}_2 \vee x_4, \\ f_5 &= x_2x_3 \vee \bar{x}_3 = x_2 \vee \bar{x}_3. \end{aligned}$$

$$\begin{aligned} f(1, x_2, x_3, x_4) &= f_3f_4 \vee f_3f_4x_2 \vee f_3f_5x_3 = f_3f_4 \vee f_3f_5x_3 \\ &= 1 \cdot (\bar{x}_2 \vee x_4) \vee 1 \cdot (x_2 \vee \bar{x}_3)x_3 \\ &= \bar{x}_2 \vee x_4 \vee x_2x_3 = \bar{x}_2 \vee x_3 \vee x_4. \end{aligned}$$

Therefore, we have the expression:

$$f(x_1, x_2, x_3, x_4) = \bar{x}_1(x_2\bar{x}_4 \vee \bar{x}_2 x_3) \vee x_1(\bar{x}_2 \vee x_3 \vee x_4).$$

∎

3.7 BINARY DECISION DIAGRAM

A **binary decision diagram** (**BDD**) is a graphical representation of a logic function, and often has a more compact representation than other methods. Thus, BDDs are very important in the design of logic networks using computers.

For example, consider the logic function $f = A \vee B\overline{C}$. Let $A = 1$, then we have $f = 1$. Let $A = 0$, if $B = 0$, then $f = 0$. Let $A = 0$ and $B = 1$. In this case, if $C = 1$ then $f = 0$, and if $C = 0$, then $f = 1$. Fig. 3.24 shows this relation as a decision graph, where Ⓐ, Ⓑ, and Ⓒ denote **non-terminal nodes**, and correspond to the input variables. Also, [0] and [1] are **terminal nodes**, and correspond to the values of the logic function. Such a decision diagram is a BDD. Fig. 3.25 shows the BDDs for the AND, the OR, and the EXOR functions of three variables. Given a BDD, deriving the truth table is easy. Conversely, to construct the BDD from the truth table in Fig. 3.26(a), first, consider the **complete binary decision tree** shown in Fig. 3.26(b).

In general, the complete binary decision tree for an n-variable function has 2^n different paths, and there is one to one correspondence between the terminal nodes of BDDs and 2^n elements of the truth table. The total number of nodes is $2^{n+1} - 1$.

As an example of a simplification method for a complete binary decision tree, consider the function $f = ABC \vee \overline{A}\,\overline{C}$. First, in the decision tree in Fig. 3.27(a), when $A = 1$ and $B = 0$, the function values under C are 0 independently of the value of C. Thus, we can merge these constants 0. Similarly, the two leftmost nodes for C have the same subtrees. By merging these nodes, we have the BDD in Fig. 3.27(b). Next, in Fig. 3.27(b), the leftmost B node and the middle C node are redundant. By removing these redundant nodes, we have the BDD in Fig. 3.27(c).

To transform the logical expression into BDD, we apply the following Shannon expansion repeatedly:

$$f(A, B, C, \ldots) = \overline{A}f(0, B, C, \ldots) \vee Af(1, B, C, \ldots).$$

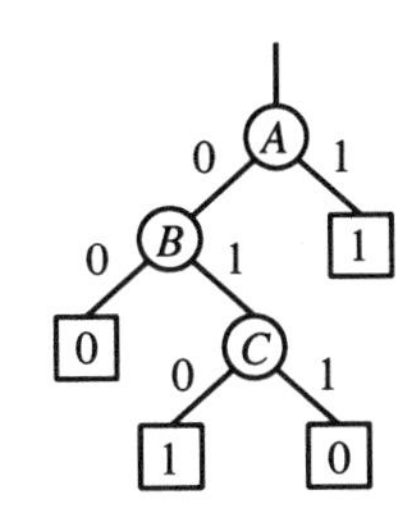

Figure 3.24 BDD for $A \vee B \cdot \overline{C}$.

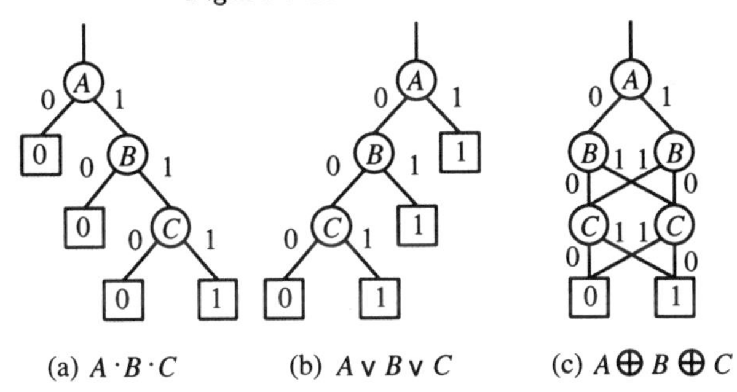

Figure 3.25 BDD for three-variable functions.

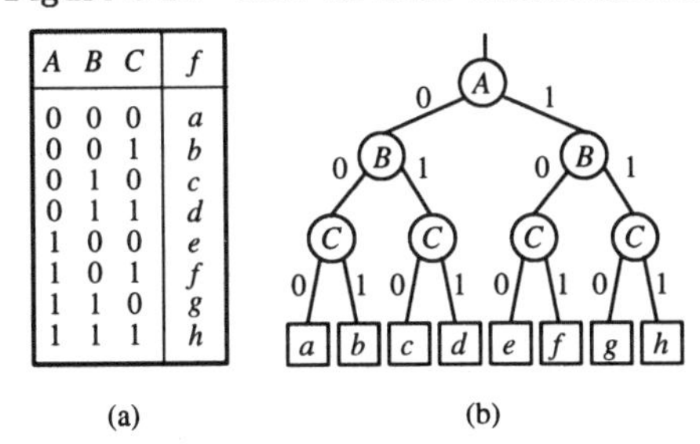

A	B	C	f
0	0	0	a
0	0	1	b
0	1	0	c
0	1	1	d
1	0	0	e
1	0	1	f
1	1	0	g
1	1	1	h

Figure 3.26 Transformation from truth table to BDD.

In this case, the BDD is constructed from the root, and this is a top-down operation. Fig. 3.28 shows a construction of a BDD for the function

$$f = \overline{C}(\overline{A}\,\overline{D} \vee AB \vee AD) \vee C(\overline{A}BD \vee \overline{B}\,\overline{D})$$

by applying the Shannon expansion. First, in f, set $A = 0$ to obtain f_0. And write f_0 as the left child of the node A. Next, in f, set $A = 1$ to obtain f_1. Write f_1 as the right child of the node A, and obtain Fig. 3.28(a). Do the

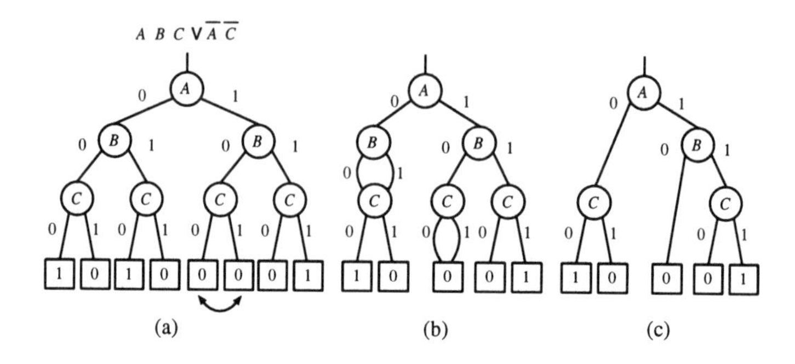

Figure 3.27 Simplification of BDD.

similar operations to variable B, to obtain four sub-functions f_{00}, f_{01}, f_{10}, f_{11} in Fig. 3.28(b). Do the similar operations to variables C and D. In this case, if the same sub-functions appear, then the nodes are shared. Fig. 3.28(c) shows the final BDD. If we apply the Shannon expansions repeatedly to an n-variable function, then all the paths terminate in constant 0 or 1. Also, any path from the root node to the terminal node will visit at most n non-terminal nodes.

ROBDD

In a BDD, if we fix the order of the variables for expansion, and delete redundant nodes, and share the subgraphs that represent the same functions, then we have a unique BDD. Such a BDD is a **Reduced Ordered Binary Decision Diagram (ROBDD)**. Given the order of the variables for this expansion, the ROBDD is unique for a function f, and is a canonical representation of f. Thus, when two logic functions are represented by ROBDDs, the decision of the equivalence is done by checking the equivalence of two graphs, which can be done very quickly. Usually, an ROBDD is simply called a BDD.

Size of BDDs

The number of nodes of a BDD (**size of a BDD**) greatly depends on the order of variables for expansion. Consider the function $x_1x_2 \vee x_3x_4 \vee x_5x_6$. Fig. 3.29 shows the BDDs with two different orders: $x_1 < x_2 < x_3 < x_4 < x_5 < x_6$, and $x_1 < x_3 < x_5 < x_2 < x_4 < x_6$. Note that the BDD with the first ordering has 8 nodes, while the BDD with the second ordering has 16 nodes. In general, the

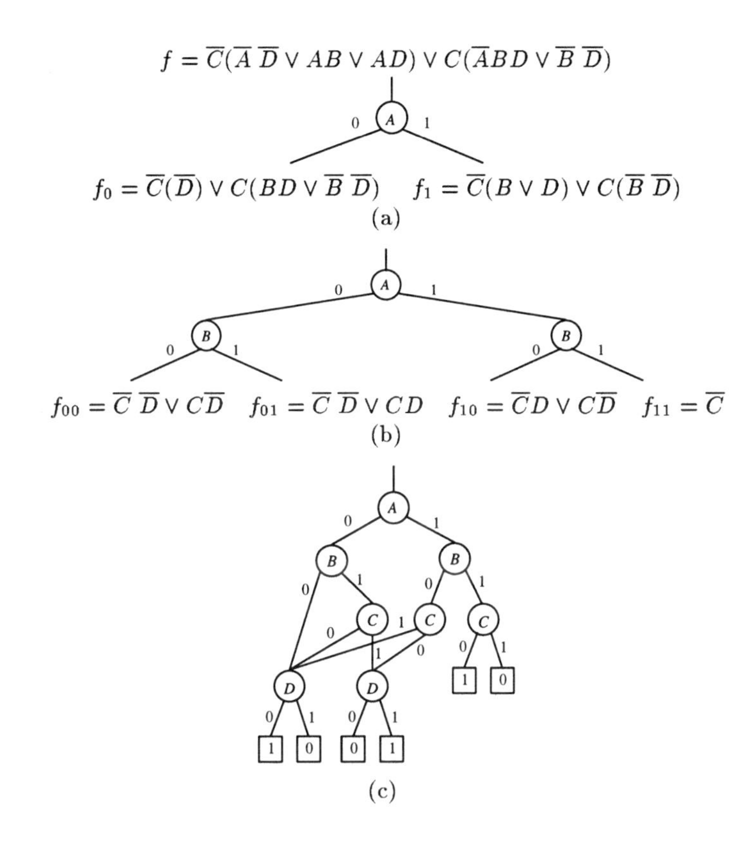

Figure 3.28 Construction of BDD.

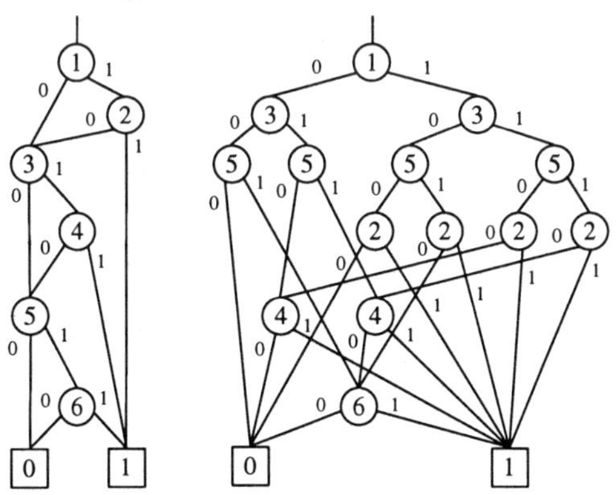

Figure 3.29 BDDs for $x_1x_2 \vee x_3x_4 \vee x_5x_6$ with different variable orderings.

size of BDD for the $2n$ variable function $x_1x_2 \vee x_3x_4 \vee \cdots \vee x_{2n-1}x_{2n}$ is $2n+2$ when the variable ordering is $x_1 < x_2 < x_3 < \cdots < x_{2n-1} < x_{2n}$, and 2^{n+1} when the variable ordering is $x_1 < x_3 < x_5 < \cdots < x_{2n-1} < x_2 < x_4 < \cdots < x_{2n}$.

To find the ordering that reduces the number of nodes in the ROBDD is important, since the large BDD require long computation time as well as large amount of memory. Various methods to reduce the number of nodes in BDDs are developed.

3.8 COMPARISON OF REPRESENTATION METHODS

This section compares four different methods to represent n-variable logic functions for computers.

Truth Table

To represent an n-variable function by a truth table, only the function part must be stored in the computers. So, 2^n bits are sufficient. In a computer with a 32-bit word, a five-variable function can be represented by a word. When, n is small, the operation can be done very quickly. Also, it has a merit that the representation is unique. However, the necessary size of memory is proportional to 2^n. Thus, when $n \geq 30$, the necessary amount of memory is too large.

Sum-Of-Products Expression (SOP)

A product term can be represented by $2n$ bits (ref. Chapter 9). In this representation, the AND, the OR, and the NOT operations are relatively easy, and various manipulation algorithms have been developed. The representation corresponds to an AND-OR two-level network. So, in order to represent a parity function or an adder with n inputs, we need product terms that is proportional to 2^n. For the **Achilles' heel function**,

$$f = x_1x_2x_3 \vee x_4x_5x_6 \vee \cdots \vee x_{3m-2}x_{3m-1}x_{3m},$$

the number of products in the SOP for $\bar{f}$ is 3^m. Also, in an SOP, the estimation of number of gates for multi-level logic networks is difficult. In general, many SOPs exist for a function, and the decision of equivalence of the function is time consuming.

Factored Expression

A **factored expression** (FCE) is defined recursively as follows:

1) A literal is an FCE.
2) A logical sum of FCEs is an FCE.
3) A logical product of FCEs is an FCE.

That is, an FCE is the logical expression where the complements are permitted only for the literals. For example, $x \vee y$ and $(x \vee y) \cdot (z \vee (\bar{w}(u \vee \bar{v})))$ are FCEs. The sizes for FCEs for function f and its complement $\bar{f}$, are the same. This is because, in an FCE, by replacing the AND with the OR operations, and by replacing the variables with their complements, we have $\bar{f}$. An FCE corresponds to a fan-out free multi-level logic network. The size of the memory for FCEs is not so large as SOPs or truth tables. Literals in an FCE corresponds to a contact of a series-parallel contact network. In general, many FCEs exist for a function.

Binary Decision Diagram (BDD)

In a BDD for f, if we interchange the terminal nodes 0 with 1, we have the BDD for $\bar{f}$. If we fix the order of the input variables, then the ROBDD is unique. Thus, a BDD is a canonical representation of a logic function. In BDDs, the AND, the OR, and the complement operations as well as the equivalence decision can be done efficiently. The size of the BDD greatly depends on the order of input variables. To represent an arbitrary n-variable logic function, we need $O(2^n/n)$ nodes (ref. Chapter 14). However, for many of the practical functions, the size of BDDs are not so large. So, in many cases, BDDs require smaller amount of memory than the truth tables or SOPs. Thus, the BDD representation of logic function is very important in logic design.

Table 3.3 Logical equation and propositional logic.

Logical Equation	Propositional Logic
$x = 1$	x is true
$\bar{x} = 1$	x is false
$x \vee y = 1$	Either x or y is true, or both x and y are true.
$xy = 1$	Both x and y are true
$\bar{x} \vee y = 1$	$x \Rightarrow y$
$xy \vee \bar{x}\bar{y} = 1$	$x \Leftrightarrow y$

3.9 LOGICAL EQUATIONS AND PROPOSITIONAL CALCULUS

A **proposition** is a declarative statement that is either true or false, but not both. With every proposition, we associate a **propositional variable** x, y, etc., which assumes the value 1 if it is true, and assumes the value 0 if it is false. Two propositions x and y can be combined to form new propositions. The **conjunction** of x and y, denoted by xy is the proposition "x and y." The **disjunction** of x and y, denoted by $x \vee y$ is the proposition "x or y or both." The negation of x, denoted by $\bar{x}$, is the proposition that is true if $x = 0$, and false if $x = 1$.

The **equivalence** of x and y, denoted by $x \Leftrightarrow y$, is the proposition "whenever one is true, the other is true, and vice versa." **Implication**, denoted by $x \Rightarrow y$, is false only when x is true and y is false.

Propositional calculus and Boolean algebra with $B = \{0, 1\}$ are isomorphic. Table 3.3 show the relation between them.

Therefore, $(x \Rightarrow y)$ and $(y \Rightarrow z)$ in propositional calculus correspond to $(\bar{x} \vee y = 1)$ and $(\bar{y} \vee z = 1)$ in logical equations, respectively. These two logical equations hold iff $(\bar{x} \vee y)(\bar{y} \vee z) = 1$. In general, given m logical equations $f_i = 1$ $(i = 1, 2, \ldots, m)$, a necessary and sufficient condition for all these equations to hold is $\bigwedge_{i=1}^{m} f_i = 1$.

Example 3.5 (Problem) Five students x_1, x_2, x_3, x_4, and x_5 are planning a trip. Let $x_i = 1$ mean student x_i goes, and $x_i = 0$ mean student x_i stays. They must satisfy all the following conditions:

1) Either x_1 or x_2, or both x_1 and x_2 go on a trip.

2) Either x_3 or x_5, but not both go on a trip.
3) Either both x_1 and x_3 go, or neither go on a trip.
4) If x_4 goes, then x_5 also goes on a trip (i.e., $x_4 \Rightarrow x_5$).
5) If x_2 goes, then both x_1 and x_4 go on a trip (i.e., $x_2 \Rightarrow x_1x_4$).

Obtain all possible combinations of person to go on a trip.

(Solution) Each condition can be replaced by a logical equation as follows:

1) $x_1 \vee x_2 = 1$.
2) $x_3\bar{x}_5 \vee \bar{x}_3x_5 = 1$.
3) $x_1x_3 \vee \bar{x}_1\bar{x}_3 = 1$.
4) $\bar{x}_4 \vee x_5 = 1$.
5) $\bar{x}_2 \vee x_1x_4 = 1$.

All the conditions hold iff $(x_1 \vee x_2)(x_3\bar{x}_5 \vee \bar{x}_3x_5)(x_1x_3 \vee \bar{x}_1\bar{x}_3)(\bar{x}_4 \vee x_5)(\bar{x}_2 \vee x_1x_4) = 1$. By expanding this equation, we have $x_1\bar{x}_2x_3\bar{x}_4\bar{x}_5 = 1$. Thus, the only combination of students that go on the trip is $\{x_1, x_3\}$, which is shown by positive literals for x_1 and x_3 in the equation. ∎

Bibliographical Notes

Many textbooks on logic design are available [126, 157, 170, 172, 201, 212, 218, 222, 295]. Textbooks treating MOS gate networks are [58, 265, 278, 424]. Early development of switching theory is shown in [285, 286, 293, 379]. For optical devices, see [7]. For magnetic bubble logic devices, see [206, 337, 338, 340]. Original papers on BDDs are [2, 48, 221]. For the implementation of BDD packages, see [32, 254, 256, 302, 303]. As for BDD variable ordering, see [57, 128, 132, 186, 334]. For a discussion of the complexities of BDDs, see [49, 56, 225, 249, 421]. Other representations of logic functions are shown in [71, 196, 363].

Exercises

3.1 Consider the combinations of binary numbers x, y, z, and w. If the number of 1s is an even number, then $f = 1$, else $f = 0$. Show the truth table for f.

3.2 Let $\boldsymbol{x} = (x_1, x_0)$ and $\boldsymbol{y} = (y_1, y_0)$ be binary numbers with 2 bits. Show the truth table for the function z where $z = 1$ iff $\boldsymbol{x} > \boldsymbol{y}$.

3.3 Consider the combinations of binary numbers x, y, z, and w. If the number of 1s is greater than two, then $f = 1$, else $f = 0$. Show the truth table for f.

3.4 Obtain the canonical SOP (sum of minterms) and the canonical POS (product of maxterms) for the following function:

$$f = xy\bar{z}\bar{w} \vee (\bar{x} \vee \bar{y})zw \vee (x \vee y)(\bar{z}w \vee z\bar{w}).$$

3.5 Prove that an arbitrary n-variable logic function has an SOP with at most 2^{n-1} products.

3.6 Obtain the Reed-Muller canonical expressions for the following logic functions:

$$\begin{aligned} f &= x \vee y, \\ g &= xy(\bar{z} \vee \bar{w}) \vee zw(\bar{x} \vee \bar{y}). \end{aligned}$$

3.7 Represent the logic function $f = x(y \oplus z)$ by the canonical SOP, the canonical POS, and the Reed-Muller canonical expression.

3.8 Prove the following relations:

(a) $\underbrace{1 \oplus 1 \oplus \cdots \oplus 1}_{n\ 1\text{'s}} = \begin{cases} 0 & \text{when } n \text{ is an even number,} \\ 1 & \text{when } n \text{ is an odd number.} \end{cases}$

(b) $x \oplus x = 0$, $x \oplus 1 = \bar{x}$.

(c) $(x \oplus y)(y \oplus z)(z \oplus x) = 0$.

(d) $x \oplus y = x \oplus z \Leftrightarrow y = z$.

(e) $f = g \oplus h \Leftrightarrow g = f \oplus h$.

(f) $(x \oplus a)(x \oplus b) = (\bar{a} \oplus b)x \oplus ab$.

(g) $(x \vee y) \oplus (y \vee z) = x\bar{y} \oplus \bar{y}z$.

(h) $xy \vee yz \vee zx = xy \oplus yz \oplus zx$.

3.9 Prove or show the counterexample for the following proposition:

$$x \oplus (y \vee z) = (x \oplus y) \vee (x \oplus z).$$

3.10 Suppose that the variables and their complements are available as inputs. Realize the following function by using AND and OR gates with three inputs:

$$f = \bar{x}_1\bar{x}_2x_4 \vee x_2\bar{x}_3\bar{x}_4 \vee x_2x_3x_4 \vee x_1\bar{x}_2\bar{x}_3x_4.$$

3.11 Show BDDs for the functions:
(a) $f = xy \oplus xz \oplus xw \oplus yz \oplus yw \oplus zw$.
(b) $g = x \oplus y \oplus z \oplus w$.

3.12 (M) Let $x_1 < x_2 < \cdots < x_{2n}$ be the order of the variables for expansion. Show that the number of nodes in the ROBDD for the following function is 2^{n+1}:

$$f = x_1x_{2n} \vee x_2x_{2n-1} \vee \cdots \vee x_nx_{n+1}.$$

3.13 Let $h(f, g)$ be an arbitrary logic function of f and g. Prove that the following relation holds, where x is a variable in f and g:

$$h(f, g) = \bar{x}h(f(|\ x = 0), g(|\ x = 0)) \vee xh(f(|\ x = 1), g(|\ x = 1)).$$

3.14 A large room has four entrances and a lamp in the center of the room. Each entrance has a switch to turn on and off the lamp. Design a contact network such that each switch can turn the light on and off. You can use transfer switches if necessary.

3.15 Prove Theorem 3.4 (page 46).

3.16 Show the truth table of the functions f and g for the MOS network in Fig. 3.30.

3.17 Show the truth table of the logic network shown in Fig. 3.31.

3.18 Show the truth table of the logic network shown in Fig. 3.32.

3.19 Show the truth table of the logic network shown in Fig. 3.33.

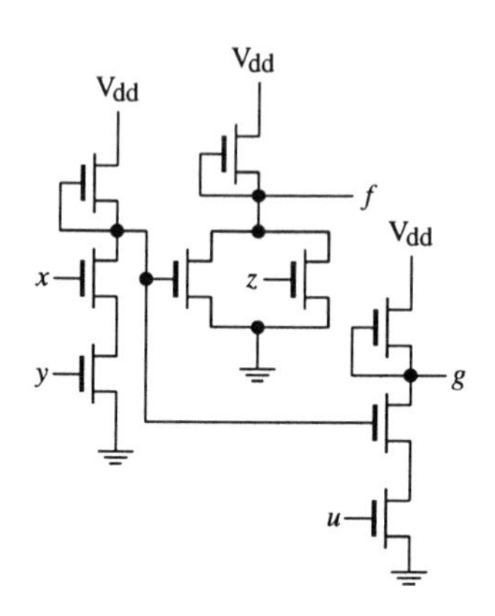

Figure 3.30

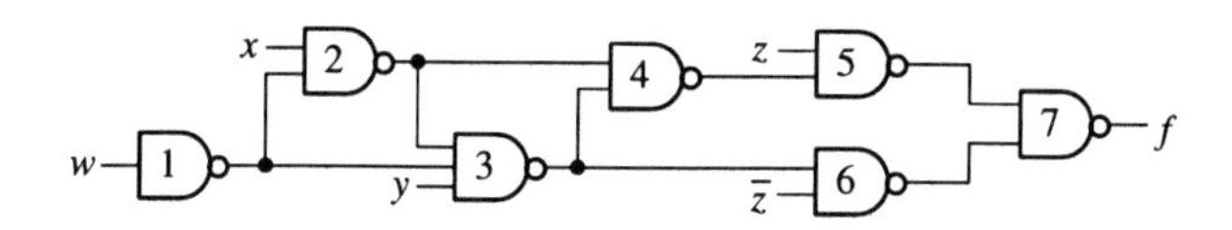

Figure 3.31

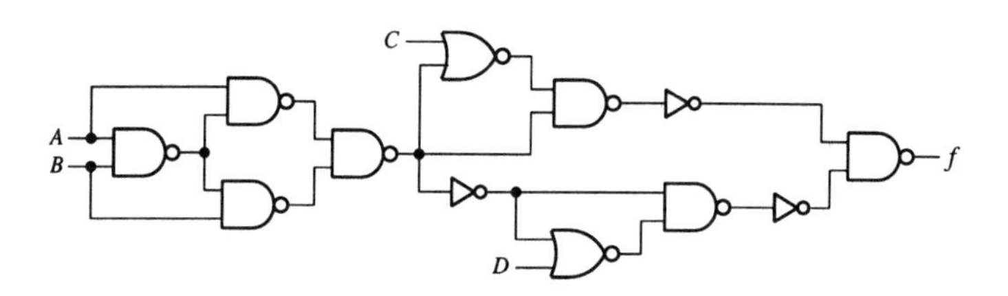

Figure 3.32

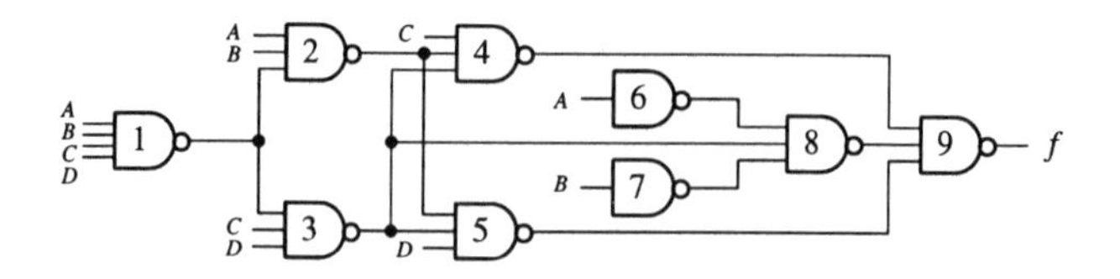

Figure 3.33

3.20 Is the following proposition function, true, false or dependent on the weather?
"It is raining in Tokyo, but not in Nagasaki," or "It is not raining in at least one of the three cities: Tokyo, Nagasaki, and Kyoto," or "It is raining in Tokyo, in Nagasaki, and in Kyoto." To analyze this proposition, use symbols as follows:
T : It is raining in Tokyo.
N : It is raining in Nagasaki.
K : It is raining in Kyoto.

3.21 Let $x\bar{y} \vee \bar{x}y = z$. Show that $x\bar{z} \vee \bar{x}z = y$.

3.22 Obtain the assignments of (x, y, z) that satisfy the following conditions: $x \vee y = z$. $x \vee yz = 1$.

3.23 Obtain the assignments of (x, y, z) that satisfy the following conditions: $\bar{x} \vee xy = 0$. $xy = xz$.

4

OPTIMIZATION OF AND-OR TWO-LEVEL LOGIC NETWORKS

Sum-of-products expressions (SOPs) correspond to AND-OR two-level logic networks, and are very important. This chapter describes two methods to simplify AND-OR two-level logic networks by using SOPs. The first method, using the Karnaugh map, is suitable for manual design, while the second method, the Quine McCluskey method, is suitable for computer implementation.

4.1 SOPS AND TWO-LEVEL LOGIC NETWORKS

A logic function has, in general, many SOPs, and the complexity of the network depends on the SOP. For example, Fig. 4.1 is the AND-OR two-level logic network for $F: x_1\bar{x}_2 \vee \bar{x}_1 x_2 \vee x_1 x_2 \vee x_3 x_4$. In this case, each product in the SOP corresponds to an AND gate, and each literal corresponds to an AND connection. If we apply the laws of Boolean algebra repeatedly to F to obtain the simplified SOP for f, we have the following:

$$\begin{aligned} f &= x_1\bar{x}_2 \vee (\bar{x}_1 \vee x_1)x_2 \vee x_3x_4 = x_1\bar{x}_2 \vee x_2 \vee x_3x_4 \\ &= x_1\bar{x}_2 \vee x_1x_2 \vee x_2 \vee x_3x_4 = x_1(\bar{x}_2 \vee x_2) \vee x_2 \vee x_3x_4 \\ &= x_1 \vee x_2 \vee x_3x_4. \end{aligned}$$

Fig. 4.2 is the network corresponding to the last SOP. As illustrated in this example, the reduction of products in an SOP corresponds to the reduction of the AND gates, and the reduction of the literals corresponds to the reduction of AND connections. Here, to simplify the SOP, we used laws of Boolean algebra by inspections. However, this method is unsuitable for SOPs with many inputs.

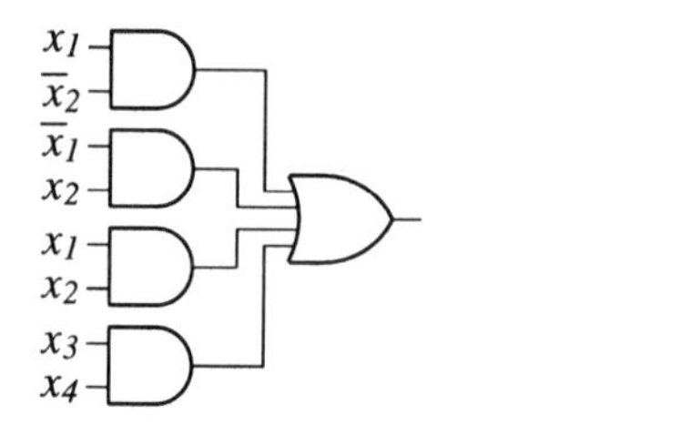

Figure 4.1 Network for $x_1\bar{x}_2 \vee \bar{x}_1 x_2 \vee x_1 x_2 \vee x_3 x_4$.

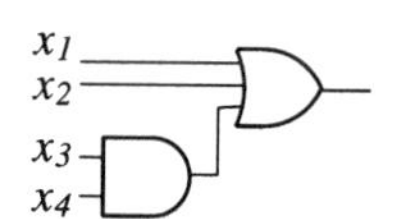

Figure 4.2 Network for $x_1 \vee x_2 \vee x_3 x_4$.

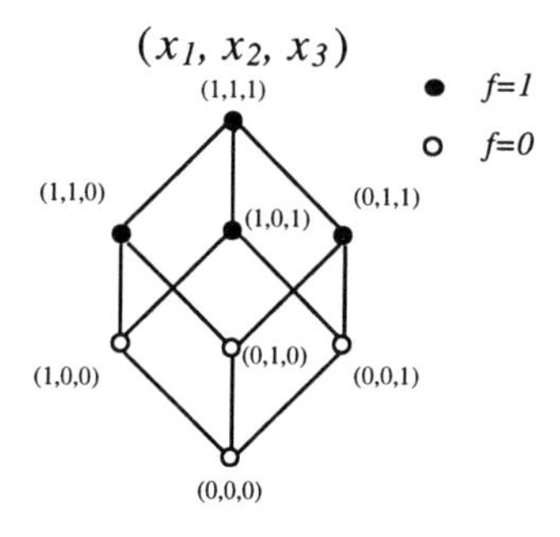

Figure 4.3 Three-dimensional cube.

In the followings, we will consider systematic methods to reduce the number of gates and literals.

4.2 *N*-DIMENSIONAL CUBE

An n-variable logic function is represented as a subset of 2^n vertices of the n-dimensional cube.

Example 4.1 Consider the three-variable function:

$$f = x_1 x_2 x_3 \vee \bar{x}_1 x_2 x_3 \vee x_1 \bar{x}_2 x_3 \vee x_1 x_2 \bar{x}_3.$$

This function is represented as a set of vertices shown in Fig. 4.3. Each vertex of a cube corresponds to a minterm. For example, a vertex (1,1,1) corresponds to the minterm $x_1x_2x_3$. Next, consider the edge that consists of two vertices (1,1,1) and (1,1,0). The logical sum of minterms for this edge is simplified to

$$x_1x_2x_3 \vee x_1x_2\bar{x}_3 = x_1x_2(x_3 \vee \bar{x}_3) = x_1x_2.$$

The edge that consists of vertices (1,1,1) and (0,1,1) is represented as

$$x_1x_2x_3 \vee \bar{x}_1x_2x_3 = x_2x_3.$$

Similarly, the edge that consists of vertices (1,1,1) and (1,0,1) is represented as

$$x_1x_2x_3 \vee x_1\bar{x}_2x_3 = x_1x_3.$$

Therefore, the given function is represented as the logical sum of three products:

$$f = x_1x_2 \vee x_1x_3 \vee x_2x_3.$$

This SOP has fewer products and fewer literals than the original SOP. In this way, the n-dimensional cube is useful for simplification of SOPs. ∎

Don't Care

In practical applications, there are combinations that never appear in the inputs, or the case where the value of the function can be either 0 or 1. In such a case, the output for the input can be either 0 or 1, and the output is **don't care.** A function with a *don't care* is an **incompletely specified function.** On the other hand, a function without a *don't care* is a **completely specified function.**

Example 4.2 Table 4.1 is a truth table for 4-input 1-output function. The value of the function is 1 iff the value of inputs is greater than four in the BCD (Binary Coded Decimal) code. Note that in the BCD code, the numbers that are greater than 9 are not defined. So, such combinations never appear as inputs. In the truth table, such a combination is represented by d.

4.3 KARNAUGH MAP

A **Karnaugh map** is an n-dimensional cube written in a plane. It is convenient for the representation of logic functions and simplification of SOPs with 3–5

Table 4.1 Truth table with *don't care.*

x_3	x_2	x_1	x_0	f
0	0	0	0	0
0	0	0	1	0
0	0	1	0	0
0	0	1	1	0
0	1	0	0	0
0	1	0	1	1
0	1	1	0	1
0	1	1	1	1
1	0	0	0	1
1	0	0	1	1
1	0	1	0	d
1	0	1	1	d
1	1	0	0	d
1	1	0	1	d
1	1	1	0	d
1	1	1	1	d

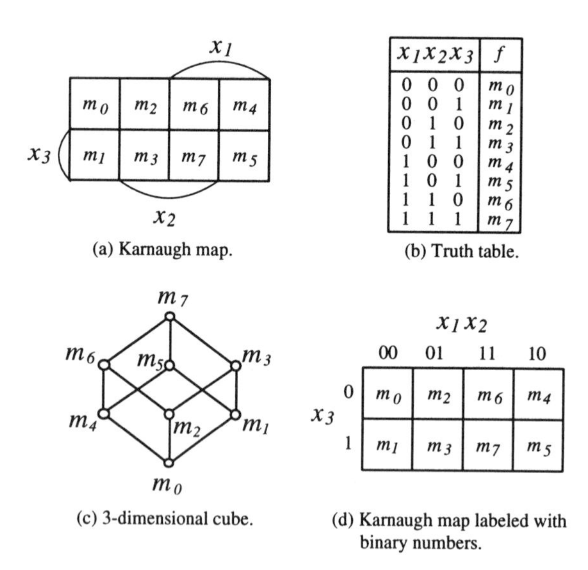

Figure 4.4 Representations by Karnaugh maps.

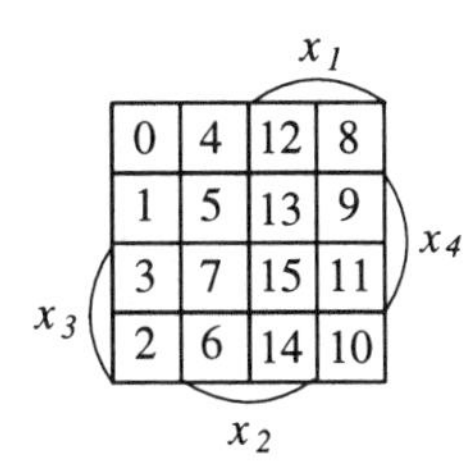

Figure 4.5 Karnaugh map for 4 variables.

0	8	24	16	17	25	9	1
2	10	26	18	19	27	11	3
6	14	30	22	23	31	15	7
4	12	28	20	21	29	13	5

Figure 4.6 Karnaugh map for 5 variables.

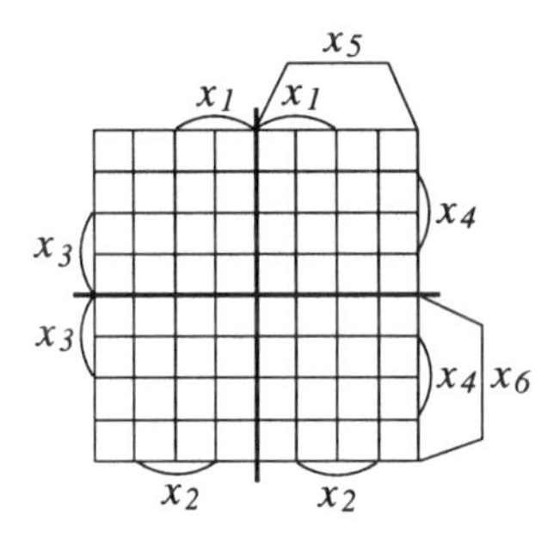

Figure 4.7 Karnaugh map for 6 variables.

variables. Fig. 4.4(a) shows a three-variable Karnaugh map. It consists of 8 **cells**, and each cell corresponds to a minterm. For example, in Fig. 4.4(a), m_0 corresponds to the minterm $\bar{x}_1\bar{x}_2\bar{x}_3$; m_1 corresponds to the minterm $\bar{x}_1\bar{x}_2x_3$; and so on. m_0, m_1,..., and m_7 correspond to values of the truth table, and they are either 0, 1 or d. Usually, 0's are not entered in the cell. So, a blank cell represents 0. In a Karnaugh map, *don't care* is represented by the symbol d or $\times$. As shown in Fig. 4.4(d), the Karnaugh maps labeled with binary numbers are also used. By comparing the Karnaugh map in Fig.4.4 (a) and the cube in Fig. 4.4(c), we find that the cell m_7 is adjacent to the cells m_3, m_5, and m_6. If we consider the vectors of (x_1, x_2, x_3), we have the following: The vector corresponding to the cell m_7 is (1,1,1), and the vector corresponding to the cell m_3 is (0,1,1). The **Hamming distance** (i.e., the number of different bits in the vectors) between two cells m_7 and m_3 is one. However, the cells m_7 and m_4 are not adjacent: They are in diagonal positions. This can be verified as follows: The vector corresponding to the cell m_4 is (1,0,0), and the vector corresponding to the cell m_7 is (1,1,1). And the Hamming distance between

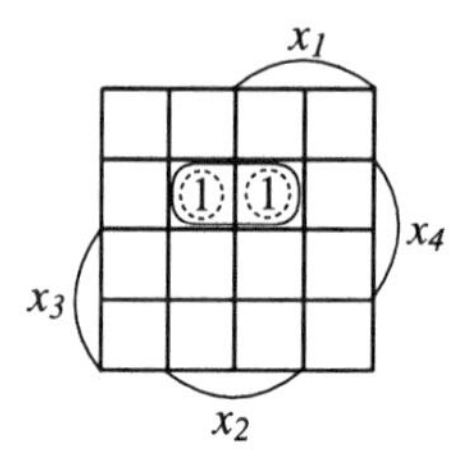

Figure 4.8 Merge of minterms.

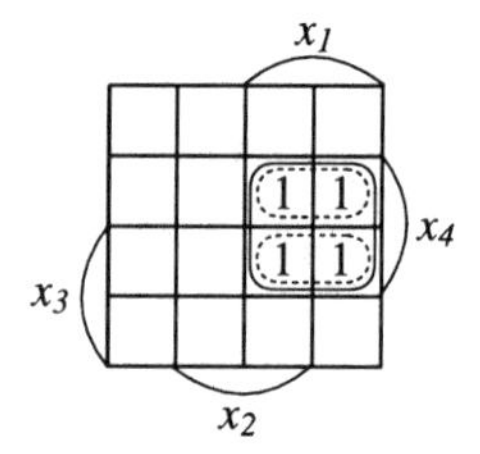

Figure 4.9 Merge of two loops.

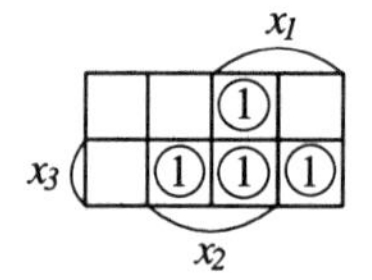

Figure 4.10 Karnaugh map for F_1.

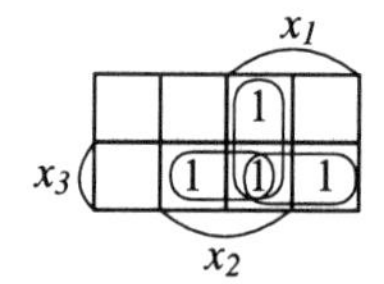

Figure 4.11 Karnaugh map for F_2.

these two vectors is two. Note that cells m_0 and m_4 are adjacent. This is clear because the Hamming distance between two vectors (0,0,0) and (1,0,0) is one. Also, it can be seen that in the cube in Fig. 4.4(c), these vertices are adjacent. Similarly, the cells m_1 and m_5 are adjacent. The Karnaugh map in Fig. 4.4(a) is obtained from Fig. 4.4(c) as follows: First, cut the edges (m_0, m_4) and (m_1, m_5) to make the cube into the plane. Then, replace each vertex with a cell. A cell containing a 1 is a **1-cell**, and a cell containing 0 is a **0-cell**. Fig. 4.5 shows a Karnaugh map with four variables. The method to attach the labels for variables may be different. However, each of the 16 different cells must have a unique label. In a Karnaugh map with four variables, two cells that are in opposite ends of the same column (row) are adjacent.

Fig. 4.6 shows a Karnaugh map with five variables, and Fig. 4.7 shows a Karnaugh map with six variables. When $n \geq 5$, a computer program is more convenient to the simplification of SOPs.

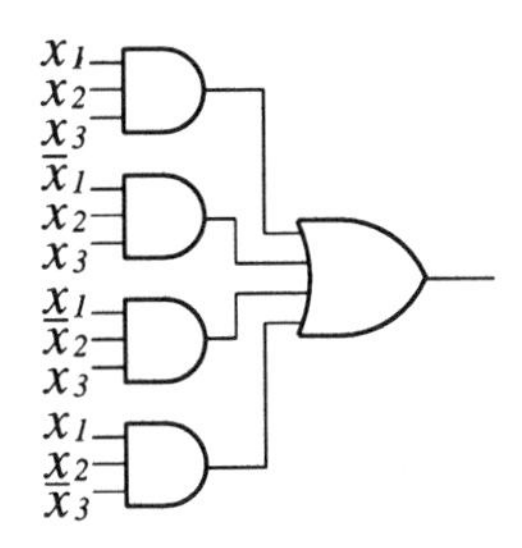

Figure 4.12 AND-OR two-level network for F_1.

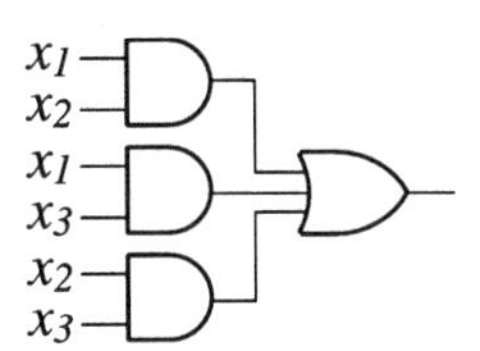

Figure 4.13 AND-OR two-level network for F_2.

Loops on Karnaugh Map

Consider two 1-cells which are adjacent in a row (or a column). In two minterms that correspond to these two 1-cells, there is a variable y which is contained as an un-complemented literal in one minterm, and as a complemented literal in the other minterm. For example, in Fig. 4.8, two minterms $\bar{x}_1x_2\bar{x}_3x_4$ and $x_1x_2\bar{x}_3x_4$, shown by dotted line loops, are adjacent. These two minterms are different only in variable x_1, and they can be merged as follows:

$$\bar{x}_1x_2\bar{x}_3x_4 \vee x_1x_2\bar{x}_3x_4 = x_2\bar{x}_3x_4(x_1 \vee \bar{x}_1) = x_2\bar{x}_3x_4.$$

In the Karnaugh map of Fig. 4.8, this merge operation corresponds to covering two 1-cells by the solid loop. The solid loop represents the products that are obtained by deleting the literal from the minterms of the 1-cells. Next, consider the map in Fig. 4.9. This map has two loops written in dotted lines, and they represent products $x_1\bar{x}_3x_4$ and $x_1x_3x_4$. Since these two loops written in dotted lines are adjacent, they can be merged to make the solid loop. The solid loop represent the product $x_1\bar{x}_3x_4 \vee x_1x_3x_4 = x_1x_4$.
In general, if two loops for 2^i 1-cells represent $P\bar{y}$ and Py, then we can merge these loops into a loop with 2^{i+1} 1-cells that represents the product P.

Example 4.3 Fig. 4.10 and Fig. 4.11, respectively, show the Karnaugh maps for two SOPs which appeared in Example 4.1:

$$F_1 : x_1x_2x_3 \vee \bar{x}_1x_2x_3 \vee x_1\bar{x}_2x_3 \vee x_1x_2\bar{x}_3, \text{ and}$$
$$F_2 : x_1x_2 \vee x_1x_3 \vee x_2x_3.$$

Also, Fig. 4.12 and Fig. 4.13 show the AND-OR two-level networks for F_1 and F_2, respectively. ∎

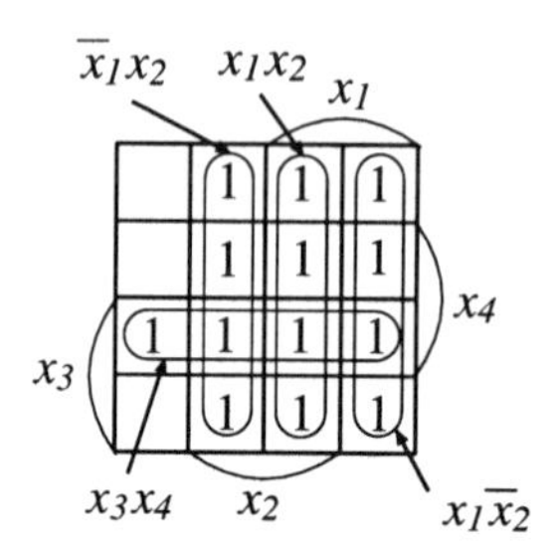

Figure 4.14 Karnaugh map for F_1.

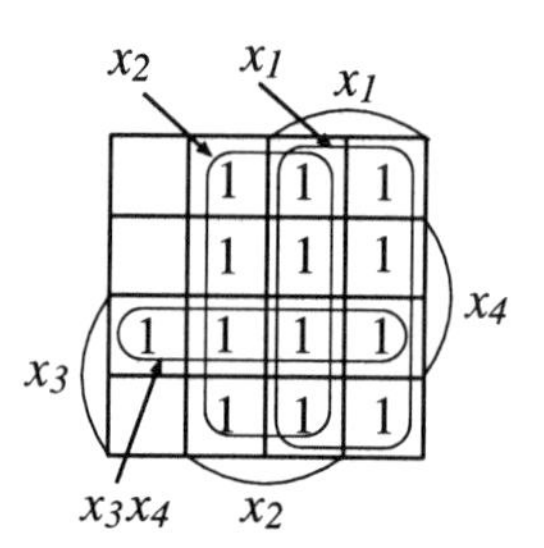

Figure 4.15 Karnaugh map for F_2.

Example 4.4 The SOP that appeared in Section 4.1,

$$F_1 : x_1\bar{x}_2 \vee \bar{x}_1x_2 \vee x_1x_2 \vee x_3x_4$$

is simplified as follows:

$$F_2 : x_1 \vee x_2 \vee x_3x_4.$$

Fig. 4.14 and Fig. 4.15 show the Karnaugh maps for F_1 and F_2, respectively. Fig. 4.1 and Fig. 4.2 show the AND-OR two-level networks for Fig. 4.14 and Fig. 4.15, respectively. Note that each loop in Fig. 4.14 corresponds to an AND gate in Fig. 4.1. Among the loops in Fig. 4.15, no AND gate is necessary for x_1 and x_2. For such loops, the variables are directly connected to the OR gate. ∎

As shown in the above examples, in a single-output AND-OR two-level logic network (SOP) that is represented by a Karnaugh map, we have the following properties:

- The fewer the loops, the less the AND gates (or the products).
- The larger the loops, the fewer the number of connections (or the number of literals).

Thus, the simplification of a single-output AND-OR two-level network (SOP) corresponds to "Cover all the 1-cells in the Karnaugh map by using as few loops and by using as large loops as possible."

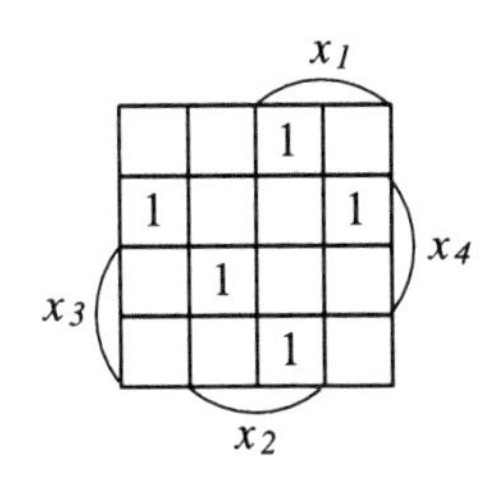

(a) Karnaugh map for f.

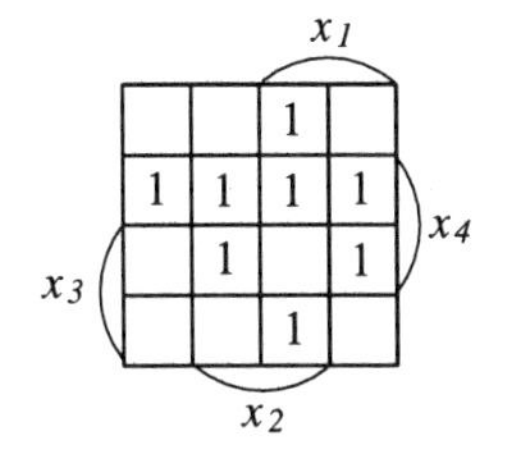

(b) Karnaugh map for g.

Figure 4.16 Relation of $f \leq g$.

4.4 PRIME IMPLICANT

As shown in the previous section, the simplification of an SOP corresponds to "Cover all the 1-cells in the Karnaugh map by using as few loops and by using as large loops as possible." In this section, we introduce the concept of the prime implicant, which is a formal definition of "as large loop as possible."

Definition 4.1 *In two logic functions f and g, if $g(\boldsymbol{x}) = 1$ for all $\boldsymbol{x}$ such that $f(\boldsymbol{x}) = 1$, then g* **contains** *f, denoted by $f \leq g$.*

Example 4.5 Between two functions f and g in Fig. 4.16, the relation $f \leq g$ holds. ∎

Definition 4.2 *If a logic function f contains a product c, then c is an* **implicant** *of f. Furthermore, if c is a minterm, then c is a* **minterm** *of f.*

Definition 4.3 *A product P is a* **sub-product** *of Q, if all the literals in P also appear in Q.*

Example 4.6 Let $c_1 = x_1x_2$ and $c_2 = x_1$. Since all the literals (x_1) in the product c_2 also appear in the product c_1, c_2 is a sub-product of c_1. ∎

The sub-product is a concept of the set of literals, which is converse relation of the inclusion relation (Fig. 4.16) in the Karnaugh map.

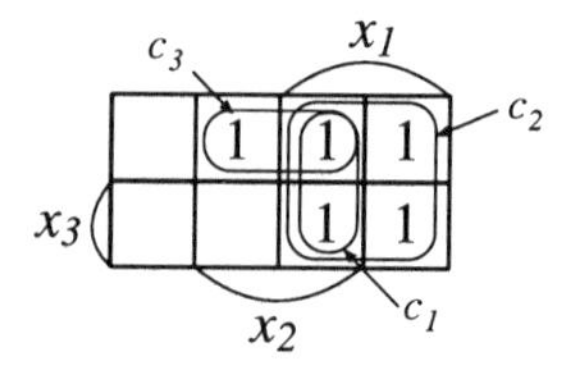

Figure 4.17 Implicant and prime implicant.

Definition 4.4 *Let P be an implicant of a logic function f. If no other implicant Q of f is a sub-product of P, then P is a* **prime implicant** *(PI) of f.*

Example 4.7 Consider the function f in Fig. 4.17. c_1 is an implicant of f. c_2, another implicant of f, is a sub-product of c_1. Thus, c_1 is not a prime implicant. The sub-products of $c_3 = x_2\bar{x}_3$ are c_3, x_2, $\bar{x}_3$ and the constant 1. However, except c_3, none of them is an implicant of f. Thus, c_3 is a prime implicant. Similarly, c_2 is also a prime implicant. ∎

Generation of Prime Implicants

Among the loops for 2^i 1-cells, the loop that corresponds to a prime implicant of f is a **prime implicant loop**. When a cell contains *don't care d*, choose the value of d so that the loop becomes as large as possible.

Example 4.8 All the prime implicants for the function in Fig. 4.18(a), are shown in Figs. 4.18(b)–4.18(d). Note that the loop that covers four cells in the corners in Fig. 4.18(b), and the loop that covers four cells in Fig. 4.18(d) are often overlooked by beginners. ∎

4.5 MINIMUM SOP

A minimum sum-of-products expression (MSOP) corresponds to an AND-OR two-level logic network with the minimum gates. Thus, to obtain an MSOP for a given logic function is very important in logic design. To derive an MSOP for a function is often refereed as a *minimization of the logic function* [212, 243].

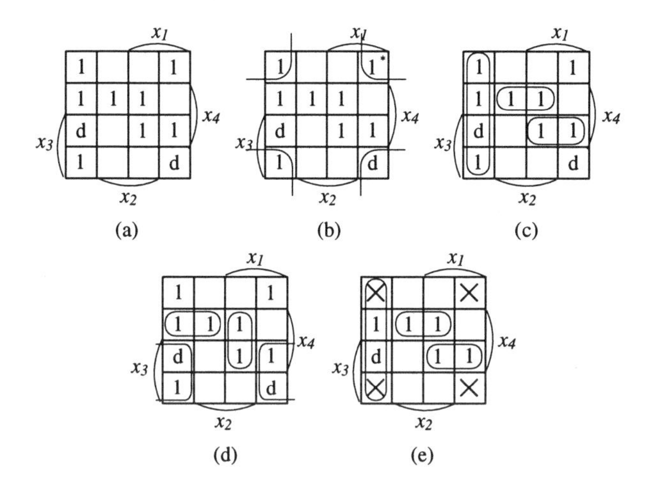

Figure 4.18 Simplification by Karnaugh maps.

However, for a completely specified logic function, functions represented by equivalent SOPs are the same.

Definition 4.5 *Consider an SOP that consists of prime implicants. If any SOP that is obtained by removing any of the prime implicants does not represent the original function f, then the original SOP is an* **irredundant sum-of-products expression** *(***ISOP***) of f.*

Definition 4.6 *Among the ISOPs for f, one with the minimum number of products is a* **minimum sum-of-products expression** *(***MSOP***). Among the MSOPs for f, one with the minimum number of literals is a* **exact minimum sum-of-products expression** *(***exact MSOP***) of f.*

Definition 4.7 *Let m be a minterm of a logic function f. If only one prime implicant of f covers m, then m is a* **distinguished minterm***. Among all the prime implicants of f, if there is only one that covers m, then it is an* **essential prime implicant***.*

When f has a distinguished minterm m, then the MSOP for f has the essential prime implicant that covers m. Some logic functions do not have distinguished minterms.

Example 4.9 (Problem) In the Karnaugh map in Fig. 4.18, obtain the distinguished minterms and the essential prime implicants.
(Solution) $x_1\bar{x}_2\bar{x}_3\bar{x}_4$ is a distinguished minterm, since only one prime implicant covers it. In Fig. 4.18(b), this cell is marked with $*$. Thus, $\bar{x}_2\bar{x}_4$ is the essential prime implicant. ∎

4.6 SIMPLIFICATION OF SOPS WITH KARNAUGH MAP

For the functions with at most four variables, MSOPs are easily obtained on the Karnaugh maps as follows:

1. Obtain all the prime implicant loops in the Karnaugh map. If a cell contains *don't care d*, set the value of d so that the loops become as large as possible, and write the prime implicant loops. Loops that consists of only d need not be written.
2. Among the 1-cells, if a 1-cell is covered by only one prime implicant loop, then mark the 1-cell with $*$. A 1-cell marked with $*$ corresponds to a distinguished minterm, and the prime implicant loop that covers the 1-cell is an **essential prime implicant loop**. The essential prime implicant must be included in the final solution.
3. Cover the uncovered 1-cells by using minimum number of prime implicant loops.
4. If there are more than one combination in the above step, choose the combination with the largest loops.

Example 4.10 (Problem) Obtain an MSOP for the function shown in Fig. 4.18(a).

(Solution)
1. There are 7 prime implicants (see Figs. (b)–(d)).
2. Only one prime implicant covers the minterm $x_1\bar{x}_2\bar{x}_3\bar{x}_4$. So, it is a distinguished minterm. In Fig. 4.18(b), the cell is marked with $*$. Encircle the 1-cells by the essential prime implicant loop $\bar{x}_2\bar{x}_4$. Write a $\times$ in the 1-cells that are covered by the essential prime implicant, as shown in Fig. 4.18(e).
3. Two loops cover the minterm $x_1\bar{x}_2x_3x_4$: $\bar{x}_2x_3$ and $x_1x_3x_4$. Since $x_1x_3x_4$ covers more uncovered minterms, we choose $x_1x_3x_4$. Similarly, two loops cover $x_1x_2\bar{x}_3x_4$: $x_1x_2x_4$ and $x_2\bar{x}_3x_4$. Since $x_2\bar{x}_3x_4$ covers more uncovered minterms, we choose $x_2\bar{x}_3x_4$ as shown in Fig. 4.18(e).

4. Two loops cover $\bar{x}_1\bar{x}_2\bar{x}_3x_4$: $\bar{x}_1\bar{x}_2$ and $\bar{x}_1\bar{x}_3x_4$. Both loops cover only one uncovered minterm. However, since $\bar{x}_1\bar{x}_2$ is the larger loop, we choose $\bar{x}_1\bar{x}_2$, as shown in Fig. 4.18(e).
5. All the minterms are covered by the chosen loops. The obtained MSOP is $\bar{x}_2\bar{x}_4 \vee x_1x_3x_4 \vee x_2\bar{x}_3x_4 \vee \bar{x}_1\bar{x}_2$. ∎

4.7 QUINE-MCCLUSKEY METHOD

SOPs with three to five variables are easily simplified by using Karnaugh maps. However, when the number of variables is six or more, the maps are too complicated to use. The **Quine-McCluskey method** (**QM method**) for minimizing SOPs is suitable for computer manipulation. It consists of two parts:

1) Generate all the prime implicants, and make the prime implicant table.
2) Obtain the minimum cover of the prime implicant table.

However, when the number of input variables is greater than 10, the necessary computation time and memory storage are excessive. So, we have to resort to the heuristic methods shown in Chapter 10.

Definition 4.8 *$x_i^{c_i}$ denotes x_i when $c_i = 1$, and $\bar{x}_i$ when $c_i = 0$. For a minterm $x_1^{c_1}x_2^{c_2}\cdots x_n^{c_n}$, the decimal representation of the binary number $c_1c_2\cdots c_n$ is a* **decimal representation of a minterm**.

Example 4.11 When $n = 3$, the decimal representations of $\bar{x}_1\bar{x}_2\bar{x}_3$, $\bar{x}_1\bar{x}_2x_3$, and $\bar{x}_1x_2\bar{x}_3$ are 0, 1, and 2, respectively. ∎

Definition 4.9 *$x_i^{c_i}$ denotes x_i when $c_i = 1$, $\bar{x}_i$ when $c_i = 0$, and 1 when $c_i = -$. For a product $x_1^{c_1}x_2^{c_2}\cdots x_n^{c_n}$, $c_1c_2\cdots c_n$ is a* **cubical representation of a product**, *or a* **cube**.

Example 4.12 When $n = 4$, the cubical representation of $x_1\bar{x}_2\bar{x}_3x_4$ is 1001, the cubical representation of $\bar{x}_1x_2$ is $01--$, and the cubical representation of the constant 1 is $----$. ∎

Definition 4.10 (**Merge** *of cubes*)
Let two cubes be

$$\begin{array}{l} \alpha = c_1 c_2 \cdots\cdots \; 1 \; \cdots\cdots c_n, \text{ and} \\ \beta = c_1 c_2 \cdots\cdots \; 0 \; \cdots\cdots c_n. \\ \qquad\qquad\qquad \uparrow \\ \qquad\qquad\qquad i \end{array}$$

Note that α and β are different only in the ith variable. In this case, the cube

$$\begin{array}{l} \gamma = c_1 c_2 \cdots\cdots \; - \; \cdots\cdots c_n \\ \qquad\qquad\qquad \uparrow \\ \qquad\qquad\qquad i \end{array}$$

is obtained by merging α and β, denoted by $\gamma = \alpha \vee \beta$. Note that γ contains minterms of α and β.

Example 4.13 Let two cubes be $\alpha = 00-1-01$ and $\beta = 00-0-01$. Then, $\alpha \vee \beta = 00---01$. ∎

Generation of All the Prime Implicants

Procedure 4.1 *(Generation of all the Prime Implicants)*

1. *Obtain the minterm expansion of the given function. Make the list of decimal representations and cubical representation for the minterms.*
2. *Make a list of cubes which are sorted in the ascending order of the number of 1's in the cubes.*
3. *By the number of 1's, partition the set of cubes into the groups. Make the 1st list. Let $I \leftarrow 1$.*
4. *Compare the cubes between the 1st group and 2nd group of the Ith list. If the pair is different in only one variable, put the check marks to these cubes in the list. Next, in the $(I+1)$th list, write the decimal representations of the minterms that are covered by the pair of cubes. After that, write the cube that covers the pair of cubes. However, if the cube is already in the list and just the order of the minterms is different, then the cube need not be written.*

   If we finish all the cubes in the 1st group, write the horizontal line under the 1st group of the $(I+1)$th list.
5. *Compare the 2nd group and 3rd group of the Ith list. If there is a pair of cubes that can be merged, then merge them and put it in the $(I+1)$th list. Repeat the above procedure until the last two groups are compared.*
6. *If the $(I+1)$th list has more than one group, then let $I \leftarrow I+1$ and go to step 4.*
7. *In the final list, the cubes without check marks are prime implicants.*

Minterm	Cube $x_1\, x_2\, x_3\, x_4$	
0	0 0 0 0	√
1	0 0 0 1	√
2	0 0 1 0	√
8	1 0 0 0	√
3	0 0 1 1	√
5	0 1 0 1	√
10	1 0 1 0	√
11	1 0 1 1	√
13	1 1 0 1	√
15	1 1 1 1	√

(a) First list.

Minterm	Cube $x_1\, x_2\, x_3\, x_4$	
0,1	0 0 0 –	√
0,2	0 0 – 0	√
0,8	– 0 0 0	√
1,3	0 0 – 1	√
1,5	0 – 0 1	*D*
2,3	0 0 1 –	√
2,10	– 0 1 0	√
8,10	1 0 – 0	√
3,11	– 0 1 1	√
5,13	– 1 0 1	*E*
10,11	1 0 1 –	√
11,15	1 – 1 1	*F*
13,15	1 1 – 1	*G*

(b) Second list.

Minterm	Cube $x_1\, x_2\, x_3\, x_4$	
0,1,2,3	0 0 – –	*A*
0,2,8,10	– 0 – 0	*C*
2,3,10,11	– 0 1 –	*B*

(c) Third list.

Figure 4.19 Generation of prime implicants.

Example 4.14 (Problem) Using Procedure 4.1, derive all the prime implicants for the function in Fig. 4.18.
(Solution) First, make the cubes for the minterms of the given function. Classify the cubes according to the number of 1's, and we have the 1st list shown in Fig. 4.19(a). Next, compare the cubes in the adjacent groups to make the 2nd list. In the 2nd list, each cube has one – (Fig. 4.19(b)). From the 2nd list, derive the 3rd list shown in Fig. 4.19(c). The cubes without check marks denote prime implicants. There are 7 prime implicants: $\bar{x}_1\bar{x}_2$, $\bar{x}_2x_3$, $\bar{x}_2\bar{x}_4$, $\bar{x}_1\bar{x}_3x_4$, $x_2\bar{x}_3x_4$, $x_1x_3x_4$, $x_1x_2x_4$. ∎

Procedure 4.2 *(Construction of the Prime Implicant Table)*

1. *Classify the prime implicants according to the numbers of literals.*
2. *List the prime implicants along the vertical axis. The group with the minimum number of literals is in the top, followed by groups with more literals. Draw a horizontal line between two groups. List the decimal representations of minterms along the horizontal axis. The minterms for don't cares need not be listed.*

Table 4.2 Example of prime implicant table.

	0	1	2	5	8	11	13	15
$A = \bar{x}_1\bar{x}_2$	×	×	×					
$B = \bar{x}_2 x_3$			×			×		
$C = \bar{x}_2\bar{x}_4$	×		×		×			
$D = \bar{x}_1\bar{x}_3 x_4$		×		×				
$E = x_2\bar{x}_3 x_4$				×			×	
$F = x_1 x_3 x_4$						×		×
$G = x_1 x_2 x_4$							×	×

3. *If a prime implicant that corresponds to the ith row contains the minterm that corresponds to the jth column, then write a $\times$ mark in the (i, j) element.*

Example 4.15 (Problem) By using Procedure 4.2, construct the prime implicant table for the function in Fig. 4.18.
(Solution) See Table 4.2. ∎

Minimum Covering Problem

Definition 4.11 *In the covering table, if there is a $\times$ mark in the (i, j) element, then the row i* **covers** *the column j. The* **minimum covering problem** *is to find a set of minimum rows that covers all the columns. In some cases, each row has a cost, and the solution where the sum of the costs is the minimum must be found.*

The set of rows that satisfies the minimum covering corresponds to an MSOP. Among MSOPs, the set of rows that contains as many rows in the upper groups as possible corresponds to an exact MSOP. A minimum covering problem is useful not only for the simplification of SOPs, but also other optimization problems.

Example 4.16 A department of a university has 11 courses: m_1–m_{11}.
Professor A can teach, m_1, m_2, m_4, and m_7.
Professor B can teach, m_1, m_3, m_4, and m_8.
Professor C can teach, m_2, m_5, m_7, and m_{10}.
Professor D can teach, m_4, m_7, m_8, and m_{11}.

Table 4.3 Covering table.

	m_1	m_2	m_3	m_4	m_5	m_6	m_7	m_8	m_9	m_{10}	m_{11}
A	×	×		×			×				
B	×		×	×				×			
C		×			×		×			×	
D				×			×	×			×
E			×			×					
F					×				×		
G						×	×				

Professor E can teach, m_3, and m_6.
Professor F can teach, m_5, and m_9.
Professor G can teach, m_6, and m_7.
In this case, obtain the minimum number of professors to teach all the courses m_1–m_{11}.
This problem is formulated as a minimum covering problem. The **covering table** is shown in Table 4.3. ∎

Procedure 4.3 *(Reduction of Covering Table)*

1. *Detection of essential rows:*
   The column with only one × mark is a **distinguished column**. *The distinguished column denotes a distinguished minterm. The row that has a × mark in the distinguished column is an* **essential row**. *The essential row denotes an essential prime implicant. Mark the distinguished column with ○. Mark the essential row with ∗, and include in the final solution. Delete essential rows, also delete the columns covered by essential rows.*
2. *Elimination of dominated rows:*
   In the (reduced) covering table, if the row P has ×'s in all the columns in which the row Q has ×'s, then the row P **dominates** *the row Q. If the row P dominates the row Q, and the the number of literals of the Q is not less than that of P, then delete Q.*
3. *Elimination of dominating columns:*
   In a (reduced) covering table, if the column i has ×'s in all the row in which the column j has ×'s, then the column i dominates the column j. If the column i dominates the column j, then delete the column i.
4. *Detection of the secondary essential rows:*
   In a reduced covering table, the column with only one × mark is the **secondary distinguished column**. *The row that contains such × mark is*

		0	1	2	5	8	11	13	15
		√		√		○			
	A	×	×	×					
	B			×			×		
*	C	×		×		×			
	D		×		×				
	E				×			×	
	F						×		×
	G							×	×

(a) Selection of essential row.

		1	5	11	13	15
△	A	×				
△	B			×		
	D	×	×			
	E		×		×	
	F			×		×
	G				×	×

(b) Deletion of dominated row.

	1	5	11	13	15
D	×	×			
E		×		×	
F			×		×
G				×	×

(c) Deletion of dominated column. (When rows A and B are deleted)

		1	11	13
		○	○	
*	D	×		
	E			×
*	F		×	
	G			×

(d) Detection of secondary essential rows.

Figure 4.20 Example of minimum cover.

the **secondary essential row**. *Mark the secondary distinguished column with* ○. *Mark the secondary essential row with* $*$, *and include it in the final solution. Delete the secondary essential rows. Also, delete the columns that are covered by the secondary essential row.*

5. *Repeat the steps 2 through 4 while the reduction is possible.*

If some columns remain after the application of the above procedure, then the reduced covering table contains at least two × *marks in each column. Such table is a* **cyclic table**. *To obtain the minimum cover of the cyclic table, we use Procedure 4.5.*

Example 4.17 (Problem) Obtain the minimum cover for the function in Fig. 4.18 by using Procedure 4.3.

(Solution) See Fig. 4.20.

1. In Fig. 4.20(a), mark the distinguished column with a ○, and mark the essential row with $*$.
2. Delete the essential row. Delete the columns (marked with √) that are covered by the essential row, and we have Fig. 4.20(b). In this case, the

row D dominates the row A, and the row F dominates the row B. Unfortunately, the numbers of literals in A and B are less than those of D and F. So, we cannot delete the rows A and B. However, this produce the cyclic table, and cannot continue the procedure. So, we will delete rows A and B.

3. By deleting rows A and B, we have Fig. 4.20(c). In this case, the column 5 dominates column 1, and column 15 dominates column 11.
4. By deleting columns 5 and 15, we have Fig. 4.20(d). Mark the secondary distinguished column with ○, and mark the secondary essential row with *.
5. By deleting columns 1 and 11, we have the table with only the column 13. To cover this column, we need either E or G. Thus, the minimum cover is either $\{C, D, F, E\}$ or $\{C, D, F, G\}$. Therefore, the following are the MSOPs:

$$f = \bar{x}_2\bar{x}_4 \vee \bar{x}_1\bar{x}_3x_4 \vee x_1x_3x_4 \vee p,$$

where $p = x_2\bar{x}_3x_4$ or $x_1x_2x_4$. However, as will be shown, these are not exact MSOPs, but MSOPs. ∎

Solution Methods for Cyclic Tables

The problem to cover all the columns by using the minimum number of rows is solved by the following recursive procedure:

Procedure 4.4 *(A naive algorithm to obtain a minimum cover)*

(a) Assume that a row is contained in the final solution, reduce the covering table, and obtain a minimum cover.
(b) Assume that a row is not contained in the final solution, reduce the covering table, and obtain a minimum cover.
(c) Compare (a) with (b), and select one with lower cost as the solution.

However, this method essentially searches all the combinations of rows, and it is quite inefficient. To improve the efficiency, we use the following techniques:

1) Choose the row so that near optimum solutions can be found at the preliminary stage of the computation.
2) Avoid searching the useless combinations.

1) is the **branching operation**. The strategies are:

	1	5	11	13	15
A	×				
B			×		
D	×	×			
E		×		×	
F			×		×
G				×	×

Figure 4.21 Example of cyclic table.

- Select the row that covers as many uncovered columns.
- Select the row that covers as many important columns as possible.

2) is the **bounding operation.** If the improving of the existing solution is impossible, then stop the search. For example, use the techniques in Exercises 4.4–4.6 to obtain lower bounds on the number of products.

Procedure 4.5 (**Branch and bound method** *for cyclic table)*
If a cyclic table appears in Procedure 4.3, apply this procedure. Do the following operations at the same time repeatedly.

1. **Branch**
   (a) Choose one row from the covering table. Let it be A.
   (b) Assume that A is included in the final solution, and apply Procedure 4.3 to obtain a minimum cover.
   (c) Assume that A is not included in the final solution, and apply Procedure 4.3 to obtain a minimum cover.
   (d) Select one with smaller cost between (b) and (c).

2. **Bound**
   (a) If the cost of the solution is equal to the lower bound on the cost of minimum cover, then stop the search.
   (b) If the solution of the current problems cannot be better than the best solutions ever found, then stop the search.

Example 4.18 In Example 4.17, for the sake of explanation, we deleted rows A and B. However, since the number of literals of A and B, are less than that of D and F, we cannot delete them when we need an exact MSOP. Therefore, in order to find an exact MSOP, we have to solve the cyclic table shown in Fig. 4.21.

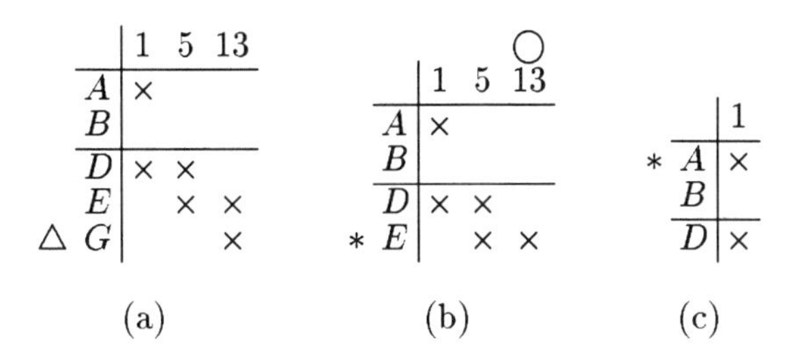

Figure 4.22 When the solution contains F.

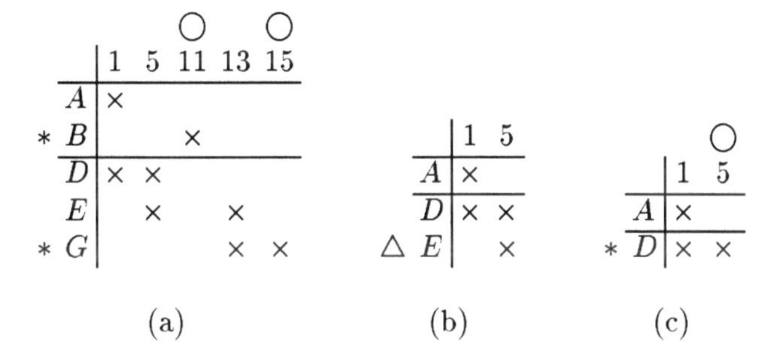

Figure 4.23 When the solution does not contain F.

1. First, assume that F in the minimum solution, and solve the cyclic table. Then, columns 11 and 15 are deleted. F is deleted (Fig. 4.22(a)).
2. E dominates G. So, delete G, which is marked with $\triangle$ (Fig. 4.22(b)).
3. Then, column 13 is a distinguished column. And the row E becomes an essential row. Delete the columns 5 and 13. Delete E (Fig. 4.22(c)).
4. Then, column 1 is a distinguished column. A becomes an essential row.
5. All the columns are deleted. $\{A, C, E, F\}$ is one of the solutions.
6. Next, assume that F is not included in the minimum solution, and solve the cyclic table.
7. Delete F, and columns 11 and 15 becomes distinguished columns. B and G becomes essential rows (Fig. 4.23(a)). Delete columns 11, 13, and 15. Delete B and G (Fig. 4.23(b)).
8. Since D dominates E, which is marked with $\triangle$, delete E. (Fig. 4.23(c)).
9. The column 5 becomes a distinguished column, and D becomes an essential row.
10. All the columns are deleted. $\{B, C, D, G\}$ is one of the solutions.
11. Note that the number of products in both solutions $\{A, C, E, F\}$ and $\{B, C, D, G\}$ are four. Also the number of literals of both solutions are

Table 4.4 Cyclic table.

	m_a	m_b	m_c	m_d	m_e
P_1	×	×		×	
P_2		×	×		
P_3	×		×		
P_4				×	×
P_5			×		×
P_6		×			×

10. Thus, the one of the exact MSOP is

$$f = \bar{x}_1\bar{x}_2 \vee \bar{x}_2\bar{x}_4 \vee x_2\bar{x}_3x_4 \vee x_1x_3x_4.$$

Note that this exact MSOP requires one fewer literal than the MSOP found in Example 4.17. ∎

Algebraic Method for Cyclic Tables

This part shows another method to solve cyclic tables, the **Petrick's method.** In the solution of a covering problem, each column of the covering table must be covered at least once by the row(s) that has × in the column. For each row P_i, let g_i be a Boolean variable showing that the row P_i is in the covering solution. The condition that the column m_j is covered by at least one row is represented as: $\bigvee_{i\in I_j} g_i = 1$, where I_j is a set of indexes of rows that covers column m_j. For example, in the cyclic table in Table 4.4, the column m_a is covered when $g_1 \vee g_3 = 1$. Similarly, for $m_b, \ldots, m_e$, we have the following conditions:

$$\begin{aligned} m_b &: g_1 \vee g_2 \vee g_6 = 1, \\ m_c &: g_2 \vee g_3 \vee g_5 = 1, \\ m_d &: g_1 \vee g_4 = 1, \text{ and} \\ m_e &: g_4 \vee g_5 \vee g_6 = 1. \end{aligned}$$

These conditions are satisfied simultaneously iff

$$(g_1 \vee g_3)(g_1 \vee g_2 \vee g_6)(g_2 \vee g_3 \vee g_5)(g_1 \vee g_4)(g_4 \vee g_5 \vee g_6) = 1.$$

By expanding the equation by using distributive law, and by deleting the products that are absorbed by other products, we have

$$g_1g_5 \vee g_1g_2g_4 \vee g_1g_2g_6 \vee g_2g_3g_4 \vee g_3g_4g_6 \vee g_1g_3g_4 \vee g_1g_3g_6 = 1.$$

Among all the products, the one with the minimum number of literals is $g_1 g_5$. That is, if $g_1 = g_5 = 1$, then the above expression becomes 1. This means P_1 and P_5 covers all the columns, and $\{P_1, P_5\}$ is the minimum cover.

In general, the condition to cover all the m columns of the covering table is shown by the **Petrick's equation**:

$$\bigwedge_{j=1}^{m} \left(\bigvee_{i \in I_j} g_i \right) = 1.$$

If we expand the left-hand side of the equation into an SOP, then each product represents the set of rows that covers all the columns. If we assume that all the cost of the rows are the same, then the product with the fewest literals corresponds to the minimum cover.

4.8 MSOPS AND THEIR APPLICATIONS

A minimum sum-of-products expression (MSOP) corresponds to the simplest AND-OR two-level logic network. In this section, we show methods to design various networks by using MSOPs.

Definition 4.12 *A (double-rail input)* **AND-OR two-level network** *satisfies the following conditions:*

1. *The level of the network is at most two, and AND gates are used in the first level, and the OR gate is used in the second level (output).*
2. *For each input, both uncomplemented variable x_i and complemented variable $\bar{x}_i$ are available.*
3. *There is no fan-in restriction.*

In many cases, "double-rail input" are omitted, and simply refereed as an AND-OR two-level network.

Definition 4.13 *The* **minimum AND-OR two-level network** *is one with the minimum number of gates. The* **exact minimum AND-OR two-level network** *is a minimum AND-OR two-level network with the minimum number of connections.*

Theorem 4.1 *The (exact) minimum AND-OR two-level network corresponds to the (exact) MSOP.*

By modifying the design method of AND-OR minimum networks, we can design the following minimum networks:

Procedure 4.6 *(Realization of an OR-AND two-level network)*

1. *Obtain the minimum AND-OR network for* $\bar{f}$.
2. *Replace each AND gate by an OR gate, and each OR gate by an AND gate.*
3. *Interchange uncomplemented variables with complemented variables.*

Procedure 4.7 *(Realization of an NAND two-level network)*

1. *Obtain the minimum AND-OR network for* f.
2. *Replace all the gates with NAND gates.*
3. *If the output gate is connected to input variables or their complement, then interchange the uncomplemented variables with complemented variables.*

Procedure 4.8 *(Realization of NOR two-level networks)*

1. *Obtain the minimum AND-OR network for* $\bar{f}$.
2. *Replace all the gates with NOR gates.*
3. *Replace the uncomplemented variables that are connected to the input gates with complemented ones, and the complemented variables that are connected to the input gates with uncomplemented ones.*

Example 4.19 (Problem) Let $f = x_1x_2 \vee x_2x_3 \vee x_1x_3 \vee x_4$. Realize the minimum network for f by an AND-OR two-level network, an OR-AND two-level network, a NAND two-level network, and a NOR two-level network.
(Solution) See Fig. 4.24. ∎

4.9 SIMPLIFICATION OF MULTI-OUTPUT NETWORKS

Consider the design of multi-output networks. The network obtained by separate simplifications of each output often requires more gates than minimum.

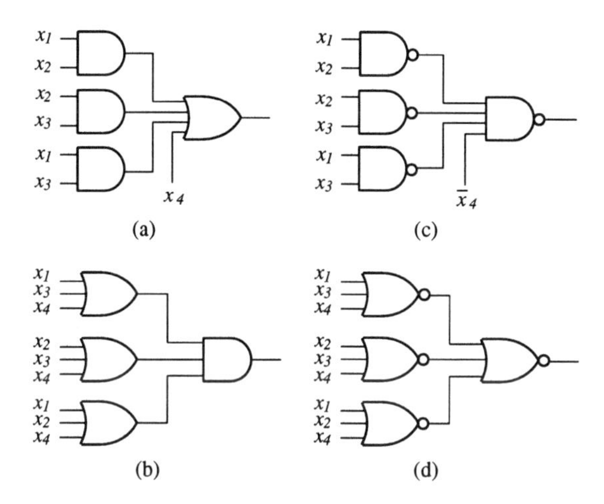

Figure 4.24 Minimum networks for $f = x_1x_2 \vee x_2x_3 \vee x_1x_3 \vee x_4$.

Table 4.5 Multi-output function.

x	y	z	f_1	f_2
0	0	0	0	0
0	0	1	0	0
0	1	0	1	0
0	1	1	1	0
1	0	0	0	1
1	0	1	0	1
1	1	0	0	0
1	1	1	1	1

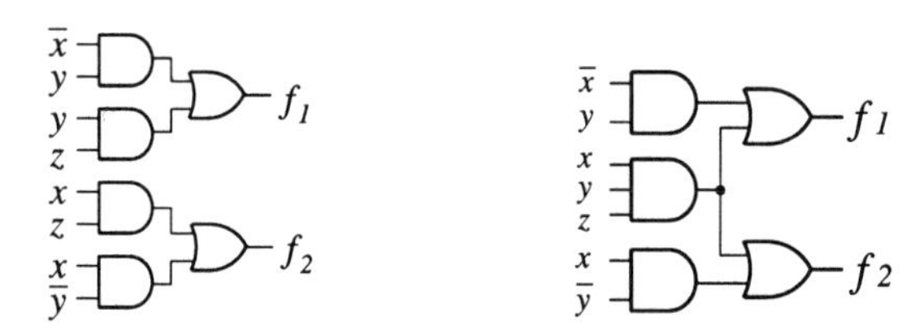

Figure 4.25 Separate simplification.

Figure 4.26 Simultaneous simplification.

In order to make the number of gates minimum, we have to consider all the outputs simultaneously.

Example 4.20 Consider the design for three-input two-output function shown in Table 4.5. If f_1 and f_2 are simplified separately, then we have the network in Fig. 4.25, that has four AND gates and two OR gates. However, if two outputs are considered simultaneously, then we have the network shown in Fig. 4.26, which has three AND gates and two OR gates. ∎

The simplification method for multi-output networks will be considered in Chapter 10.

Bibliographical Notes

Standard textbooks that explain on SOP minimization are [171, 212, 245, 277]. Many papers have been published for each topic: Classical Prime implicant (PI) generation [243, 263, 320, 405]; other PI generation [87, 386]; logic functions with many PIs [112]; average numbers of PIs for the functions with the given number of true minterms [251]; functions with maximal number of PIs [183]; a class of functions whose SOPs are difficult to minimize [305, 372]; minimum cover [138, 231]; cyclic table [138, 313]; lower bounds on the number of PIs in MSOPs [185, 250]; and classical methods to simplify SOPs for multi-output functions [13, 244].

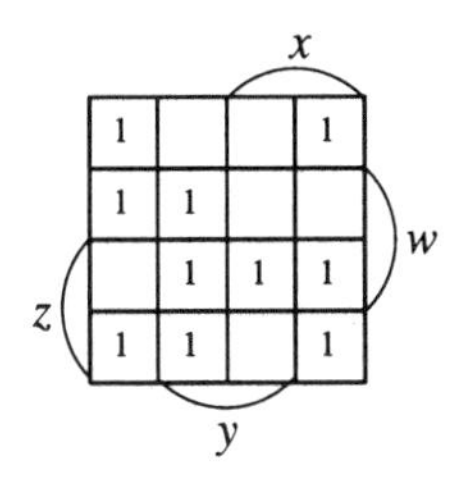

Figure 4.27

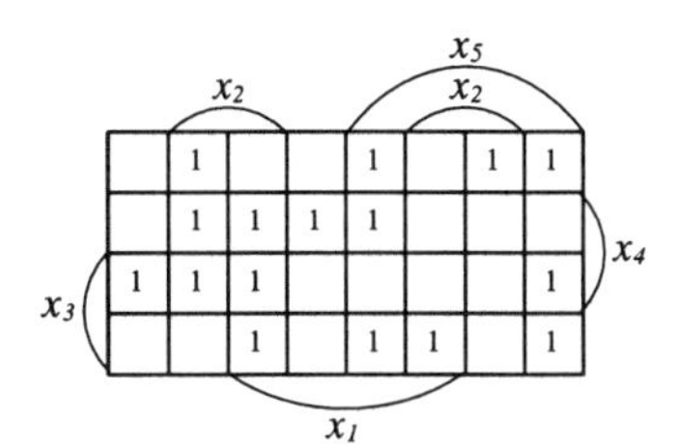

Figure 4.28

Exercises

4.1 Prove or show a counterexample for each of the following statements.

a) If the MSOP for the function f is unique, then all the prime implicants for f are essential prime implicants.

b) If the MSOP for the function f is unique, then the minimum POS of f is unique.

c) Let p be an arbitrary prime implicant of f. There is an MSOP that has p.

d) The MSOP for a completely specified function f is unique.

e) The number of products in an MSOP for a completely specified function f is equal to the number of sums in an MPOS for f.

4.2 Obtain the MSOP for the function in Fig. 4.27.

4.3 Show that $xy \vee yz \vee zx = xy \vee x\bar{y}z \vee \bar{x}yz$.

4.4 Show that an MSOP for $f = (x_1 \vee x_2 \vee x_3)(\bar{x}_1 \vee \bar{x}_2 \vee \bar{x}_3)$ has at least three products.

4.5 (M) Show that an MSOP for $f = (x_1 \vee x_2 \vee \cdots \vee x_n)(\bar{x}_1 \vee \bar{x}_2 \vee \cdots \vee \bar{x}_n)$ has at least n products.

4.6 Show that an MSOP for $f = \overline{xyz \vee xyw \vee yzw \vee xzw}$ has at least 6 products.

4.7 Simplify the following SOP, and obtain the minimum AND-OR two-level network:

$$xyz \vee xy\bar{z} \vee \bar{x}yz \vee \bar{x}\bar{y}z \vee \bar{x}\bar{y}\bar{z}.$$

4.8 Convert the following POS into an SOP, and obtain the minimum AND-OR two-level network:

$$f = (x_1 \vee x_2 \vee x_3 \vee x_4)(\bar{x}_1 \vee \bar{x}_2 \vee \bar{x}_3 \vee \bar{x}_4).$$

4.9 Design the minimum AND-OR two-level network and a minimum OR-AND two-level network, and compare the complexities for the networks:

$$f = xz \vee \bar{x}\bar{y}\bar{z}\bar{w} \vee yw.$$

4.10 In a BCD code (x_3, x_2, x_1, x_0), design a NOR two-level network that realizes z, where $z = 1$ iff the number 1's in the code is an odd number. Use *don't cares* to simplify the network.

4.11 Design a network by using ANDs, ORs, and inverters that compares two binary numbers $X = (x_2, x_1, x_0)$ and $Y = (y_2, y_1, y_0)$, and produces $Z = (z_2, z_1, z_0)$ where

$$\begin{array}{l} X > Y \text{ iff } z_0 = 1, \\ X = Y \text{ iff } z_1 = 1, \text{ and} \\ X < Y \text{ iff } z_2 = 1. \end{array}$$

4.12 Realize a multiplication network that produces the product $Z = (z_3, z_2, z_1, z_0)$ of two binary numbers $X = (x_1, x_0)$ and $Y = (y_1, y_0)$. Use an AND-OR two-level network.

4.13 (M) Let f be an Achilles' heel function (Section 3.8) of n variables. Show that the number of products in an MSOP for $\bar{f}$ is 3^m.

4.14 Obtain the minimum cover of Table 4.3.

4.15 In the Karnaugh map of Fig. 4.28,

a) Obtain all the prime implicants.
b) Obtain an MSOP.

4.16 To graduate from a university, a student must satisfy the following conditions on six courses x, y, z, u, v, and w:

a) Always finish x and y.
b) If z is not finished, then finish both x and u.
c) If u is not finished, then finish either y or v.
d) If u is finished, then w is unnecessary.

Represent the condition of graduation by an SOP with x, y, z, u, v, and w. Simplify the SOP.

4.17 Derive all the ISOPs for

$$f = (x_1 \vee x_2 \vee x_3 \vee x_4)(\bar{x}_1 \vee \bar{x}_2 \vee \bar{x}_3 \vee \bar{x}_4)$$

by using the Petrick's equation.

4.18 Prove the following relation:

$$x_1x_2x_3 \vee x_1x_2x_4 \vee x_1x_3x_4 \vee x_2x_3x_4 \leq (x_1 \vee x_3)(x_2 \vee x_3)(x_2 \vee x_4)$$

5

LOGIC FUNCTIONS WITH VARIOUS PROPERTIES

In this chapter, we present properties of self-dual functions, monotone functions, unate functions, linear functions, symmetric functions, threshold functions, and majority functions. If the given logic function has a special property, then the function is often realized by using fewer elements. Also, the analysis of the networks realizing such a function is relatively easy. In addition, we consider the universal set of logic functions, which can realize an arbitrary logic function. Finally, we will introduce equivalence classes of logic functions.

5.1 SELF-DUAL FUNCTION

Definition 5.1 *The* **dual** *of a function* $f(x_1, x_2, \ldots, x_n)$ *is* $\bar{f}(\bar{x}_1, \bar{x}_2, \ldots, \bar{x}_n)$, *denoted by* f^d. f^d *is obtained first by replacing each literal* x_i *with* $\bar{x}_i$ *and then by complementing the function.*

For example, the dual function of $(x \cdot y) \vee (z \cdot w)$ is $\overline{(\bar{x} \cdot \bar{y}) \vee (\bar{z} \cdot \bar{w})} = (x \vee y) \cdot (z \vee w)$. When a logical expression is represented by AND, OR and NOT operations, the dual function is obtained by first appropriately appending parenthesis, and then by interchanging the AND with OR operations. This is clear by the De Morgan's theorem.

Definition 5.2 *A* **self-dual function** *is the function such that* $f = f^d$.

Example 5.1 The dual function of $f = xy \vee yz \vee zx$ is $f^d = \overline{\bar{x}\bar{y} \vee \bar{y}\bar{z} \vee \bar{z}\bar{x}} = (x \vee y)(y \vee z)(z \vee x) = xy \vee yz \vee zx = f$. Thus, f is a self-dual function. ∎

Table 5.1 Truth table for a self-dual function.

xyz	$f(x,y,z)$	$f(\bar{x},\bar{y},\bar{z})$	$\bar{f}(\bar{x},\bar{y},\bar{z})$
000	f_0	f_7	$\bar{f}_7$
001	f_1	f_6	$\bar{f}_6$
010	f_2	f_5	$\bar{f}_5$
011	f_3	f_4	$\bar{f}_4$
100	f_4	f_3	$\bar{f}_3$
101	f_5	f_2	$\bar{f}_2$
110	f_6	f_1	$\bar{f}_1$
111	f_7	f_0	$\bar{f}_0$

To analyze more properties of self-dual functions, consider the truth table for a general three-variable function shown in Table 5.1. In this table, $f(x,y,z)$ denotes the original function, $f(\bar{x},\bar{y},\bar{z})$ denotes the function that is obtained by replacing all the variables with their complements in f, and $\bar{f}(\bar{x},\bar{y},\bar{z})$ is obtained by complementing the function $f(\bar{x},\bar{y},\bar{z})$, i.e., the dual function of f. The elements in $f(x,y,z)$ and $f(\bar{x},\bar{y},\bar{z})$ are symmetric with respect to the central horizontal line of the truth table. For example, f_0 which is the element in the 1st line of the column for $f(x,y,z)$, is in the lowest line in the column for $f(\bar{x},\bar{y},\bar{z})$.

In the case of self-dual function, $f(x,y,z)=\bar{f}(\bar{x},\bar{y},\bar{z})$, and so we have the relations $f_0 = \bar{f}_7$, $f_1 = \bar{f}_6$, $f_2 = \bar{f}_5$, and $f_3 = \bar{f}_4$. Thus, the values for f_0, f_1, f_2, and f_3 completely specify the function. From this, we know that there are $2^4 = 16$ self-dual functions of three variables. Also, the number of input combinations that make $f = 1$ is four, the half of the total input combinations. By generalizing this, we have the following:

Theorem 5.1 *There are $2^{2^{n-1}}$ different self-dual functions of n variables.*

Theorem 5.2 *Let f be a self-dual function of n variables, and let $|f|$ be the number of inputs $\boldsymbol{a}$ for which $f(\boldsymbol{a}) = 1$, then $|f| = 2^{n-1}$.*

Theorem 5.3 *A function which is obtained by assigning a self-dual function to a variable of a self-dual function is also self-dual.*

(Proof) Let $f(x_1, x_2, \ldots, x_n)$ and $g(y_1, y_2, \ldots, y_m)$ be self-dual functions. In this case, we will show that $h = f(x_1, x_2, \ldots, g(y_1, y_2, \ldots, y_m), \ldots, x_n)$ is also self-dual. By definition, we have $h^d = \bar{f}(\bar{x}_1, \bar{x}_2, \ldots, g(\bar{y}_1, \bar{y}_2, \ldots, \bar{y}_m), \ldots, \bar{x}_n)$. Since g is self-dual, we have $\bar{g}(\bar{y}_1, \bar{y}_2, \ldots, \bar{y}_m) = g(y_1, y_2, \ldots, y_m)$. Thus, $h^d = \bar{f}(\bar{x}_1, \bar{x}_2, \ldots, \bar{g}(y_1, y_2, \ldots, y_m), \ldots, \bar{x}_n)$. Since, f is self-dual, we have $\bar{f}(\bar{x}_1, \bar{x}_2, \ldots, \bar{x}_n) = f(x_1, x_2, \ldots, x_n)$. Thus, we have $h^d = f(x_1, x_2, \ldots, g(y_1, y_2, \ldots, y_m), \ldots, x_n) = h$. Hence, h is also self-dual. □

Definition 5.3 *A* **self-anti-dual function** *is the function such that* $f(x_1, x_2, \ldots, x_n) = f(\bar{x}_1, \bar{x}_2, \ldots, \bar{x}_n)$.

For example, $f(x, y) = x \oplus y$ is a self-anti-dual function, since $f = (\bar{x} \oplus 1) \oplus (\bar{y} \oplus 1) = \bar{x} \oplus \bar{y} = f(\bar{x}, \bar{y})$.

5.2 MONOTONE FUNCTION AND UNATE FUNCTION

A monotone increasing function is a function that can be represented by AND and OR gates only. A unate function is a generalization of a monotone function.

Definition 5.4 *Let* $\boldsymbol{a}$ *and* $\boldsymbol{b}$ *be Boolean vectors. If* f *satisfies* $f(\boldsymbol{a}) \geq f(\boldsymbol{b})$, *for any vectors such that* $\boldsymbol{a} \geq \boldsymbol{b}$, *then* f *is a* **monotone increasing function** *or a* **positive function**.

Theorem 5.4 f *is a monotone increasing function iff* f *is a constant or represented by an SOP without complemented literals.*

(Proof) (Monotone increasing $\Rightarrow$ SOP without complemented literals)
Let f be a monotone increasing function. Then, we have $f(1, x_2, \ldots, x_n) \geq f(0, x_2, \ldots, x_n)$. From this, we have

$$f(1, x_2, \ldots, x_n) \vee f(0, x_2, \ldots, x_n) = f(1, x_2, \ldots, x_n). \tag{5.1}$$

From the Shannon's expansion theorem, we have

$$f(x_1, x_2, \ldots, x_n) = x_1 f(1, x_2, \ldots, x_n) \vee \bar{x}_1 f(0, x_2, \ldots, x_n).$$

By applying (5.1) to this equation, we have

$$f(x_1, x_2, \ldots, x_n) = x_1(f(1, x_2, \ldots, x_n) \vee f(0, x_2, \ldots, x_n)) \vee \bar{x}_1 f(0, x_2, \ldots, x_n)$$

$$= x_1 f(1, x_2, \ldots, x_n) \vee (x_1 \vee \bar{x}_1) f(0, x_2, \ldots, x_n)$$
$$= x_1 f(1, x_2, \ldots, x_n) \vee f(0, x_2, \ldots, x_n).$$

Note that the last expression does not contain $\bar{x}_1$. By applying the same operations to other variables, we have an SOP without literals $\bar{x}_2, \ldots, \bar{x}_n$.

(SOP without complemented literals $\Rightarrow$ Monotone increasing)
Since f is represented by an SOP without literals $\bar{x}_i$, it is written as $f(x_1, x_2, \ldots, x_n) = x_i g_1 \vee g_2$. Let

$$\boldsymbol{c} = (x_1, x_2, \ldots, 0, \ldots, x_n) : (\text{vector with } x_i = 0),$$
$$\boldsymbol{d} = (x_1, x_2, \ldots, 1, \ldots, x_n) : (\text{vector with } x_i = 1).$$

We have $f(\boldsymbol{c}) = g_2$ and $f(\boldsymbol{d}) = g_1 \vee g_2$. Thus, $f(\boldsymbol{c}) \leq f(\boldsymbol{d})$ holds. Let $\boldsymbol{a}, \boldsymbol{b} \in B^n$ such that $\boldsymbol{a} \leq \boldsymbol{b}$. Consider the sequence of binary vectors such that $\boldsymbol{a} \leq \boldsymbol{a}_1 \leq \boldsymbol{a}_2 \leq \cdots \leq \boldsymbol{b}$ and in $\boldsymbol{a}, \boldsymbol{a}_1, \boldsymbol{a}_2, \ldots, \boldsymbol{b}$, the number of 1's in the vector increases one by one. Then, we have $f(\boldsymbol{a}) \leq f(\boldsymbol{a}_1) \leq f(\boldsymbol{a}_2) \leq \cdots \leq f(\boldsymbol{b})$. Hence, f is a monotone increasing function. □

As shown in the above theorem, a monotone increasing function is represented by AND and OR operators only. The constant functions, 0 and 1, are also monotone increasing functions.

Example 5.2 The monotone increasing functions of two variables are: 0, x, y, xy, $x \vee y$, and 1. ∎

Enumeration of the monotone increasing functions is not so simple.

Theorem 5.5 *A function that is obtained by assigning a monotone increasing function to an arbitrary variable of a monotone increasing function is also a monotone increasing function.*

(Proof) A monotone increasing function f is represented by an SOP without complemented literals. That is, f is represented by AND and OR operators only. The function that is obtained by assigning a monotone increasing function to a variable of f is also represented by AND and OR operators only, so it is also a monotone increasing function. □

Definition 5.5 *Let $\boldsymbol{a}$ and $\boldsymbol{b}$ be Boolean vectors. If f satisfies $f(\boldsymbol{a}) \leq f(\boldsymbol{b})$ for any vectors such that $\boldsymbol{a} \geq \boldsymbol{b}$, then f is a* **monotone decreasing function** *or a* **negative function**.

Example 5.3 The monotone decreasing functions of two variables are: 0, $\bar{x}$, $\bar{y}$, $\bar{x}\bar{y}$, $\bar{x} \vee \bar{y}$, and 1. A monotone decreasing function other than constant functions is realized by a MOS gate. For example, Fig. 3.10 and Fig. 3.11 show MOS gates that realize NAND and NOR, respectively. ∎

Theorem 5.6 *f is a monotone decreasing function iff f is a constant or represented by an SOP with complemented literals only. A monotone decreasing function is obtained by complementing a monotone increasing function.*

Definition 5.6 *If a function f is a constant or is represented by an SOP using either uncomplemented or complemented literals for each variable, then f is a* **unate function**.

Unate functions represents a class of functions that include monotone increasing functions and monotone decreasing functions.

Example 5.4 There are 16 functions of two variables. Among them, non-unate functions are $x \oplus y$ and $x \oplus \bar{y}$. The other 14 two-variable functions are all unate: 0, x, $\bar{x}$, y, $\bar{y}$, xy, $\bar{x}y$, $x\bar{y}$, $\bar{x}\bar{y}$, $x \vee y$, $\bar{x} \vee y$, $x \vee \bar{y}$, $\bar{x} \vee \bar{y}$, and 1. ∎

Example 5.5 The following functions are represented by SOPs where only either complemented or uncomplemented literals appear for each variable. Thus, they are unate functions:

$$f_1(x, y, z) = x \vee yz, \; f_2(x, y, z) = \bar{x}\bar{y} \vee \bar{z}.$$

In the following SOP, both x and $\bar{x}$ appear. So, one may think it is not a unate function.

$$f_3(x, y) = x \vee \bar{x}\bar{y}.$$

However, this function is also represented as $f_3(x, y) = x \vee \bar{y}$. Thus, it is a unate function. $f_4(x, y, z) = xy \vee \bar{x}z$ is not a unate function. ∎

5.3 LINEAR FUNCTION

When a logic function is represented by a Reed-Muller canonical expression, sometimes the degrees of all the products are less than two. This section considers such functions.

Definition 5.7 *If a logic function f is represented as:*

$$f = a_0 \oplus a_1 x_1 \oplus a_2 x_2 \oplus \cdots \oplus a_n x_n, \tag{5.2}$$

where a_i =0 or 1, then f is a **linear function**.

Let $\boldsymbol{x}$ and $\boldsymbol{y}$ be Boolean vectors, and a be a constant ($a \in \{0, 1\}$). Then, the "linear function" in ordinary mathematics satisfies the following:

$$f(\boldsymbol{x} \oplus \boldsymbol{y}) = f(\boldsymbol{x}) \oplus f(\boldsymbol{y}), \tag{5.3}$$
$$f(a\boldsymbol{x}) = af(\boldsymbol{x}). \tag{5.4}$$

Let $a_0 = 0$ in (5.2). Then, the linear functions satisfy the above conditions.

Theorem 5.7 *There are 2^{n+1} linear functions of n variables.*

Example 5.6 There are 8 linear functions $f(x, y)$ on two variables: 0, 1, x, $x \oplus 1 = \bar{x}$, y, $1 \oplus y = \bar{y}$, $x \oplus y$, and $\bar{x} \oplus y$. ∎

If $a_1 = a_2 = \cdots = a_n = 1$ in Definition 5.7, then f is a **parity function** of n variables. A parity function decides whether the number of 1's in the inputs is an even number or an odd number.

Theorem 5.8 *The function that is obtained by assigning a linear function to an arbitrary variable of a linear function is also a linear function.*

(Proof) Let f and g be linear functions as follows: $f = a_0 \oplus a_1 x_1 \oplus a_2 x_2 \oplus \cdots \oplus a_n x_n$, $g = b_0 \oplus b_1 y_1 \oplus b_2 y_2 \oplus \cdots \oplus b_m y_m$. Then, we have

$$\begin{aligned} f_1 &= f(x_1, x_2, \ldots, x_{i-1}, g, x_{i+1}, \ldots, x_n) \\ &= a_0 \oplus a_1 x_1 \oplus a_2 x_2 \oplus \cdots \oplus a_i(b_0 \oplus b_1 y_1 \oplus b_2 y_2 \oplus \cdots \oplus b_m y_m) \oplus \cdots \oplus a_n x_n. \end{aligned}$$

Note that this expression does not contain the products whose degrees are greater than one. Thus, f_1 is also a linear function. □

Theorem 5.9 *A linear function is either a self-dual function or a self-anti-dual function.*

(Proof) Since, $\bar{x}_i = 1 \oplus x_i$, we have

$$\begin{aligned} f(\bar{x}_1, \bar{x}_2, \ldots, \bar{x}_n) &= a_0 \oplus a_1(x_1 \oplus 1) \oplus a_2(x_2 \oplus 1) \oplus \cdots \oplus a_n(x_n \oplus 1) \\ &= a_0 \oplus a_1 x_1 \oplus a_2 x_2 \oplus \cdots \oplus a_n x_n \oplus (a_1 \oplus a_2 \oplus \cdots \oplus a_n). \end{aligned}$$

Thus, if $a_1 \oplus a_2 \oplus \cdots \oplus a_n = 1$, then

$$f(\bar{x}_1, \bar{x}_2, \ldots, \bar{x}_n) = f(x_1, x_2, \ldots, x_n) \oplus 1 = \bar{f}(x_1, x_2, \ldots, x_n),$$

so f is self-dual. On the other hand, if $a_1 \oplus a_2 \oplus \cdots \oplus a_n = 0$, then

$$f(\bar{x}_1, \bar{x}_2, \ldots, \bar{x}_n) = f(x_1, x_2, \ldots, x_n),$$

so f is self-anti-dual. □

5.4 SYMMETRIC FUNCTION

Functions that appears in arithmetic circuits often have symmetries. When logic function have some symmetries, they are often realized by using fewer elements.

Definition 5.8 *A function f is a* **totally symmetric function** *if any permutation of the variables in f does not change the function. A totally symmetric function is also called a* **symmetric function**.

Definition 5.9 *In a function $f(x_1, \ldots, x_i, \ldots, x_j, \ldots, x_n)$, if the function $f(x_1, \ldots, x_j, \ldots, x_i, \ldots, x_n)$ that is obtained by interchanging variables x_i with x_j, is equal to the original function, then f is* **symmetric with respect to x_i and x_j**. *If any permutation of subset S of the variables does not change the function f, then f is a* **partially symmetric function**.

Definition 5.10 *The* **elementary symmetric functions** *of n variables are*

$$\begin{aligned} S_0^n &= \bar{x}_1 \bar{x}_2 \cdots \bar{x}_n, \\ S_1^n &= x_1 \bar{x}_2 \cdots \bar{x}_n \vee \bar{x}_1 x_2 \bar{x}_3 \cdots \bar{x}_n \vee \cdots \vee \bar{x}_1 \bar{x}_2 \cdots \bar{x}_{n-1} x_n, \\ &\cdots\cdots\cdots\cdots\cdots\cdots, \text{ and} \\ S_n^n &= x_1 x_2 \cdots x_n. \end{aligned}$$

S_i^n =1 iff exactly i out of n inputs are equal to one. Let $A \subseteq \{0, 1, \ldots, n\}$. A symmetric function S_A^n is defined as follows:

$$S_A^n = \bigvee_{i \in A} S_i^n.$$

Example 5.7 $f(x_1, x_2, x_3) = x_1x_2x_3 \vee x_1\bar{x}_2\bar{x}_3 \vee \bar{x}_1x_2\bar{x}_3 \vee \bar{x}_1\bar{x}_2x_3$ is a totally symmetric function. $f = 1$ when all the variables are one, or when only one variable is one. Thus, f can be written as $S_1^3 \vee S_3^3 = S_{\{1,3\}}^3$. ∎

Theorem 5.10 *An arbitrary n-variable symmetric function f is uniquely represented by elementary symmetric functions $S_0^n, S_1^n, \ldots, S_n^n$ as follows:*

$$f = \bigvee_{i \in A} S_i^n = S_A^n, \text{ where } A \subseteq \{0, 1, \ldots, n\}.$$

Theorem 5.11 *Let f and g be totally symmetric functions of n variables. Then, $f \vee g$, fg, $f \oplus g$, and $\bar{f}$ are also totally symmetric functions. Let $f = S_A^n$ and $g = S_B^n$. Then, $f \cdot g = S_{A \cap B}^n$, $f \vee g = S_{A \cup B}^n$, $f \oplus g = S_{A \oplus B}^n$, and $\bar{f} = S_{\bar{A}}^n$.*

Consider a three-variable symmetric function S_A^3. Since $A \subseteq \{0, 1, 2, 3\}$, there are $2^4 = 16$ different ways to specify A. Thus, the total number of symmetric functions of three variables is 16. By generalizing this, we have the following:

Theorem 5.12 *There are 2^{n+1} symmetric functions of n variables.*

Example 5.8 There are 8 symmetric functions of two variables: $S_\phi^2 = 0$, $S_{\{0\}}^2 = \bar{x}_1\bar{x}_2$, $S_{\{1\}}^2 = x_1 \oplus x_2$, $S_{\{2\}}^2 = x_1x_2$, $S_{\{0,2\}}^2 = x_1 \oplus \bar{x}_2$, $S_{\{0,1\}}^2 = \bar{x}_1 \vee \bar{x}_2$, $S_{\{1,2\}}^2 = x_1 \vee x_2$, and $S_{\{0,1,2\}}^2 = 1$. ∎

5.5 THRESHOLD FUNCTION

The **3-input majority function** has three variables, and the output is 1 iff two or more of the variables are 1. A threshold function is a generalization of a majority function, and considered as a model of a **neuron**.

Definition 5.11 *Let $(w_1, w_2, \ldots, w_n)$ be an n-tuple of real numbers called* **weights**, *and t be a real number called* **threshold** *(Fig. 5.1). A* **threshold function** *is a function such that:*

$$f = \begin{cases} 1 & \text{when } w_1x_1 + w_2x_2 + \cdots w_nx_n \geq t, \\ 0 & \text{otherwise.} \end{cases}$$

Example 5.9 Let $n = 3$, $w_1 = 2$, $w_2 = 1$, $w_3 = 1$, and $t = 1.5$ in Definition 5.11. Then, we have

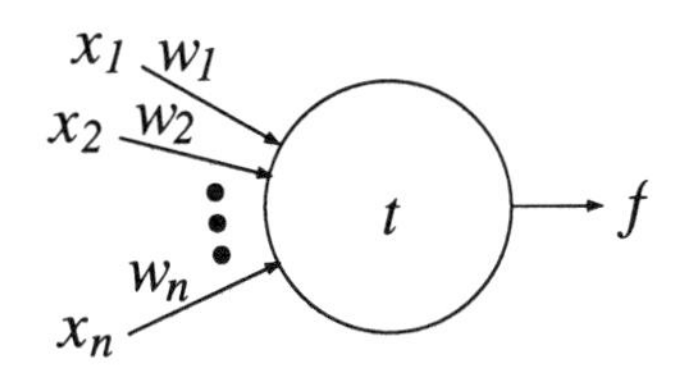

Figure 5.1 Threshold function.

$$f = \begin{cases} 1 & \text{when } 2x_1 + x_2 + x_3 \geq 1.5, \\ 0 & \text{otherwise.} \end{cases}$$

The logic function satisfying this condition is represented by $f = x_1 \vee x_2x_3$. Let $n = 2$. In this case, when $w_1 = w_2 = 1$, and $t = 0.5$, $f = x_1 \vee x_2$; when $w_1 = w_2 = 1$, and $t = 1.5$, $f = x_1x_2$; when $w_1 = w_2 = -1$, and $t = -0.5$, $f = \overline{x_1 \vee x_2}$; and when $w_1 = w_2 = -1$, and $t = -1.5$, $f = \overline{x_1x_2}$. ∎

Example 5.10 There are 16 functions of two variables. Among them, only two functions are not threshold functions: $x_1 \oplus x_2$ and $x_1 \oplus \bar{x}_2$. All other functions are threshold functions. ∎

Example 5.11 All the monotone increasing function with up to three variables are threshold functions. However, the four-variable function $x_1x_2 \vee x_3x_4$ is not a threshold function. ∎

Theorem 5.13 *A threshold function is a unate function.*

(Proof) By the definition of a threshold function, if $w_i = 0$, then f is independent of x_i. Thus, we have to only consider the variables where $w_i \neq 0$. If $w_i > 0$, then $f(x_1, \ldots, 1, \ldots, x_n) \geq f(x_1, \ldots, 0, \ldots x_n)$. If $w_i < 0$, then $f(x_1, \ldots, 1, \ldots, x_n) \leq f(x_1, \ldots, 0, \ldots x_n)$. Thus, when $w_i > 0$, f is represented by an SOP that does not contain $\bar{x}_i$. Also, when $w_i < 0$, f is represented by an SOP that does not contain x_i. By the definition of the unate function, it is clear that the theorem holds. □

Definition 5.12 *A* **majority function** *is a threshold function, where* $n = 2m + 1$, $t = m + 1$ *and* $w_1 = w_2 = \cdots = w_n = 1$.

A majority function f is equal to 1 iff the inputs have more 1's than 0's.

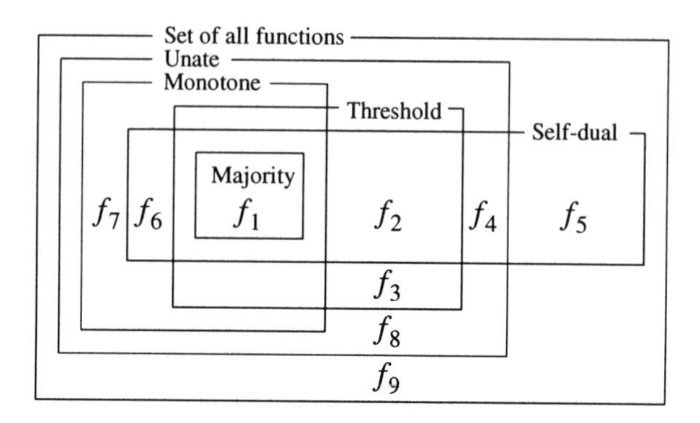

Figure 5.2 Relations among various functions.

Example 5.12 Let $m = 1$ in Definition 5.12. Then we have the majority function of three variables:

$$M(x_1, x_2, x_3) = x_1x_2 \vee x_2x_3 \vee x_3x_1.$$

This function is 1 iff two or more inputs are 1. It is a self-dual function, and can also represented as

$$M(x_1, x_2, x_3) = (x_1 \vee x_2)(x_2 \vee x_3)(x_3 \vee x_1).$$

∎

Theorem 5.14 *A majority function is totally symmetric, monotone increasing, and self-dual.*

(Proof) Let $n = 2m + 1$. The n-variable majority function f is represented by $f = S^n_{\{m+1,\ldots,n\}}$. Thus, f is a totally symmetric function. Also, note that

$$\begin{aligned}
\bar{f}(\bar{x}_1, \bar{x}_2, \ldots, \bar{x}_n) &= \bar{S}^n_{\{m+1,\ldots,n\}}(\bar{x}_1, \bar{x}_2, \ldots, \bar{x}_n) \\
&= \bar{S}^n_{\{0,\ldots,m\}}(x_1, x_2, \ldots, x_n) \\
&= S^n_{\{m+1,\ldots,n\}}.
\end{aligned}$$

Thus, f is a self-dual function. f is monotone increasing since all the weights are positive. □

Fig 5.2 shows relations among various functions.

5.6 UNIVERSAL SET OF LOGIC FUNCTIONS

If an arbitrary logic function is represented by a given set of logic functions, then the set is **universal** or **complete**. For example, consider the set consisting of the 2-input NAND functions, which is denoted by $\{\overline{x \cdot y}\}$. Then, it is universal. That is, an arbitrary logic function is realized by 2-input NAND gates only. The theory of universality is useful in selecting basic logic elements.

Definition 5.13 *Let $\mathcal{F} = \{f_1, f_2, \ldots, f_m\}$ be a set of logic functions. If an arbitrary logic function is realized by a loop-free combinational network using the logic elements that realize function $f_i (i = 1, 2, \ldots, m)$, then $\mathcal{F}$ is universal.*

Example 5.13 Both $\{\overline{xy}\}$ and $\{x \vee y,\ xy,\ \overline{x}\}$ are universal. $\{x \vee y,\ xy\}$ is not universal. ∎

Definition 5.14 *A function such that $f(0, 0, \ldots, 0) = 0$ is a* **0-preserving function**. *A function such that $f(1, 1, \ldots, 1) = 1$ is a* **1-preserving function**.

Theorem 5.15 *A function that is obtained by assigning a 0-preserving function to an arbitrary variable of a 0-preserving function is also a 0-preserving function.*

(Proof) Let $f(x_1, x_2, \ldots, x_n)$ and $g(y_1, y_2, \ldots, y_m)$ be 0-preserving functions. Let Y be the function that is obtained by assigning g to a variable x_i of f. Then, we have

$$\begin{aligned} Y &= f(x_1, x_2, \ldots, g(y_1, y_2, \ldots, y_m), \ldots, x_n) \\ &= Y(x_1, x_2, \ldots, x_n, y_1, y_2, \ldots, y_m). \end{aligned}$$

Thus, $Y(0, 0, \ldots, 0) = f(0, 0, \ldots, 0, \ldots, 0) = 0$. That is, Y is also a 0-preserving function. □

Theorem 5.16 *A function that is obtained by assigning a 1-preserving function to an arbitrary variable of a 1-preserving function is also a 1-preserving function.*

Theorem 5.17 *Let*

$\mathcal{M}_0$ be the set of 0-preserving functions,
$\mathcal{M}_1$ be the set of 1-preserving functions,
$\mathcal{M}_2$ be the set of self-dual functions,
$\mathcal{M}_3$ be the set of monotone increasing functions, and
$\mathcal{M}_4$ be the set of linear functions.

Then, the set of functions $\mathcal{F}$ is universal iff

$$\mathcal{F} \not\subseteq \mathcal{M}_i \ (i = 0, 1, 2, 3, 4).$$

(Proof for necessity) Let $f = \overline{xy}$, then $f \notin \mathcal{M}_i$ $(i = 0, 1, 2, 3, 4)$. Each of the sets $\mathcal{M}_i (i = 0, 1, 2, 3, 4)$ is **closed under the composition of the functions.** That is, assigning a function in $\mathcal{M}_i$ to a variable of $\mathcal{M}_i$, also produces a function in $\mathcal{M}_i$. So, if $\mathcal{F} \subseteq \mathcal{M}_i$, then the function such that $f \notin \mathcal{M}_i$ cannot be realized. Thus, $\mathcal{F}$ is not universal. □

The proof for the sufficiency of Theorem 5.17 is rather complicated. First, we will prove Lemmas 5.1–5.6.

Lemma 5.1 *The complement $\bar{x}$ is realized by any non-monotone increasing function and constants 0 and 1.*

(Proof) Let $f(x_1, x_2, \ldots, x_n)$ be a non-monotone increasing function. Then, there exist two Boolean vectors $\boldsymbol{a}$ and $\boldsymbol{b}$ $(\boldsymbol{a} \leq \boldsymbol{b})$ where only one bit position is different, such that $f(\boldsymbol{a}) = 1$ and $f(\boldsymbol{b}) = 0$. Assigning the following values to the variable $x_i (i = 1, \ldots, n)$ of the f, f realizes $\bar{x}$: when $a_i = 0$ and $b_i = 0$, assign $x_i = 0$, when $a_i = 0$ and $b_i = 1$, assign $x_i = x$, and when $a_i = 1$ and $b_i = 1$, assign $x_i = 1$. □

Example 5.14 $f(x, y, z) = x\bar{y} \vee z$ is a non-monotone increasing function. Note that $\boldsymbol{a} = (1, 0, 0)$, $\boldsymbol{b} = (1, 1, 0)$, $f(\boldsymbol{a}) = 1$, and $f(\boldsymbol{b}) = 0$. By assigning $x = 1$ and $z = 0$, we have $f(1, y, 0) = \bar{y}$, the complement of y. ▮

Lemma 5.2 *The AND and OR functions are realized by any non-linear function, complement, and constants 0 and 1.*

(Proof) Let f be a non-linear function. When f is represented by a Reed-Muller canonical form, the expression has at least one product whose degree is more than one. Let

$$x_{i_1} x_{i_2} \cdots x_{i_s} \ (s \geq 2) \tag{5.5}$$

be such a product with the lowest degree. For simplicity, assume that $i_1 = 1$, $i_2 = 2, \ldots$, and $i_s = s$. Next, consider the function such that

$$g(x_1, x_2) = f(\underbrace{x_1, x_2, 1, \ldots, 1}_{s \text{ variables}}, 0, \ldots, 0).$$

Then, it can be represented as $g(x_1, x_2) = c_0 \oplus c_1 x_1 \oplus c_2 x_2 \oplus c_{12} x_1 x_2$, where $c_{12} = 1$. This is because, among the products in the Reed-Muller canonical form for f, all the products that contain literals other than $x_1, x_2, \ldots, x_s$ disappear. By the assumption that (5.5) has the lowest degree, except for the products in (5.5), all the products whose degrees are more than one disappears. Next, by setting $h(x_1, x_2) = g(x_1 \oplus c_2, x_2 \oplus c_1)$, we have

$$\begin{aligned} h(x_1, x_2) &= c_0 \oplus c_1(x_1 \oplus c_2) \oplus c_2(x_2 \oplus c_1) \oplus (x_1 \oplus c_2)(x_2 \oplus c_1) \\ &= d_0 \oplus x_1 x_2, \end{aligned}$$

where $d_0 = c_0 \oplus c_1 c_2$. If $d_0 = 1$, then h denotes the NAND function. If $d_0 = 0$, then h denotes the AND function. From the NAND and a complement, we can synthesize the AND function. Also, from the AND and a complement, we can synthesize the OR function. □

Example 5.15 $f(x, y, z, w) = 1 \oplus x \oplus y \oplus xyz \oplus xyzw$ is a non-linear function. Among the products whose degree are more than one, the product with the lowest degree is xyz. Let $z = 1$, and $w = 0$. Then, we have $g(x, y) = 1 \oplus x \oplus y \oplus xy$. $h(x, y) = g(x \oplus 1, y \oplus 1) = 1 \oplus (x \oplus 1) \oplus (y \oplus 1) \oplus (x \oplus 1)(y \oplus 1) = xy$. This shows that we can realize the AND function. ∎

Lemma 5.3 *Constants 0 and 1 are realized by a non-self-dual function and the complement.*

(Proof) Let f be a non-self-dual function. Then, there exists a vector $\boldsymbol{a} = (a_1, a_2, \ldots, a_n)$ such that $f(\boldsymbol{a}) = f(\bar{\boldsymbol{a}}) = c$, where $c = 0$ or 1. For $a_i = 1$, assign $x_i = x$. For $a_i = 0$, assign $x_i = \bar{x}$. Then, we have $f(x_1, x_2, \ldots, x_n) \equiv c$, and we can realize a constant 0 or a constant 1 function. By using the complement, we can realize the other constant. □

Example 5.16 $f(x, y) = xy$ is a non-self-dual function. Let $\boldsymbol{a} = (0, 1)$. Then, we have $\bar{\boldsymbol{a}} = (1, 0)$, $f(\boldsymbol{a}) = f(\bar{\boldsymbol{a}}) = 0$, and $f(x, \bar{x}) = 0$. Thus, the constant 0 function is realized. ∎

Lemma 5.4 *If f is non 1-preserving and non 0-preserving, then the complement can be realized from f.*

(Proof) Since $f(1,\ldots,1) = 0$ and $f(0,\ldots,0) = 1$, we have $f(x,\ldots,x) = \bar{x}$. Thus, we have the complement. □

Lemma 5.5 *If f is 1-preserving and non 0-preserving, then the constant 1 is realized from f.*

(Proof) Since $f(1,\ldots,1) = 1$ and $f(0,\ldots,0) = 1$, we have $f(x,\ldots,x) = 1$. Thus, we have the constant 1. □

Lemma 5.6 *If f is 0-preserving and non 1-preserving, then the constant 0 is realized from f.*

(Proof) Since $f(1,\ldots,1) = 0$ and $f(0,\ldots,0) = 0$, we have $f(x,\ldots,x) = 0$. Thus, we have the constant 0. □

Here, we have completed the preparation of the proof for the theorem.
(Outline of the proof for the sufficiency of Theorem 5.17) From Lemmas 5.1–5.6, we can realize AND, OR, and NOT. Thus, an arbitrary logic function is realized from $\mathcal{F}$. □

Example 5.17 Let $f_1 = \bar{x}\bar{y}$, $f_2 = x\bar{y}$, $f_3 = x \vee \bar{y}$, $f_4 = x \oplus y$, $f_5 = 1$, $f_6 = 0$, $f_7 = xy \vee yz \vee zx$, $f_8 = x \oplus y \oplus z$, $f_9 = \bar{x}$, $f_{10} = xy$, and $f_{11} = x$. Table 5.2 shows the inclusion relation of f_i and $\mathcal{M}_j$. When $f_i \notin \mathcal{M}_j$, element (i,j) has a symbol $\times$. In order for the given set of functions $\mathcal{F}$ to be universal, all the columns must contain $\times$. Also, to obtain the **minimal universal set** of the functions, we have to obtain the minimal covering of Table 5.2. For example, $\{f_1\}$, $\{f_2, f_3\}$, $\{f_2, f_5\}$, $\{f_3, f_4\}$, $\{f_3, f_6\}$, $\{f_4, f_5, f_7\}$, $\{f_5, f_6, f_7, f_8\}$, and $\{f_9, f_{10}\}$ are minimal sets. ∎

5.7 EQUIVALENCE CLASSES OF LOGIC FUNCTIONS

It is useful to categorize functions into equivalence classes. For example, if we have a minimum network for each of the representative functions of the equivalence classes, then the minimum network for an arbitrary function can be obtained by a table look-up and a simple transformation. This method is useful for logic functions with three to five variables.

Table 5.2 Table showing functional properties.

	$\mathcal{M}_0$	$\mathcal{M}_1$	$\mathcal{M}_2$	$\mathcal{M}_3$	$\mathcal{M}_4$
f_1	×	×	×	×	×
f_2		×	×	×	×
f_3	×		×	×	×
f_4		×	×	×	
f_5	×		×		
f_6		×	×		
f_7					×
f_8				×	
f_9	×	×		×	
f_{10}			×		×
f_{11}					

Definition 5.15 *For a given logic function f, if a function g is derived from f by the combination of the following three operations (including the combination that do not use some of the operations), then the function g is* **NPN-equivalent** *to f.*

(1) Negation of some variables in f.
(2) Permutation of some variables in f.
(3) Negation of f.

This relation is the **NPN-equivalence relation**. *The set of functions that are NPN-equivalent to the given function f forms an NPN-equivalence class, and f is a* **representative function** *of the equivalence class. The function that is obtained by the operations (1) and (2) is* **NP-equivalent** *to f. This relation is an* **NP-equivalence relation**. *Similarly,* **P-equivalence** *(permutation operation of the variables only) and* **N-equivalence** *(only the negation operation of the variables) can be defined.*

For example, in the case of three-variable functions, the number of the P-equivalence classes is 80. A table of minimum NAND networks for the representative functions for all the equivalence classes is available. A permutation of the variables does not change the number of gates in the minimum network. Thus, in the case of three-variable functions, the design of minimum NAND networks can be done by a table look-up. Also, when both un-complemented literals and complemented literals are available as inputs, the negation of the

Table 5.3 Classification of two-variable functions.

# of variables	All functions	P	NP	NPN
0	0	0	0	0
	1	1	1	
1	x, y	x	x	x
	$\bar{x}$, $\bar{y}$	$\bar{x}$		
2	xy	xy	xy	xy
	$\bar{x}y$ $x\bar{y}$	$\bar{x}y$		
	$\bar{x}\bar{y}$	$\bar{x}\bar{y}$		
	$x \vee y$	$x \vee y$	$x \vee y$	
	$\bar{x} \vee y$ $x \vee \bar{y}$	$\bar{x} \vee y$		
	$\bar{x} \vee \bar{y}$	$\bar{x} \vee \bar{y}$		
	$x \oplus y$	$x \oplus y$	$x \oplus y$	$x \oplus y$
	$\bar{x} \oplus y$	$\bar{x} \oplus y$		

P: Representative function of P-equivalence classes.
NP: Representative function of NP-equivalence classes.
NPN: Representative function of NPN-equivalence classes.

input variables does not change the number of gates in the minimum network. Thus, we need only to obtain 22 networks for the representative functions of NP-equivalence classes. For example, when we classify the AND-OR two-level logic networks, we can use NP-equivalence classes. In ECL networks, the gates realize both NOR and OR functions. In this case, we can consider NPN-equivalence classes. Note that the P-equivalence relation is a refinement of the NP-equivalence relation. And the NP-equivalence relation is a refinement of the NPN-equivalence relation.

Example 5.18 There are 16 functions of two variables. If we classify them by the P-equivalence relation, we have 12 equivalence classes. If we classify them by the NP-equivalence relation, we have 6 equivalence classes. And if we classify them by the NPN-equivalence relation, we have 4 equivalence classes. Table 5.3 shows the representative functions. ∎

In the case of P-equivalence, the network that realizes a given function is obtained by a permutation of the input variables on a network of a representative function. For example, the network for the function $x\bar{y}$ is obtained from the

Table 5.4 Number of equivalence classes under various equivalence relations.

	0	1	2	3	4
All functions	2	4	16	256	65,536
P-equivalence class	2	4	12	80	3,984
NP-equivalence class	2	3	6	22	402
NPN-equivalence class	1	2	4	14	222

network for the representative function $\bar{x}y$, by interchanging x and y. In classifying NAND networks, we use P-equivalence. In the case of NP-equivalence, an equivalent logic function is obtained by a permutation and/or the negation of the input variables. For example, the function $x\bar{y}$ is obtained from the network for the representative function xy, by replacing the input y with its complement. NP-equivalence class is useful where both true and complemented literals are available for each variables (**double-rail input logic**). In the case of the NPN-equivalence, the equivalent logic function can be obtained by a permutation and/or negation of the input variables, and/or the negation of the output function. For example, the function $x \vee \bar{y}$ is obtained from the network for the representative function xy, by replacing input x with its complemented literal $\bar{x}$, and then complement the output function. The NPN-equivalence is useful for double-rail input logic, where each logic element realizes both a function and its complement. Table 5.4 shows the numbers of equivalence classes up to $n = 4$. When n is sufficiently large, the number of equivalence classes are approximated as follows:

The number of P-equivalence classes is $\dfrac{2^{2^n}}{n!}$.

The number of NP-equivalence classes is $\dfrac{2^{2^n}}{2^n \cdot n!}$.

The number of NPN-equivalence classes is $\dfrac{2^{2^n}}{2^{n+1} \cdot n!}$.

Enumeration of the exact number of equivalence classes for n-variable functions requires combinatorial mathematics. Tables in [157] and [277] show the number of equivalence classes up to n=6.

Bibliographical Notes

Details on the topics in this chapter can be found as follows: Self-dual function [325, 407]; monotone and unate function [246]; liner function, symmetric function [157, 260, 315, 381, 412]; threshold function [275]; parametron [397]; universal set of logic elments [126, 182, 199, 268, 316]; equivalence classes of logic functions [157, 158, 277, 297, 298, 387]; table of minimum logic networks [16, 79, 164, 211, 388]; the enumeration of unate functions is difficult, but functions realized by tree or cascade networks are enumerated [53, 54, 162, 341].

Exercises

5.1 Which of the following are self-dual functions?

$$f = xy,\ g = \bar{x},\ h = xyz \vee \bar{x}y \vee \bar{x}\bar{y}z.$$

5.2 Prove that $(f^d)^d = f$.

5.3 Let $f(x_1, x_2, \ldots, x_n)$ be an arbitrary n-variable function. Show that the following function is self-dual:

$$h(x_1, x_2, \ldots, x_n, x_{n+1}) = \bar{x}_{n+1} f^d \vee x_{n+1} f.$$

5.4 Show that there are $2^{2^{n-1}}$ self-dual functions of n variables.

5.5 Enumerate the self-dual functions that depend on three variables.

5.6 Let a network F which consists of AND, OR, and NOT gates realize a function f. The network that is obtained from F by replacing all the AND gates with the OR gates and vice versa in F, will realize f^d. Prove this.

5.7 Prove that the minimal product-of-sums expressions of f is obtained by the following procedure:

1. Obtain the MSOP for f^d, the dual function of f.
2. Interchange the logical sum with logical product in the MSOP.

5.8 Let a network consisting of NAND gates only (does not contain constant 0 nor 1) realizes the parity function $f = x_1 \oplus x_2 \oplus \cdots \oplus x_n$. Consider the network that is obtained by replacing all the NAND gates with NOR gates. Discuss whether the network realizes f or not.

5.9 Let a logic function $f(x, y)$ satisfy the conditions: $f(x, 1) \geq f(x, 0)$ and $f(1, y) \geq f(0, y)$. Then, prove the following:

$$f(x, y) = f(0, 0) \vee yf(0, 1) \vee xf(1, 0) \vee xyf(1, 1).$$

5.10 (M) Let f be a monotone increasing function of n variables. Prove the following:

1) All the prime implicants of f are essential prime implicants.
2) There is a unique MSOP for f.

5.11 (M) Prove that the number of the monotone increasing functions of n variables is at least 2^N, where $N = {}_nC_{\lfloor n/2 \rfloor}$, and $\lfloor k \rfloor$ denotes the maximum integer not greater than k.

5.12 Which of the following are unate functions?

$$f = \bar{x}y \vee \bar{y} \vee xz, \; g = x\bar{y} \vee y, \; h = x\bar{y}\bar{z} \vee \bar{y}z \vee w.$$

5.13 Let f be a parity function of n variables. Show that the number of true minterms is 2^{n-1}.

5.14 Which of the following are symmetric functions?

$$\begin{aligned} f(x,y) &= (x \vee \bar{y})(\bar{x} \vee y), \\ g(x,y,z) &= xz \vee yz \vee yx, \\ h(x,y,z) &= xy \vee x\bar{y}z \vee \bar{x}yz. \end{aligned}$$

5.15 Enumerate the symmetric functions of three variables.

5.16 Prove that there are 2^{n+1} symmetric functions of n variables.

5.17 By using elementary symmetric functions, represent the function: $f = xy \oplus yz \oplus zx$.

5.18 Enumerate the monotone increasing symmetric functions of n variables.

5.19 Let n be an odd number. Prove that the number of self-dual symmetric function of n variables is $2^{(n+1)/2}$.

5.20 (M) Let $f(x_1, x_2, \ldots, x_n) = f(x_2, x_1, x_3, \ldots, x_n) = f(x_2, x_3, \ldots, x_n, x_1)$. Prove that f is a totally symmetric function.

5.21 Obtain the truth table, draw the Karnaugh map, and design the AND-OR two-level network with the minimum number of gates for the following function: f=1 iff $x + y + z + w$= 2 or 3.

5.22 (M) Prove that the MSOP for $S^n_{\{0,1,\ldots,n-2\}}$ $(n \geq 4)$ contains at least $\frac{n(n-1)}{2}$ products.

5.23 (M) Prove that the number of the prime implicants for $S^{3m}_{\{m,m+1,\ldots,2m\}}$ is $\dfrac{(3m)!}{(m!)^3}$. Also prove that its MSOP contains at least $\dfrac{(3m)!}{m!(2m)!}$ products.

5.24 Obtain the number of prime implicants and essential prime implicants for elementary symmetric function S^n_i.

5.25 (M) Let f be an n-variable threshold function. Prove that both $f \vee x_j$ $(1 \leq j \leq n+1)$ and $f \cdot x_j$ $(1 \leq j \leq n+1)$ are threshold functions.

5.26 Which of the following are threshold functions? $f = xy \vee \bar{z},\ g = x\bar{y} \vee \bar{x}y,\ h = xy \vee xzw, u = xy \vee \bar{x}yz,\ v = xy \vee \bar{x}\bar{y}$.

5.27 Obtain the SOPs for the following threshold functions:

$$f : w_1 = 1,\ w_2 = -2,\ w_3 = 4,\ t = 3.$$
$$g : w_1 = 1,\ w_2 = -4,\ w_3 = 3,\ t = 1.$$

5.28 Prove that a threshold function whose weights are all positive is a monotone increasing function.

5.29 Let $f(x_1, x_2, \ldots, x_n)$ be a threshold function with weights $(w_1, w_2, \ldots, w_n)$ and threshold t. Then, f^d, the dual function of f, is also a threshold function. In this case, obtain the threshold and weights for f^d.

5.30 (M) Show that $f = xy \vee zw$ is not a threshold function.

5.31 Let $M(x, y, z) = xy \vee yz \vee zx$. Does the following equations hold? Prove it, or show a counterexample.

a) $M(x_1, y_1, z_1) \vee M(x_2, y_2, z_2) = M(x_1 \vee x_2,\ y_1 \vee y_2,\ z_1 \vee z_2)$.

b) $M(x_1, y_1, z_1) \cdot M(x_2, y_2, z_2) = M(x_1x_2,\ y_1y_2,\ z_1z_2)$.

5.32 (M) Let T_i be the function such that

$$T_i(x_1, x_2, \ldots, x_n) = \begin{cases} 1 & \text{if } \sum_{j=1}^{n} x_j \geq i, \\ 0 & \text{otherwise.} \end{cases}$$

Obtain the number of prime implicants and essential prime implicants for T_i.

5.33 (M) Prove that the number of products in the MSOP for the n-variable majority function ($n = 2m + 1$) is ${}_nC_m$.

5.34 Let $M(x, y, z) = xy \vee yz \vee zx$. Prove the following equations:

$$
\begin{aligned}
&M(x, y, 1) = x \vee y, \\
&M(x, y, 0) = xy, \\
&M(x, 1, 0) = x, \\
&M(x, y, z) = M(y, z, x) = M(y, x, z), \\
&M(x, x, y) = M(x, y, \bar{y}) = x, \\
&M(x, y, z) = \overline{M}(\bar{x}, \bar{y}, \bar{z}), \\
&M(x, y, z) = xy \oplus yz \oplus zx, \\
&M(xy, z, w) = M(x, z, w)M(y, z, w), \\
&M(x \vee y, z, w) = M(x, z, w) \vee M(y, z, w), \\
&M(M(x, y, z), u, w) = M(M(x, u, w), y, M(z, u, w)).
\end{aligned}
$$

5.35 (M) Let $M_5(x_1, x_2, x_3, x_4, x_5)$ be a majority function of five variables. Prove the following:

$$
\begin{aligned}
&M_5(x_1, x_2, x_3, x_4, x_5) = \\
&\quad M(M(x_1, x_2, x_3),\ M(M(x_1, x_3, x_4),\ x_5,\ M(x_2, x_3, x_4)),\ x_5).
\end{aligned}
$$

5.36 By using gates that realize $\bar{x}y \vee \bar{z}$ and constant 1, realize AND, OR, and NOT.

5.37 By using gates that realizes $(\bar{x} \vee \bar{y})\bar{z}$, realize AND, OR, and NOT.

5.38 By using gates that realize $(\bar{x} \vee y)\bar{z}$ and constant 0, realize AND, OR, and NOT.

5.39 By using gates that realize $x \oplus y$, $xy \oplus yz \oplus zx$, and constant 1, realize AND, OR, and NOT.

5.40 By using gates that realize $x \oplus y \oplus z$, $xy \oplus yz \oplus zx$, and constants 0, and 1, realize AND, OR, and NOT.

5.41 In order to make a universal set, what kind of function should be added to the function $f = xy \vee \bar{x}\bar{z}$?

5.42 In order to make a universal set, what kind of function should be added to the function $f = xy \oplus z$?

5.43 (M) Show all the representative functions of the NPN-equivalence classes for three-variable functions.

5.44 Show that f and f^d, the dual function of f, belong to the same NPN-equivalence class.

5.45 (M) Enumerate all the three-variable functions that belong to the following equivalence classes:

a) An NP-equivalence class containing by $xy \vee yz \vee zx$.
b) An NPN-equivalence class containing by $xy \vee yz \vee zx$.
c) An NP-equivalence class containing by $x \oplus y \oplus z$.

5.46 Among the n-variable logic functions, how many functions are there whose number of true minterms is 2^{n-1}.

5.47 Draw the BDD for $S^3_{\{1,2\}}$.

5.48 (M) By using Lemmas 5.1–5.6, prove the sufficiency of Theorem 5.17.

5.49 Let g be a self-dual function, and h be a self-anti-dual function. Then, prove that $f = g \oplus h$ is a self-dual function.

5.50 Let g and h be self-dual functions. Then, prove that $f = g \oplus h$ is a self-anti-dual function.

5.51 Let f be a logic function. If $f \geq f^d$ or $f^d \geq f$, then f is **dual-comparable.** Show that a threshold function is dual-comparable. Also show a dual-comparable function that is not a threshold function.

5.52 Fig. 5.2 shows 9 functions $f_1, f_2, \ldots, f_9$ having different functional properties. Which of the following functions corresponds to f_i $(i = 1, 2, \ldots, 9)$?

$$g_1 = x_1x_2 \vee x_2x_3 \vee x_3x_4.$$
$$g_2 = x_1 \oplus x_2 \oplus x_3.$$
$$g_3 = \bar{x}_1x_2.$$

$$
\begin{aligned}
g_4 &= M(x_1, x_2, M(x_3, x_4, x_5)). \\
g_5 &= x_1 x_2 \vee x_2 \bar{x}_4 \vee x_2 x_3. \\
g_6 &= M(x_1, x_2, x_3). \\
g_7 &= \bar{x}. \\
g_8 &= x_1 \oplus x_2. \\
g_9 &= M(\bar{x}_1, x_2, M(x_3, x_4, x_5)).
\end{aligned}
$$

6

SEQUENTIAL NETWORKS

Sequential networks are the logic networks with memory. An example is a single input single output network that produces 1 iff three consecutive 1's appear in the inputs. Sequential networks are represented by state diagrams or state tables. Flip-flops are used for memory elements. A design of a sequential network is done as follows: First, minimize the number of states. Second, assign a binary code to each state. Third, allocate flip-flops. And, finally, realize the networks for flip-flops and outputs.

6.1 INTRODUCTION TO SEQUENTIAL NETWORKS

In combinational networks, the values of outputs are uniquely determined only by the values of present inputs. However, in **sequential networks**, the values of outputs depend not only on the values of present inputs, but also on the values of past inputs. A sequential network usually consists of a combinational network and the memory elements as shown in Fig. 6.1, where $x_1, x_2, \ldots, x_n$ are **input variables**, $y_1, y_2, \ldots, y_s$ are **state variables**, and $z_1, z_2, \ldots, z_m$ are **output variables.**

In Fig. 6.1, $x_i(t)$ and $y_i(t)$ denote values of x_i and y_i at the time t, respectively. Let Δt_i be the time for the signal to propagate the combinational network and the memory element to produce the new value of y_i. The value of Δt_i depends on parameters of the individual network, such as supply voltage, temperature, etc. $y_i(t + \Delta t_i)$, the value of y_i at the time $t + \Delta t_i$, is a function of $x_j(t)$ $(j = 1, 2, \ldots, n)$ and $y_k(t)$ $(k = 1, 2, \ldots, s)$. In a **synchronous sequential**

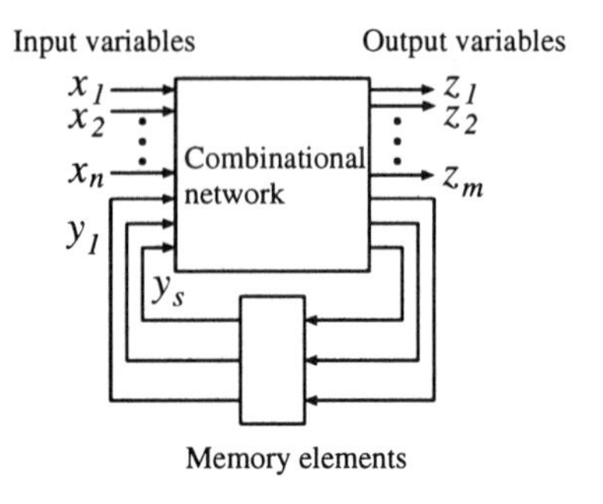

Figure 6.1 Model of sequential network.

network, the **clock pulse** synchronizes the changes of state variables, and makes Δt_i equal for all i. In a synchronous sequential network, Δt_i is the unit of the time. Thus, when the value of the input variables $x_i(t)$ $(i = 1, 2, \ldots, n)$ at time t_i are given, $y_i(t+1)$ $(i = 1, 2, \ldots, s)$, the values of the state variables at time $t+1$, and $z_j(t)$ $(j = 1, 2, \ldots, m)$, the values of the output variables at time t, are determined by the logic functions:

$$\begin{aligned} y_i(t+1) &= f_i(x_1(t), x_2(t), \ldots, x_n(t), y_1(t), \ldots, y_s(t)), \\ z_j(t) &= g_j(x_1(t), x_2(t), \ldots, x_n(t), y_1(t), \ldots, y_s(t)). \end{aligned}$$

When a synchronous sequential network is considered, we ignore the delay of interconnections. A sequential network without clock pulse is an **asynchronous sequential networks**, which will be treated in Chapter 8.

6.2 FLIP-FLOPS

Memory elements used in sequential networks are usually called **latches** or **flip-flops** (FFs for short). Latches or FFs are 1-bit memory elements, and have various types. This section introduces latches and FFs.

6.2.1 SR-Latch

A logic network requires memory elements to have more than one state. Here, we consider the basic networks that have more than one state. An SR-latch consists of two NOR gates, G_1 and G_2, as shown in Fig. 6.2(a). This network

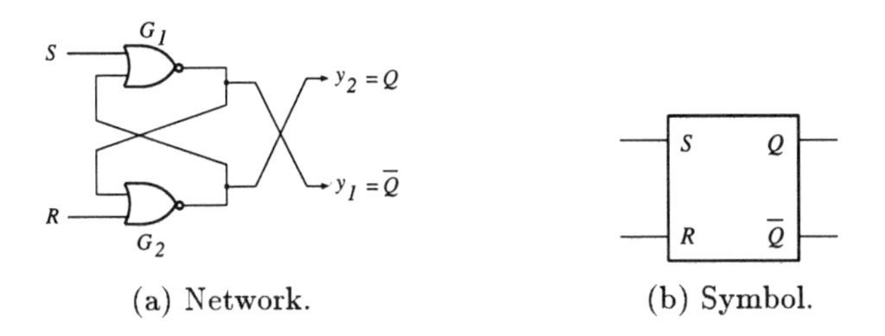

(a) Network. (b) Symbol.

Figure 6.2 SR-latch.

has two input terminals S and R, and they are connected to gates G_1 and G_2, respectively. Let the output functions for G_1 and G_2 be y_1 and y_2, respectively. When $S = R = 1$, $y_1 = y_2 = 0$. While keeping $S = 1$, if we set $R = 0$, then the output becomes $y_2 = 1$, since both inputs of G_2 are 0. Thus, both inputs for G_1 are 1, and the values of y_1 remains 0. Next, while keeping $R = 0$, if we change the value of S from 1 to 0, then the values of $y_1 = 0$ and $y_2 = 1$ will not change.

Next, while keeping $S = 0$, if we change the value of R from 0 to 1, then one of the inputs of G_2 becomes 1, and the values of y_2 changes to $y_2 = 0$. In this case, since both inputs of G_1 are 0, and y_1 changes to $y_1 = 1$. Next, while keeping $S = 0$, if we change the value of R from 1 to 0, then the outputs remain unchanged: $y_1 = 0$, $y_2 = 1$. Fig. 6.3 shows the state diagram representing operation of this network. Here, we assume that the values of S and R will not change at the same time. Then, we notice that the network has two stable states for the inputs $S = R = 0$. Also, for $S = 1$ and $R = 0$, the latch goes to the state Q_1, independently of the previous state. For $S = 0$, $R = 1$, the latch goes to the state Q_0, independently of the previous state. When the state of the latches is in Q_0 or Q_1, if $S = R = 0$, then the latch keeps the present state. If we do not use the input combination $S = R = 1$ in the stable states, then the state of a latch is represented by a logic variable $Q = y_2$ since $y_2 = \bar{y}_1$.

The input $S = 1$ makes $Q = 1$, so this input *sets* the latch. On the other hand, the input $R = 1$ makes $Q = 0$, so this input *resets* the latch.

Thus, the operations of the SR-latch are summarized as follows:

When $S = 1$ and $R = 0$, latch is set, and $Q = 1$.
When $S = 0$ and $R = 1$, latch is reset, and $Q = 0$.
When $S = R = 0$, latch keeps the present state.

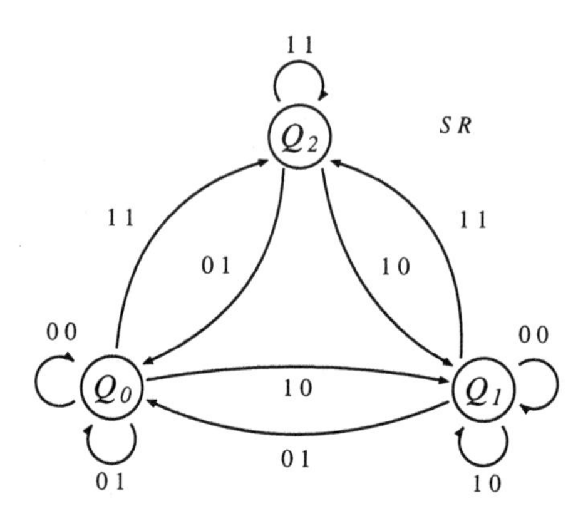

Figure 6.3 State diagram for SR-latch.

The input combination $S = R = 1$ is forbidden.

When we have to change the output value of a latch, we have to keep the values of S and R until the new value appears at the output. Suppose that $Q = 0$ and $S = R = 0$. If the value of S is changed to $S = 1$ for a very short time, and set to $S = 0$ again, then the output of the latch may not change to $Q = 1$. In order to operate the latch correctly, we have to keep the input value until the output values becomes stable. This time is at least two times of the delay time of the two NOR gates. Fig. 6.2(b) shows the symbol of an SR-latch.

6.2.2 Clocked SR-Latch

In a synchronous sequential network, state variables are defined only when the value of the clock pulse C is equal to 1. Fig. 6.4 shows a clocked SR-latch. This latch changes its state only when $C = 1$. When we use this latch in the sequential network model in Fig. 6.1, we have the following problem: Let T be the period of the clock pulse. Let τ be the time such that $C = 1$, then the time such that $C = 0$ is $T - \tau$. When $C = 0$, the output of the latch is fed to the combinational network, and the next state are computed. If $T - \tau$ is too short, then the next states are decided before the combinational network completes the computation. Thus, the network might go to the erroneous state. If τ is too short, then the inputs to the latches disappear before the latches change their states. If τ is too long, the output of the combinational network goes into the latches, and their output reenter the latches again. Thus, the network

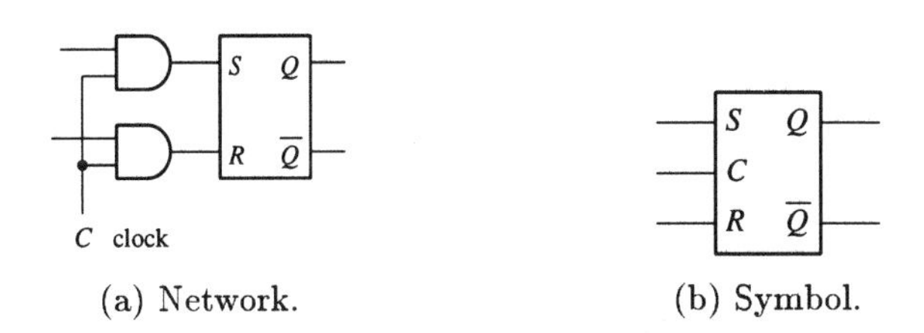

(a) Network. (b) Symbol.

Figure 6.4 Clocked SR-latch.

might make state transitions more than once. Thus, restrictions exist for the period of the clock pulse T and width of the pulse τ. Since, τ has the upper and lower bounds, this latch is difficult to use.

6.2.3 Master-Slave SR Flip-Flop

The clocked latches are hard to use due to the restrictions for the period T and the width τ of the clock pulse. The master-slave SR flip-flop (MS SR·FF) shown in Fig. 6.5 using two clocked latches has no such restrictions. When the value of the clock pulse is 1 ($C = 1$), the value of the input is fed to the first level latch (**master**). When the value of the clock pulse is 0 ($C = 0$), the value of the first level latch is transferred to the second level latch (**slave**). In the network shown in Fig. 6.1, suppose that the MS SR·FFs are used. When $C = 1$, the master latch and the slave latch are separated. So, even if the length for $C = 1$ is very large, only one state transition occurs. Thus, we can keep the output of the slave latch for enough time, so that the master latches may be excited. Also, when $C = 0$, the data is transferred from the master latch to the slave latch. And the values are supplied to the combinational network to compute the next states. Note that in this case, the master latch is separated from the input, and the master latch remain unchanged until $C = 1$.

In an MS FF, the time for reading the data to the FF, and the time for computing the next state by using the output of the FF are disjoint. Thus, we can realize a reliable network that does not depend on the width of the clock pulses. (However, in Fig. 6.5, if the width of the clock pulse is shorter than the delay time of the inverter that complements the clock pulse, then the FF may malfunction). The logic operations of an SR·FF are summarized as follows:

When (S, R)=(0, 0), keep the present value.
When (S, R)=(0, 1), reset.

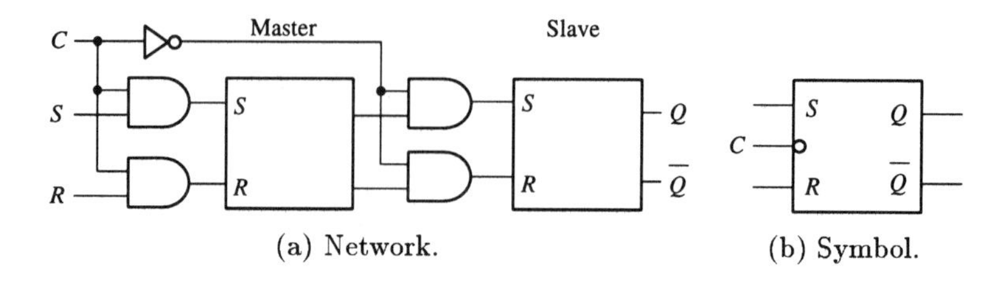

(a) Network. (b) Symbol.

Figure 6.5 Master-slave SR flip-flop.

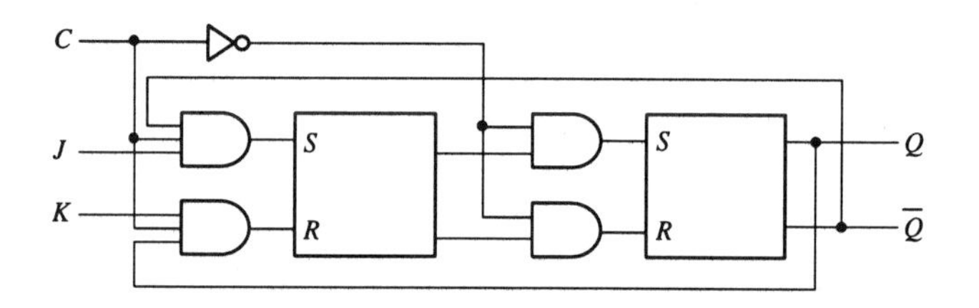

Figure 6.6 Master-slave JK flip-flop.

Table 6.1 Operation of SR flip-flop.

S	R	$Q(t+1)$
0	0	$Q(t)$
0	1	0
1	0	1
1	1	–

Table 6.2 Operation of JK flip-flop.

J	K	$Q(t+1)$
0	0	$Q(t)$
0	1	0
1	0	1
1	1	$\overline{Q}(t)$

When (S, R)=(1, 0), set.
(S, R)=(1, 1) is forbidden.

Table 6.1 also shows these operations.

6.2.4 Master-Slave JK Flip-Flop

By modifying an MS SR·FF, we have a master-slave JK flip-flop (MS JK·FF) shown in Fig. 6.6. In the SR·FF, the combination $S = R = 1$ is forbidden. However, in the JK·FF, the inputs $S = R = 1$ will change the state of the

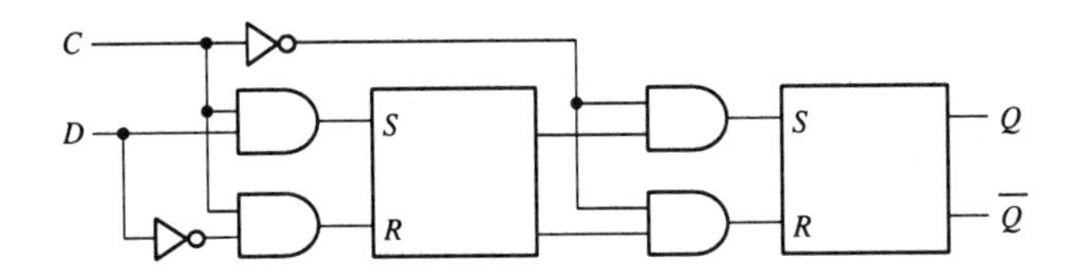

Figure 6.7 Master-slave D flip-flop.

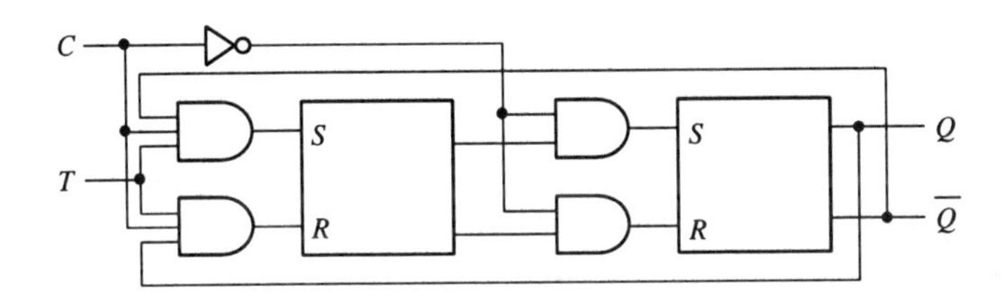

Figure 6.8 Master-slave T flip-flop.

FF. When $J = K = 0$, the inputs S and R of the master latch are both 0, and present state of the FF is hold. When $J = 1$ and $K = 0$, the S input of the master latch is $\overline{Q}$, and R input is 0. Thus, the FF is *set* regardless of the previous state. When $J = 0$ and $K = 1$, S input of the master latch is 0, and the R input is Q. Thus, the FF is *reset* regardless of the previous state. When $J = K = 1$, the S input of the master latch is $\overline{Q}$ and R input is Q, Thus, the state of the FF is complemented. Table 6.2 summarizes the operations of a JK·FF.

6.2.5 Master-Slave D Flip-Flop

In an MS SR·FF, let $D = S = \overline{R}$, and we have the master-slave D flip-flop (MS D·FF) shown in Fig. 6.7. In a D·FF, when $D = 1$, we have $S = 1$ and $R = 0$. Thus FF is *set*. When $D = 0$, we have $S = 0$ and $R = 1$. Thus, FF is *reset*. Table 6.3 summarizes the operation of the D·FF. This table shows that the state of the FF after applying the clock pulse is equal to the value of D just before the clock pulse. In other words, this FF *delays* the values of input variables D for one clock period.

Table 6.3 Operation of D flip-flop.

D	$Q(t+1)$
0	0
1	1

Table 6.4 Operation of T flip-flop.

T	$Q(t+1)$
0	$Q(t)$
1	$\overline{Q}(t)$

6.2.6 Master-Slave T Flip-Flop

In an MS JK·FF, let $T = J = K$, and we have the master-slave T flip-flop (MS T·FF) shown in Fig. 6.8. In a T·FF, when $T = 1$, we have $S = \overline{Q}$ and $R = Q$. Thus, the value of the FF is complemented. When $T = 0$, $S = R = 0$. Thus, the value of the FF is unchanged. Table 6.4 summarizes the logic operations of T·FF. In a T·FF, when $T = 1$, the state of FF changes in every clock pulse. On the other hand, when $T = 0$, the state of the FF remains unchanged. T·FFs are useful for the design of counters. However, if the T·FF fails into the erroneous state by an erroneous pulse, this FF always stays in the wrong state unless another erroneous pulse correct the state.

6.2.7 Edge-Triggered Flip-Flop

An **edge-triggered flip-flop** reads the input values in the (rising or falling) **edge** of the clock pulse. In the next moment, the input is isolated from the FF, and keeps isolated until the edge of the next clock pulse triggers. Some textbooks call latch as *flip-flop*. However, in this book, latch means the simple network such as shown in Fig. 6.2. A flip-flop means MS type or edge-triggered type FF. In practice, an ordinary D·FF is sometimes called a *register*. Also, the *latch* sometimes means the logic element that passes the data through when the control input is 0, while keeps the data when the control input is 1.

6.2.8 Inputs for Flip-Flops

Table 6.5 shows the input excitation conditions for four types of FFs. For example, when SR·FF are used,
if $Q(t) = 0$ and $Q(t+1) = 0$, then set $S = 0$ and $R = -$(*don't care*);
if $Q(t) = 0$ and $Q(t+1) = 1$, then set $S = 1$ and $R = 0$;
if $Q(t) = 1$ and $Q(t+1) = 0$, then set $S = 0$ and $R = 1$;
if $Q(t) = 1$ and $Q(t+1) = 1$, then set $S = -$ and $R = 0$.

Table 6.5 Input excitation conditions for flip-flops.

$Q(t)$	$Q(t+1)$	S	R	J	K	D	T
0	0	0	–	0	–	0	0
0	1	1	0	1	–	1	1
1	0	0	1	–	1	0	1
1	1	–	0	–	0	1	0

Similarly, we can obtain the input excitation conditions for JK·FF, D·FF, and T·FF.

SR Flip-Flop

For example, let the state transition function of the flip-flop Q be represented as

$$Q(t+1) = f = f_0\overline{Q}(t) \vee f_1 Q(t).$$

From this, the input for the SR·FF is obtained as

$$\begin{aligned} S &= \overline{Q}(t)Q(t+1) = \overline{Q}(t)(f_0\overline{Q}(t) \vee f_1Q(t)) = f_0\overline{Q}(t), \\ DC\_S &= Q(t)Q(t+1) = Q(t)(f_0\overline{Q}(t) \vee f_1Q(t)) = f_1Q(t), \\ R &= Q(t)\overline{Q}(t+1) = Q(t)(\bar{f}_0\overline{Q}(t) \vee \bar{f}_1Q(t)) = \bar{f}_1Q(t), \\ DC\_R &= \overline{Q}(t)\overline{Q}(t+1) = \overline{Q}(t)(\bar{f}_0\overline{Q}(t) \vee \bar{f}_1Q(t)) = \bar{f}_0\overline{Q}(t), \end{aligned}$$

where DC_S and DC_R denotes *don't care* conditions for S and R, respectively. Note that S and R cannot be 1 at the same time.

JK Flip-Flop

Similarly, the input for the JK·FF is obtained as

$$\begin{aligned} J &= \overline{Q}(t)Q(t+1) = \overline{Q}(t)(f_0\overline{Q}(t) \vee f_1Q(t)) = f_0\overline{Q}(t), \\ DC\_J &= Q(t), \\ K &= Q(t)\overline{Q}(t+1) = Q(t)(\bar{f}_0\overline{Q}(t) \vee \bar{f}_1Q(t)) = \bar{f}_1Q(t), \\ DC\_K &= \overline{Q}(t). \end{aligned}$$

By using DC_J, J is simplified to $J = f_0$. Also, by using DC_K, K is simplified to $K = \bar{f}_1$.

D Flip-Flop

The input for the D·FF is

$$D = Q(t+1) = f_0\overline{Q}(t) \vee f_1 Q(t).$$

There is no *don't care* conditions.

T Flip-Flop

The input for the T·FF is

$$\begin{aligned} T &= Q(t+1)\overline{Q}(t) \vee \overline{Q}(t+1)Q(t) \\ &= (f_0\overline{Q}(t) \vee f_1 Q(t))\overline{Q}(t) \vee (\bar{f}_0\overline{Q}(t) \vee \bar{f}_1 Q(t))Q(t) \\ &= f_0\overline{Q}(t) \vee \bar{f}_1 Q(t). \end{aligned}$$

There is no *don't care* conditions. T is also represented as $T = Q(t+1) \oplus Q(t)$.

6.3 REPRESENTATION OF SEQUENTIAL NETWORKS

A **state diagram** represents the function of a sequential network. This section introduces it by using a vending machine.

Example 6.1 Consider the automatic vending machine that sells soda at 150 yen. This machines accept both a 100 yen coin and a 50 yen coin, and produces change if necessary. The state diagram in Fig. 6.9 has three internal states Q_0, Q_1, and Q_2. Q_0 is the initial state, i.e., the state without money, Q_1 and Q_2 are the states that machine has received 50 yen and 100 yen, respectively. The arrows denote the transition from one state to another state. The symbols $x_1\ x_2/z_1\ z_2$ which are attached to the arrows denote: $x_1 = 1$ (the machine received a 100 yen coin), $x_2 = 1$ (the machine received a 50 yen coin), $z_1 = 1$ (the machine gives soda), and $z_2 = 1$ (the machine pays change).
The symbols 0 1/0 0 on the arrow from Q_0 to Q_1 denotes: "In Q_0 (the state without money), if a machine receive a 50 yen coin, then the internal state goes to Q_1 (the state where the machine received 50 yen)." The symbols 1 0/1 1 on the arrows from Q_2 to Q_0 denote: "In Q_2 (the state with 100 yen), if a machine receive a 100 yen coin, then the machine gives soda as well as pays 50

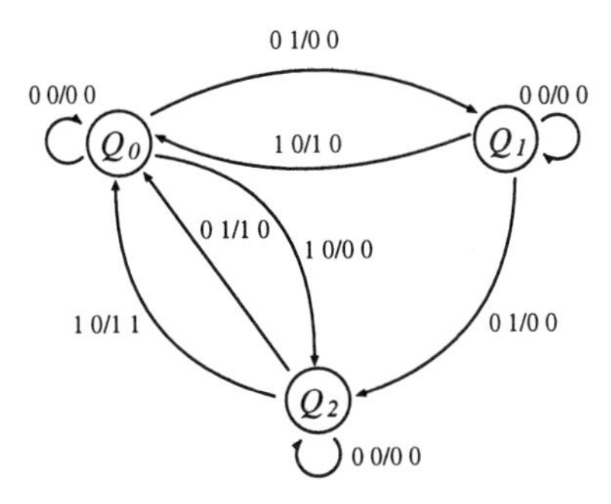

Figure 6.9 State diagram for a vending machine.

Table 6.6 State table for a vending machine.

Present state	Input x_1x_2 0 0	0 1	1 0
Q_0	Q_0, 00	Q_1, 00	Q_2, 00
Q_1	Q_1, 00	Q_2, 00	Q_0, 10
Q_2	Q_2, 00	Q_0, 10	Q_0, 11

yen change, and goes to Q_0 (the state without money)." When the number of states is not so large, a state diagram is an easy-to-understand representation of a sequential network. ∎

An alternative representation of a sequential network is a **state table.** For example, Table 6.6 is a tabular representation of the state diagram in Fig. 6.9.

6.4 STATE ASSIGNMENT AND STATE TABLE

6.4.1 State Assignment

In a state table, internal states are represented by symbols. To realize as a sequential network as shown in Fig. 6.1, the internal states must be represented

Table 6.7 State assignment.

Internal state	y_1	y_2
Q_0	0	0
Q_1	0	1
Q_2	1	0

Table 6.8 Transition table.

$y_1 y_2$	$x_1 x_2$		
	0 0	0 1	1 0
0 0	00, 00	01, 00	10, 00
0 1	01, 00	10, 00	00, 10
1 0	10, 00	00, 10	00, 11

by two-valued logic variables. Thus, each internal state is represented by a binary vector of state variables $(y_1, y_2, \ldots, y_s)$. For example, the state diagram shown in Fig. 6.9 has three internal states Q_0, Q_1, and Q_2, and each of them is represented by a binary vector (y_1, y_2) shown in Table 6.7. A **state assignment** assigns each state to a binary vector. s state variables represent at most 2^s internal states. The complexity of network depends on the method of state assignment. In addition to the internal states, when the inputs and outputs are represented by symbols, we assign them the vectors of the input variables and output variables, respectively. The problem to find a state assignment that simplifies the network is important. This problem is a **state assignment problem** or **encoding problem**. We will consider this encoding problem in Sections 7.3 and 10.8.

6.4.2 Transition Table

After the state assignment, a state table is represented by logic variables $x_1, x_2, \ldots, x_n$, $y_1, y_2, \ldots, y_s$, and $z_1, z_2, \ldots, z_m$. This is called a **transition table**. By using the state assignment in Table 6.7, the state table in Table 6.6 is converted to the transition table in Table 6.8. A transition table shows the values of input variables and state variables at time t, and corresponding values of state variables at time $t+1$ and output variables at time t.

The transition table gives the values of the state variables $y_1(t+1), y_2(t+1), \ldots, y_s(t+1)$, and the output variables $z_1(t), z_2(t), \ldots, z_m(t)$ as logic functions:

$$\begin{aligned} y_i(t+1) &= f_i(x_1(t), x_2(t), \ldots, x_n(t), y_1(t), y_2(t), \ldots, y_s(t)), \\ z_j(t) &= g_j(x_1(t), x_2(t), \ldots, x_n(t), y_1(t), y_2(t), \ldots, y_s(t)). \end{aligned}$$

In this case, f_i $(i = 1, 2, \ldots, s)$ are **state transition functions**, and g_j $(j = 1, 2, \ldots, m)$ are **output functions**. From here, for simplicity, $x_i(t)$, $y_i(t)$, and $z_j(t)$ are denoted by x_i, y_i, and z_j, respectively, and $y_i(t+1)$ is denoted by y_i'.

6.5 REALIZATION OF SEQUENTIAL NETWORKS

6.5.1 Design Procedure

Given a state table and the type of the flip-flips, we can design the sequential network as follows:

Algorithm 6.1 *(Realization of Sequential Networks)*

1. *Derive the state table from the word description.*
2. *Simplify the state table (Chapter 7 considers this problem).*
3. *Assign the codes to the state, and derive the transition table. Also, derive the output table showing output functions.*
4. *Select the flip-flops.*
5. *Derive the excitation functions for flip-flops.*
6. *Realize the logic networks for excitation functions and output functions.*

6.5.2 An Example by Using Various Types of FFs

Example 6.2 Let us realize the automatic vending machine for soda in Section 6.3, by using various types of FFs. From the state table shown in Table 6.8, the map for the state transition function is given in Fig. 6.10, and the map for the output function is given in Fig. 6.11. In the maps, $\times$ denotes the condition that never happen during operations: $y_1 = y_2 = 1$ and $x_1 = x_2 = 1$. These conditions can be considered as *don't cares.* From these maps, we have the state transition functions y_1' and y_2' as follows:

$$\begin{aligned} y_1' &= x_1\bar{y}_1\bar{y}_2 \vee x_2y_2 \vee (\bar{x}_1\bar{x}_2)y_1 \\ &= (x_1\bar{y}_2 \vee x_2y_2)\bar{y}_1 \vee (\bar{x}_1\bar{x}_2)y_1 \text{ (By considering don't care conditions.)} \\ y_2' &= (\bar{y}_1x_2)\bar{y}_2 \vee (\bar{x}_1\bar{x}_2)y_2. \end{aligned}$$

The output functions z_1 and z_2 are:

$$\begin{aligned} z_1 &= y_1(x_1 \vee x_2) \vee x_1y_2, \\ z_2 &= x_1y_1. \end{aligned}$$

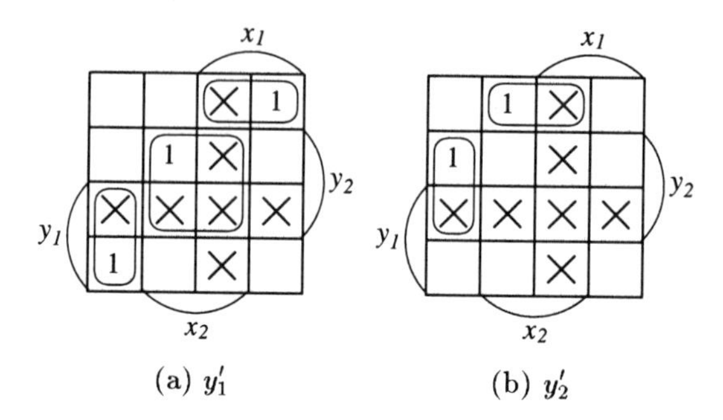

(a) y_1' (b) y_2'

Figure 6.10 Map for state transition functions.

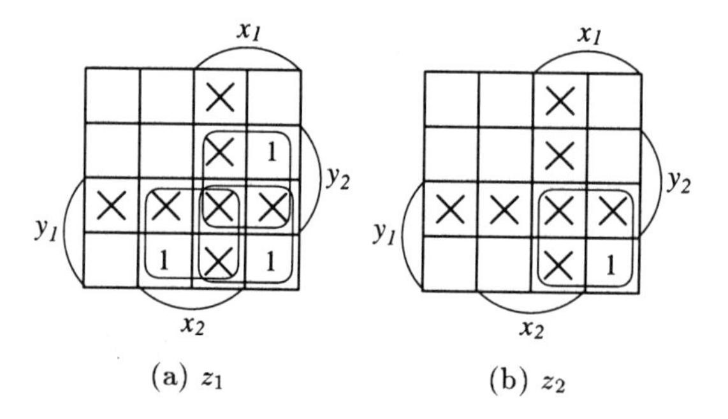

(a) z_1 (b) z_2

Figure 6.11 Map for output functions.

When SR·FFs are used

The input equation for S(Set) is $S = f_0\overline{Q}(t)$, and the don't care condition for S is $f_1Q(t)$. Therefore, the S input for y_1 FF is obtained as follows: First, to the $\bar{y}_1$ part in Fig. 6.12(a), copy the 1's in the $\bar{y}_1$ part of Fig. 6.10(a). Next, to the y_1 part in Fig. 6.12(a), copy the 1's in the y_1 part of Fig. 6.10(a), and replace these 1's with d's. Where d denotes don't care condition for S, and can be either 0 or 1. Also, copy ×'s in Fig. 6.10 to Fig. 6.12. Thus, we have the equation:

$$S_1 = x_1\bar{y}_1\bar{y}_2 \vee x_2y_2.$$

The input equation for R(Reset) input is $R = \bar{f}_1Q(t)$, and the don't care condition for R is $\bar{f}_0\overline{Q}(t)$. Thus, the R input for y_1 FF is obtained as follows: First, to the y_1 part in Fig. 6.12(b), copy 0's in the y_1 part of Fig. 6.10(a), and replace these 0's with 1's. Next, to the $\bar{y}_1$ part in Fig. 6.12(b), copy the 0's in the $\bar{y}_1$ part in Fig. 6.10(a), and replace these 0's with d's. From these, we have

$$R_1 = y_1(x_1 \vee x_2).$$

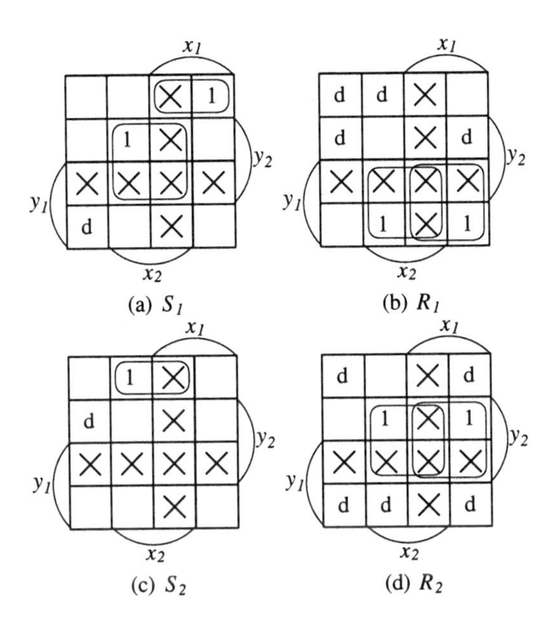

Figure 6.12 Map showing the inputs for SR·FF.

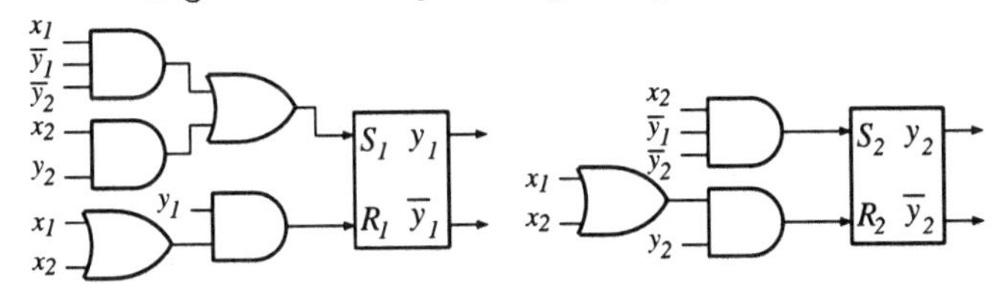

Figure 6.13 Excitation networks for SR·FF.

To obtain the inputs for y_2 FF, we can do the similar operations to the map in Fig. 6.10(b). In this case, we have to consider them with respect to y_2 and $\bar{y}_2$. The maps for S_2 and R_2 are shown in Figs. 6.12(c) and 6.12(d), respectively. Thus, we have

$$\begin{aligned} S_2 &= x_2\bar{y}_1\bar{y}_2, \\ R_2 &= (x_1 \vee x_2)y_2. \end{aligned}$$

In this way, we have the excitation network for SR·FF as shown in Fig. 6.13.

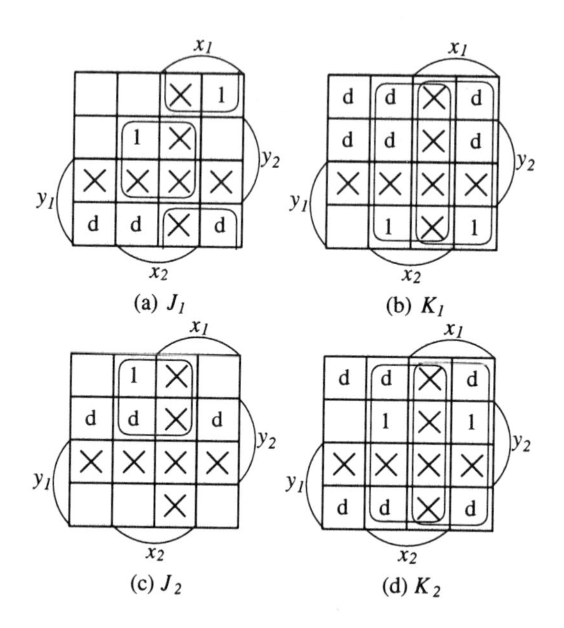

Figure 6.14 Map showing inputs for JK·FF.

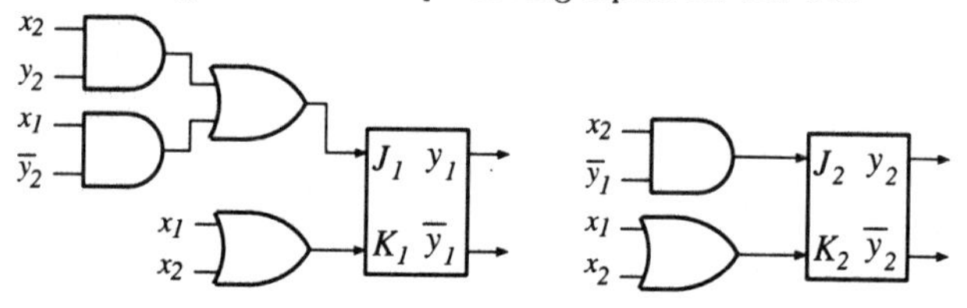

Figure 6.15 Excitation networks for JK·FF.

When JK·FFs are used

The equation for J input is $J = f_0\overline{Q}(t)$, and the don't care condition for J input is $Q(t)$. Therefore, the J input for y_1 FF is obtained as follows: First, to the $\bar{y}_1$ part of Fig. 6.14(a), copy the 1's in the $\bar{y}_1$ part of Fig. 6.10(a). Next, in all the cells in y_1 of Fig. 6.14(a), enter d's. Thus, we have

$$J_1 = x_1\bar{y}_2 \vee x_2y_2.$$

The equation for the K input is $K = \bar{f}_1Q(t)$, and the don't care condition for K is $\overline{Q}(t)$. Therefore, the K input for y_1 FF is obtained as follows: First, to the y_1 part in Fig. 6.14(b), copy the 0's in the y_1 part in Fig. 6.10(a), and replace

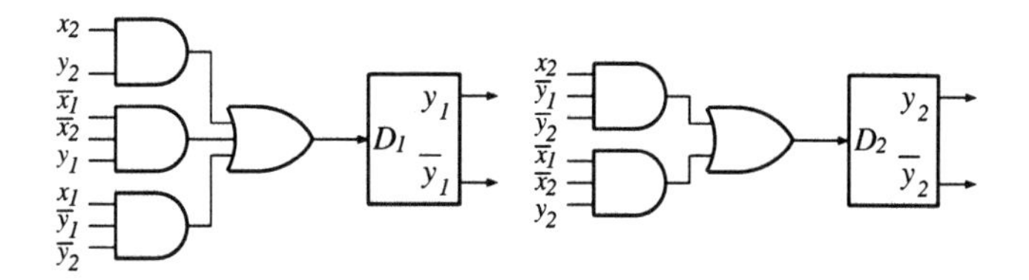

Figure 6.16 Excitation networks for D·FF.

them with 1's. Next, in all the cells in $\bar{y}_1$, enter d's. Thus, we have

$$K_1 = x_1 \vee x_2.$$

Similarly, we have

$$\begin{aligned} J_2 &= x_2\bar{y}_1, \\ K_2 &= x_1 \vee x_2. \end{aligned}$$

From these, we have the excitation network for JK·FF as shown in Fig. 6.15. The excitation network for JK·FFs is simpler than that for SR·FFs, since the maps for JK·FFs have more don't care conditions.

When D·FFs are used

Since the input for D·FF is equal to the state transition function, we can directly obtain the input conditions from Fig. 6.10:

$$\begin{aligned} D_1 &= x_2y_2 \vee \bar{x}_1\bar{x}_2y_1 \vee x_1\bar{y}_1\bar{y}_2, \\ D_2 &= x_2\bar{y}_1\bar{y}_2 \vee \bar{x}_1\bar{x}_2y_2. \end{aligned}$$

From these, the excitation network for D·FFs is obtained as Fig. 6.16.

When T·FFs are used

The input for T·FF is $T = f_0\overline{Q}(t) \vee \bar{f}_1Q(t)$. Thus, the T input for the y_1 FF is obtained as follows: First, to the $\bar{y}_1$ part in Fig. 6.17(a), copy the 1's in the $\bar{y}_1$ part of Fig. 6.10(a). Next, to the y_1 part in Fig. 6.17(a), copy the 0's in the y_1 part of Fig. 6.10(a) and replace these 0's with 1's. In this way, we have

$$T_1 = x_1\bar{y}_2 \vee x_2y_2 \vee x_2y_1.$$

Similarly, we have

$$T_2 = x_1y_2 \vee x_2\bar{y}_1.$$

From these, the excitation network for T·FF is obtained as shown in Fig. 6.18.

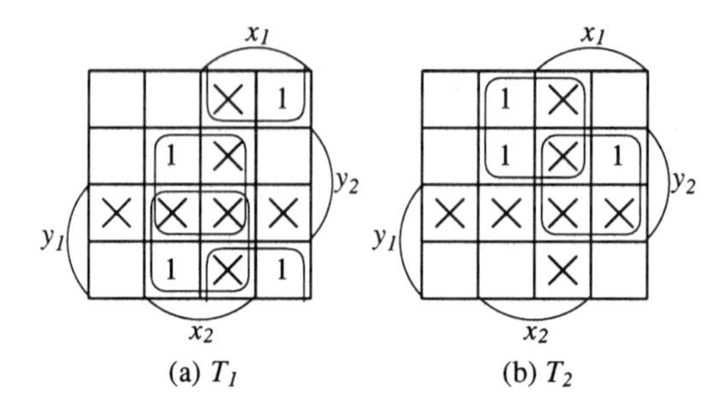

Figure 6.17 Map for the inputs of T·FF.

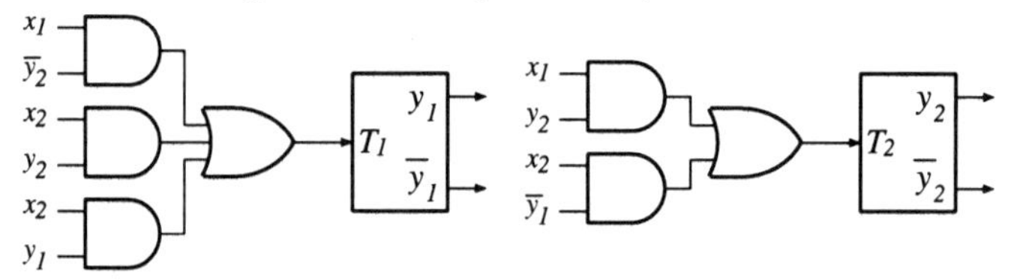

Figure 6.18 Excitation networks for T·FF.

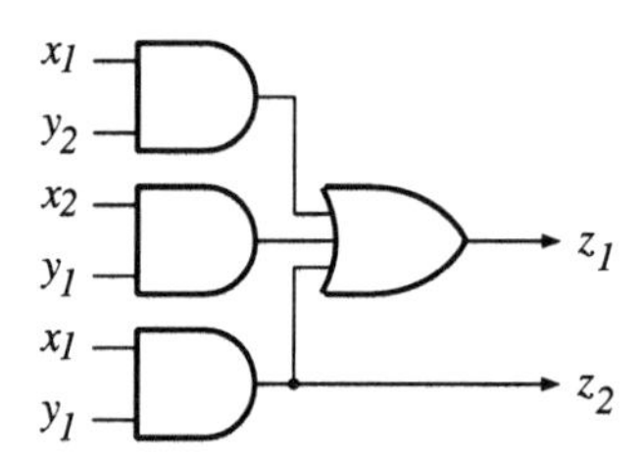

Figure 6.19 Output network.

Output network

The networks for output functions are independent of the type of FFs, and they are obtained from Fig. 6.11:

$$\begin{aligned} z_1 &= x_1 y_1 \vee x_1 y_2 \vee x_2 y_1, \\ z_2 &= x_1 y_1. \end{aligned}$$

Since an AND gate can be shared in z_1 and z_2, we have the network in Fig. 6.19. ∎

Bibliographical Notes

Detailed discussions on sequential networks can be found in [10, 70, 127, 212].

Table 6.9 Transition table.

		$x=0$		$x=1$	
y_1	y_2	y_1'	y_2'	y_1'	y_2'
0	0	1	0	0	1
0	1	0	1	1	0
1	0	1	0	1	1
1	1	0	0	0	1

Table 6.10 Transition table.

y_1	y_2	y_3	y_1'	y_2'	y_3'
0	0	0	0	0	1
0	0	1	0	1	0
0	1	0	0	1	1
0	1	1	1	0	0
1	0	0	1	0	1
1	0	1	0	0	0

Exercises

6.1 Design the following octal counter by using AND, OR, NOT gates and D·FFs: When $X = 1$, count as 0, 1, 2, 3, 4, 5, 6, 7, 0, $\cdots$. When $X = 0$, the state will not change.

6.2

(a) Draw the state diagram of the following quaternary counter: When $U = 1$, count as 0, 1, 2, 3, 0, 1, 2, 3, 0, $\cdots$. When $U = 0$, count as 0, 1, 3, 2, 0, 1, 3, 2, 0, $\cdots$.

(b) Draw the state diagram of the following quaternary up-down counter: When $U = 1$, count as 0, 1, 2, 3, 0, 1, 2, $\cdots$. When $U = 0$, count as 0, 3, 2, 1, 0, 3, 2, $\cdots$.

6.3 Design the serial binary adder by using a T·FF. Let the inputs be x_1 and x_2, and let y_1 be the previous output. Then, this network produce mod-2 addition $x_1 \oplus x_2 \oplus y_1$.

6.4 By using JK·FFs, design a sequential network for Table 6.9.

6.5 By using JK·FFs, design a sequential network for Table 6.10.

6.6 By using JK·FFs, design the **Gray code** octal counter as follows: Count as 000, 001, 011, 010, 110, 111, 101, 100, 000, $\cdots$.

6.7 Derive the transition table for Fig. 6.20. Design the network with this transition table by using JK·FFs.

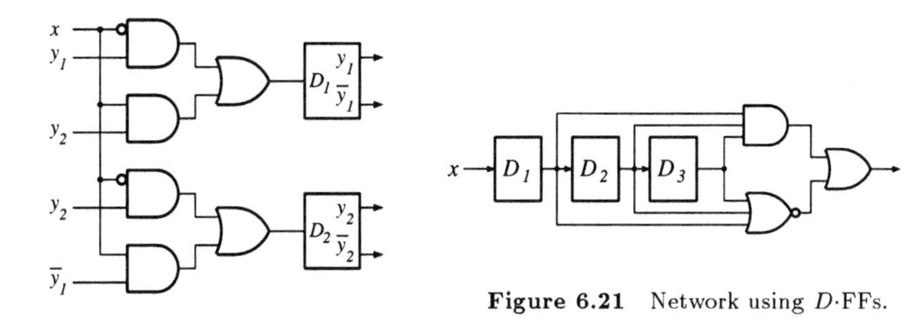

Figure 6.20 Network using D·FFs.

Figure 6.21 Network using D·FFs.

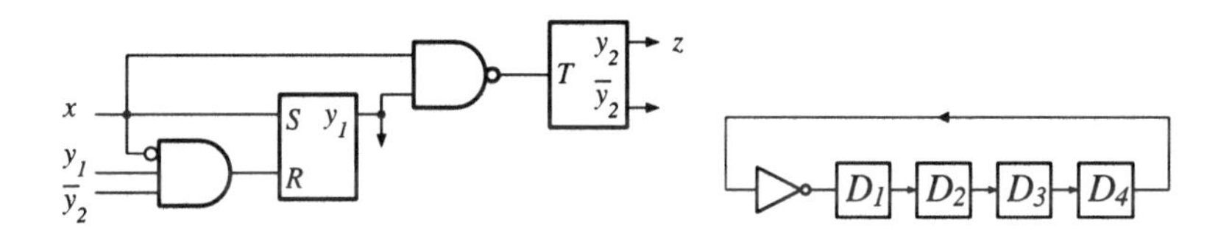

Figure 6.22 A network using two kinds of FFs.

Figure 6.23 Network using D·FFs.

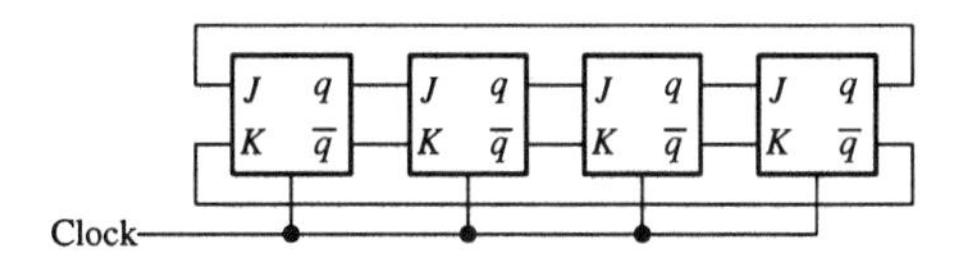

Figure 6.24 Network using JK·FFs.

6.8 Derive the transition table for Fig. 6.21. Describe the operation in words.

6.9 Derive the transition table for Fig. 6.22.

6.10 Derive the transition diagram for Fig. 6.23.

6.11 Derive the transition diagram for Fig. 6.24.

6.12 Derive the transition table for Fig. 6.25.

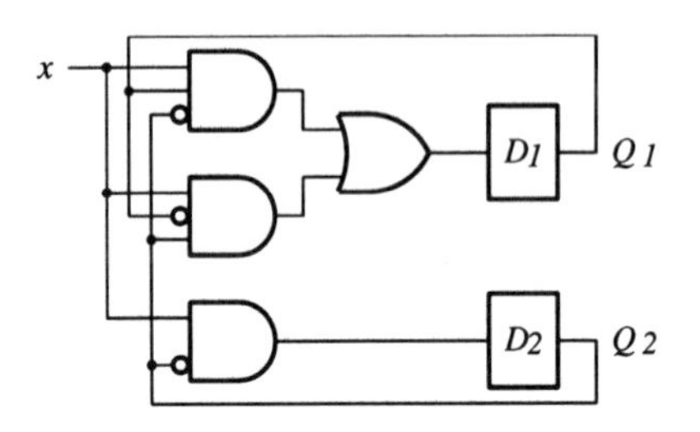

Figure 6.25 Network using D·FFs.

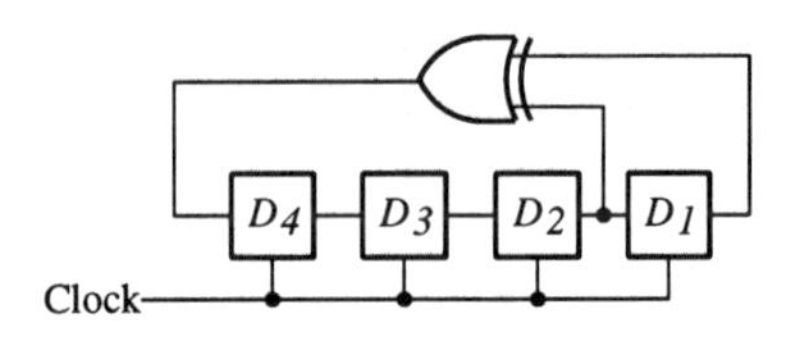

Figure 6.26 Network using D·FFs.

6.13 Derive the transition diagram for Fig. 6.26.

6.14 For any given transition table, show that the excitation network for the JK·FFs is not more complex than that for the RS·FFs.

7

OPTIMIZATION OF SEQUENTIAL NETWORKS

To realize sequential networks from state tables, we first minimize the number of states. Section 7.1 shows the minimization method for state tables without *don't cares.* Section 7.2 shows the minimization method for state tables with *don't cares.* Next, we assign a binary code to each state. This process is called state assignment. In this case, the problem is to minimize the number of flip-flops or to simplify the combinational networks. Section 7.3 shows these problems. Optimization of sequential networks often makes networks simpler and faster.

7.1 OPTIMIZATION OF COMPLETELY SPECIFIED SEQUENTIAL MACHINES

In Section 6.3, we showed a method to derive a state table from the given specification. However, as shown in the next example, the state table is not unique for a given specification.

Example 7.1 State table 7.1(a) has two states: Q_0 and Q_1. On the other hand, State table 7.1(b) has three states: Q_0, Q_1, and Q_2. However, these two state tables behave exactly the same ways as far as inputs and outputs are concerned: In the rows for Q_0, the corresponding entries in both state tables are the same. Thus, they behave the same ways in this state. In the rows for Q_1, assume that the input is $x = 0$. In table 7.1(a), the next state is Q_0, while in table 7.1(b), the next state is Q_2. However, the entries of the rows for Q_0 and Q_2 are the same. Thus, the state table that is obtained from Table 7.1(b) by replacing the label Q_2 with Q_0 behaves as it was. In the rows for Q_1, assume

Table 7.1 State tables with the same behavior.

(a)

Present state	Next state, output	
	$x = 0$	$x = 1$
Q_0	Q_0, 0	Q_1, 0
Q_1	Q_0, 1	Q_1, 1

(b)

Present state	Next state, output	
	$x = 0$	$x = 1$
Q_0	Q_0, 0	Q_1, 0
Q_1	Q_2, 1	Q_1, 1
Q_2	Q_0, 0	Q_1, 0

that the input is $x = 1$. The two state tables behave the same ways since corresponding entries are the same.
Therefore, we can conclude that state tables 7.1(a) and (b), behave exactly the same ways. In other words, when the same input sequences are applied to two tables, they produce exactly the same output sequences. ∎

As shown in Example 7.1, two different state tables behave exactly the same manner. This section shows a method to derive the table with the minimum number of states satisfying a given specification. A reduction of the number of states in a state table usually simplifies the network. Thus, the minimization of number of states is desirable in the design of sequential networks. Let R be the number of states, then the table can be realized by using $\lceil \log_2 R \rceil$ flip-flops, where $\lceil x \rceil$ denotes the minimum integer greater than or equal to x.

Definition 7.1 *A* **finite state machine** *(FSM) represents the behavior of a sequential network. A* **Moore machine** *is defined by a 5-tuple $M = (S, I, O, \delta, \lambda)$, where*

S : *set of states,*
I : *set of inputs,*
O : *set of outputs,*
$\delta : S \times I \rightarrow S$ *is the state transition function,*
$\lambda : S \rightarrow O$ *is the output function,*

and S, I, O are non-empty finite sets. A **Mealy machine** *is the same as the Moore machine except for the output function:*

$\lambda : S \times I \rightarrow O.$

Table 7.2 Original state table.

Present state	Next state, output	
	$x = 0$	$x = 1$
Q_0	Q_4, 0	Q_3, 1
Q_1	Q_5, 0	Q_3, 0
Q_2	Q_4, 0	Q_1, 1
Q_3	Q_5, 0	Q_1, 0
Q_4	Q_2, 0	Q_5, 1
Q_5	Q_1, 0	Q_2, 0

Table 7.3 Minimized state table.

Present state	Next state, output	
	$x = 0$	$x = 1$
Q_0	Q_4, 0	Q_1, 1
Q_1	Q_5, 0	Q_1, 0
Q_4	Q_0, 0	Q_5, 1
Q_5	Q_1, 0	Q_0, 0

Definition 7.2 *In an FSM, for each pair consisting of a state and an input, if the next state and output are defined, then the FSM is* **completely specified,** *otherwise the FSM is* **incompletely specified.**

Definition 7.3 *Let two states of FSM be s_i and s_j. Two states s_i and s_j are* **distinguishable** *if they produce different output sequences for the same input sequence. Such an input sequence is a* **distinguishing sequence** *of (s_i, s_j). If there is no distinguishing sequence for (s_i, s_j), then s_i and s_j are* **equivalent.** *If there is a distinguishing sequence with length k for (s_i, s_j), then (s_i, s_j) is* **k-distinguishable.**

In the case of a completely specified FSM, the set of equivalent states forms an equivalence class. And, this equivalence relation partitions all the states into equivalent sets of states. By merging equivalent states, we have the FSM with the minimum number of states. To obtain the FSM with the minimum number of states, start with a partition of states with only one block. Then, first refine this partition into 1-distinguishable states, second into 2-distinguishable states, and so on.

Theorem 7.1 *Two states s_i and s_j are in the same block in the partition Π_{k+1} iff the state table satisfies the following two conditions:*

1) *s_i and s_j are in the same block in the partition Π_k.*
2) *For any inputs, the states that go from s_i and s_j are in the same block in Π_k.*

Procedure 7.1 *(Minimization of FSM)*
In a completely specified state table,

1. *Let* Π_1 *be a partition obtained by grouping the states with the same output values for the input sequence of length one.*
2. *By using Theorem 7.1, derive* Π_{i+1} *from* Π_i*.*
3. *Repeat Step 2 until* $\Pi_i = \Pi_{i+1}$*.*
4. *The FSM for* Π_i *has the minimum number of states.*

Example 7.2 Let us minimize the number of states in the FSM shown in Table 7.2. Let Π_k be the partition obtained by the distinguishing sequence with length k. Π_0 consists of only one block consisting all the states. It corresponds to the partition where no input is applied:

$$\Pi_0 = \{[Q_0, Q_1, Q_2, Q_3, Q_4, Q_5]\}.$$

When $x = 1$ is applied as an input: For Q_0, Q_2, and Q_4, the outputs are 1, for other cases, the outputs are 0. When $x = 0$ is applied as an input: All the outputs are the same. Thus, we have

$$\Pi_1 = \{[Q_0, Q_2, Q_4], [Q_1, Q_3, Q_5]\}.$$

Next, suppose that $x = 11$ is applied as an input sequence. The output sequences for the states Q_1, Q_3, and Q_5 are 00, 00, and 01, respectively. Thus, $[Q_1, Q_3]$ and $[Q_5]$ are distinguishable. However, other input sequences of length two, 00, 01, or 10, cannot distinguish the states. Thus, we have the following:

$$\Pi_2 = \{[Q_0, Q_2, Q_4], [Q_1, Q_3], [Q_5]\}.$$

Next, suppose that $x = 111$ is applied as an input sequence. The output sequence for the states Q_0, Q_2, and Q_4 are 100, 100, and 101, respectively. Thus, $[Q_0, Q_2]$ and $[Q_4]$ are distinguishable. However, other input sequences of length three, $000, 001, \ldots, 110$, cannot distinguish the states. Thus, we have the following:

$$\Pi_3 = \{[Q_0, Q_2], [Q_4], [Q_1, Q_3], [Q_5]\}.$$

However, any input sequence of length four, such as $x = 1111$, do not distinguish states in $[Q_0, Q_2]$ nor $[Q_1, Q_3]$. Thus, we stop the operation. In this way, we know that states Q_0 and Q_2 are equivalent, and states Q_1 and Q_3 are equivalent. Table 7.3 shows the FSM with the minimum number of states. ∎

Example 7.3 Let us minimize the number of states in the FSM shown in Table 7.4. Let Π_k be the partition obtained by the distinguishing sequences

Table 7.4 Original state table.

Present state	Next state, output	
	$x = 0$	$x = 1$
Q_0	$Q_2, 0$	$Q_1, 1$
Q_1	$Q_2, 1$	$Q_0, 1$
Q_2	$Q_4, 0$	$Q_1, 1$
Q_3	$Q_5, 0$	$Q_0, 0$
Q_4	$Q_0, 0$	$Q_6, 1$
Q_5	$Q_3, 1$	$Q_2, 1$
Q_6	$Q_4, 1$	$Q_2, 1$

Table 7.5 Minimized state table.

Present state	Next state, output	
	$x = 0$	$x = 1$
Q_0	$Q_0, 0$	$Q_1, 1$
Q_1	$Q_0, 1$	$Q_0, 1$
Q_3	$Q_5, 0$	$Q_0, 0$
Q_5	$Q_3, 1$	$Q_0, 1$

with length k. Π_0 corresponds to the case where no input sequence is applied. It consists of only one partition:

$$\Pi_0 = \{[Q_0, Q_1, Q_2, Q_3, Q_4, Q_5, Q_6]\}.$$

After applying the input sequences of length one, we have

$$\Pi_1 = \{[Q_0, Q_2, Q_4], [Q_3], [Q_1, Q_5, Q_6]\}.$$

Next, in Π_1, for the input $x = 0$, from state Q_5 and Q_6, the FSM goes to states Q_3 and Q_4, respectively. Since Q_3 and Q_4 belong to the different blocks, by Theorem 7.1 we have the following partition:

$$\Pi_2 = \{[Q_0, Q_2, Q_4], [Q_3], [Q_1, Q_6], [Q_5]\}.$$

Next, in Π_2, we apply input sequences of length three. Since

$$\Pi_3 = \{[Q_0, Q_2, Q_4], [Q_3], [Q_1, Q_6], [Q_5]\} = \Pi_2,$$

we have the FSM with the minimum number of states as shown in Table 7.5. ∎

7.2 OPTIMIZATION OF INCOMPLETELY SPECIFIED SEQUENTIAL MACHINES

In the previous section, we considered a minimization for states in a completely specified sequential machine. However, in the case of incompletely specified sequential machines, we cannot use the notion of *equivalence relation* of states, but we have to introduce the notion of *compatible relation*.

Table 7.6 Incompletely specified state table.

Present state	Next state, output			
	I_1	I_2	I_3	I_4
A	–	–	$E, 1$	–
B	$C, 0$	$A, 1$	$B, 0$	–
C	$C, 0$	$D, 1$	–	$A, 0$
D	–	$E, 1$	$B,-$	–
E	$B, 0$	–	$C,-$	$B, 0$

Definition 7.4 *Two states s_i and s_j are* **compatible** *if the following conditions hold:*

a) $i = j$, or
b) For any input I_k, the outputs of the FSM are consistent. For any input sequence applied to the pair of states (s_i, s_j), FSM goes to the compatible states.

If all the state pairs in a set are compatible, then the set is **compatible state set**.

Example 7.4 In Table 7.6, state pairs (A, C) and (B, C) are both compatible. However, the state pair (A, B) is incompatible, since the outputs are different when the input I_3 is applied to the states A and B. This shows that even if (A, C) is compatible and (B, C) is compatible, (A, B) is incompatible. Thus, unlike an equivalence relation the transitive law does not hold in a compatible relation. In Table 7.6, states B and C are compatible. For the input I_2, they go to A and D, respectively. In this case, the state pair (B, C) **implies** the state pair (A, D). ∎

To obtain the compatible state set, we will use the merger graph.

Definition 7.5 *The* **merger graph** *is an undirected graph satisfying the following conditions:*

a) Each node corresponds to a state of the FSM.
b) When two states are compatible, there is an edge between corresponding nodes. Each edge has implied state pair(s) as a label.

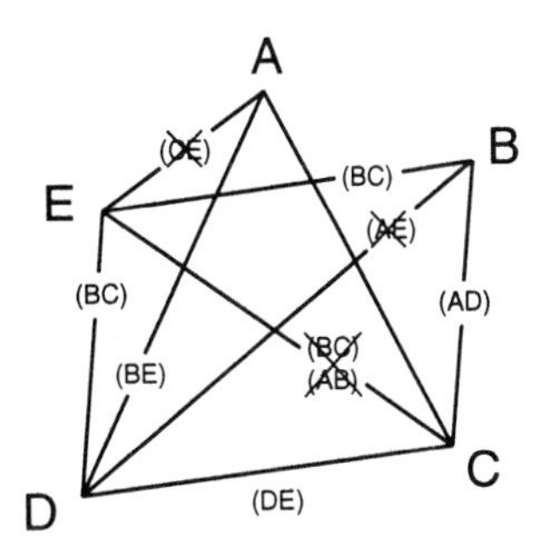

Figure 7.1 Merger graph.

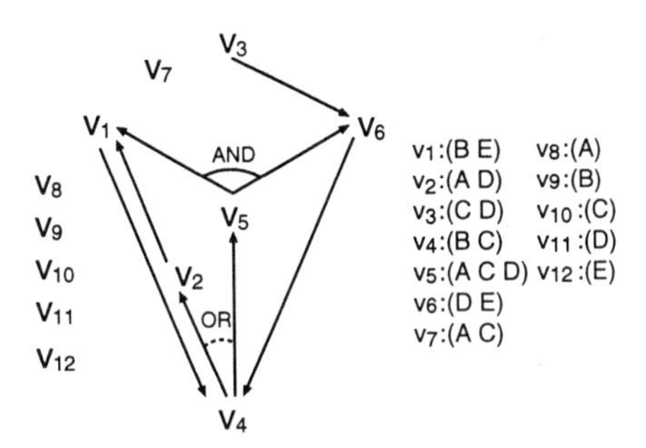

Figure 7.2 State compatibility graph.

Example 7.5 Fig. 7.1 is the merger graph for the incompletely specified state table in Table 7.6. Each edge has the condition for the pair of states to be compatible. For example, since the state pair (C, E) implies the state pairs (B, C) and (A, B), the edge CE has labels (B, C) and (A, B). However, (A, B) is incompatible, and we know that (C, E) is also incompatible. Thus, we erase the edge CE by using the $\times$ symbol. Similarly, (A, E) and (B, D) are incompatible, so we erase edges AE and BD by the $\times$ symbols. A compatible state set corresponds to the **complete subgraph**, i.e., a subgraph that has edges between all the pairs of nodes. For example, (A, C, D), (A, C), (A, D), $\dots$, (A), (B), (C), (D), (E) are complete subgraphs. ∎

To show the dependence of compatible state sets, we use the state compatibility graph.

Definition 7.6 *The* **state compatibility graph** *is a directed graph that satisfies the following conditions:*

a) Each node corresponds to a compatible state set of the merger graph.
b) When a state set implies another state set S.
Attach OR directed edges from the nodes corresponding to the original state set to the nodes corresponding to the state sets containing S.
c) When a state set implies two or more state sets: $S_0, \dots, S_{k-1}$, and S_k.
Attach AND directed edges from the node corresponding the original state set to the nodes that corresponds to $S_0, \dots,$ and S_k.

Example 7.6 Fig. 7.2 shows the state compatibility graph for the merger graph in Fig. 7.1. ∎

To obtain the FSM with the minimum number of states, we obtain the set V with the minimum number of nodes in the state compatibility graph satisfying the following conditions:

a) **Covering property**: Each state of the FSM is contained by at least one compatible set in V.
b) **Closed property**: If $v_i \in V$, then $V_i \subset V$, where V_i is the set of implied compatible sets. V_i satisfies the following conditions: For the OR directed edges that emerge from v_i, V_i contains the nodes for at least one edge. For the AND directed edges that emerge from v_i, V_i contains the nodes for all the AND directed edges.

To obtain V with the minimum number of nodes satisfying the above conditions, we represent each condition by a logical expression. Next, find the assignment of values to v_i's that makes all the equations 1 and that the weight is the minimum (i.e. the number of variables such that $v_i = 1$ is the minimum).

Example 7.7 From the state compatibility graph in Fig. 7.2, the following relations are derived:

a) Covering property.
   1. State A: $v_2 \vee v_5 \vee v_7 \vee v_8 = 1$;
   2. State B: $v_1 \vee v_4 \vee v_9 = 1$;
   3. State C: $v_3 \vee v_4 \vee v_5 \vee v_7 \vee v_{10} = 1$;
   4. State D: $v_2 \vee v_3 \vee v_5 \vee v_6 \vee v_{11} = 1$;
   5. State E: $v_1 \vee v_6 \vee v_{12} = 1$.
b) Closed property.
   6. From $v_1 \Rightarrow v_4$, we have: $\bar{v}_1 \vee v_4 = 1$;
   7. From $v_2 \Rightarrow v_1$, we have: $\bar{v}_2 \vee v_1 = 1$;
   8. From $v_3 \Rightarrow v_6$, we have: $\bar{v}_3 \vee v_6 = 1$;
   9. From $v_4 \Rightarrow (v_2 \vee v_5)$, we have : $\bar{v}_4 \vee v_2 \vee v_5 = 1$;
   10. From $v_5 \Rightarrow v_1 v_6$, we have: $(\bar{v}_5 \vee v_1)(\bar{v}_5 \vee v_6) = 1$;
   11. From $v_6 \Rightarrow v_4$, we have: $\bar{v}_6 \vee v_4 = 1$.

The condition that satisfies all the above at the same time is $\eta = 1$, where $\eta = (v_2 \vee v_5 \vee v_7 \vee v_8)(v_1 \vee v_4 \vee v_9)(v_3 \vee v_4 \vee v_5 \vee v_7 \vee v_{10})(v_2 \vee v_3 \vee v_5 \vee v_6 \vee v_{11})(v_1 \vee v_6 \vee v_{12})(\bar{v}_1 \vee v_4)(\bar{v}_2 \vee v_1)(\bar{v}_3 \vee v_6)(\bar{v}_4 \vee v_2 \vee v_5)(\bar{v}_5 \vee v_1)(\bar{v}_5 \vee v_6)(\bar{v}_6 \vee v_4)$. When $v_1 = v_2 = v_4 = 1$ and the other v_i are all 0, $\eta = 1$, and the weight is

Table 7.7 Minimized incompletely specified state table.

Present state	Next state, output			
	I_1	I_2	I_3	I_4
$(A,D) \rightarrow \alpha$	–	γ ,1	γ ,1	–
$(B,C) \rightarrow \beta$	β ,0	α ,1	β / γ ,0	α ,0
$(B,E) \rightarrow \gamma$	β ,0	α ,1	β ,0	β / γ ,0

Table 7.8 Example of state assignment.

State	Assignment 1	Assignment 2	Assignment 3
	y_1 y_2	y_1 y_2	y_1 y_2
Q_0	0 0	0 0	0 0
Q_1	0 1	1 1	1 0
Q_2	1 0	1 0	1 1
Q_3	1 1	0 1	0 1

the minimum. This corresponds to $V = \{(BC), (AD), (BE)\}$, and we have the minimum FSM for Table 7.7. The solution with the minimum weight corresponds to the products with the minimum un-complemented literals in the SOP for η. The FSMs with the minimum number of states are not unique. As shown in Table 7.7, a state in the original FSM may be associated with as two or more state sets in the simplified FSM (e.g., B is associated with β and γ). ∎

7.3 STATE ASSIGNMENT

After optimization of the state table, a state assignment is done. This assigns a binary code to a state. Different assignments produce networks with different complexities. To represent R different states, we need a binary code with at least $\lceil \log_2 R \rceil$ bits. A code with $\lceil \log_2 R \rceil$ bits is the **minimum length code.**

Example 7.8 To represent four states, we need two bits. There are 4!=24 different state assignment using a minimum length code. Table 7.8 shows three examples. ∎

As a figure-of-merit function for state assignments, we can use the complexity of the networks or the delay time. In most state assignments, only the complexity of the network is considered. The complexity of the network depends on the logic elements or types of flip-flops. The optimum state assignment can be obtained only when the network is very simple. In general, we can obtain only sub-optimal solutions. Two **heuristic methods** that obtain relatively good state assignments in a short time are known:

1. A method that simplifies logical expressions for multi-valued input two-valued output functions.
2. A method that consider the state transition of FSMs.

The first method is treated in Chapter 10. The second method uses the strategy that reduces the Hamming distance of codes of the state pairs that have frequent state transitions. When the Hamming distance of two codes is small, the network for the state transition tends to be simple.

To represent R states, we can use an R bit code whose weight is 1. Such a code is a **1-hot code.** For example, when $R = 4$, the state assignment is

$$\begin{aligned} Q_0 &= (1,0,0,0), \\ Q_1 &= (0,1,0,0), \\ Q_2 &= (0,0,1,0), \\ Q_3 &= (0,0,0,1). \end{aligned}$$

The 1-hot codes require more flip-flops than the minimum length codes, but the combinational networks are often simpler, and they are useful for high-speed networks.

Bibliographical Notes

Detailed treatments on optimization of FSMs can be found in [140, 146, 159, 247, 309]; for state assignments, see [5, 95, 418].

Table 7.9 State table.

	$x=0$	$x=1$
Q_0	Q_2, 0	Q_1, 1
Q_1	Q_2, 0	Q_3, 1
Q_2	Q_4, 0	Q_5, 1
Q_3	Q_4, 1	Q_0, 1
Q_4	Q_0, 0	Q_1, 1
Q_5	Q_4, 0	Q_3, 1

Table 7.10 State table.

	$x=0$	$x=1$
Q_0	Q_7, 0	Q_6, 1
Q_1	Q_2, 0	Q_4, 0
Q_2	Q_1, 0	Q_7, 0
Q_3	Q_4, 1	Q_7, 0
Q_4	Q_7, 0	Q_3, 1
Q_5	Q_4, 1	Q_2, 0
Q_6	Q_0, 1	Q_7, 0
Q_7	Q_3, 0	Q_5, 1

Exercises

7.1 Minimize the number of states in Table 7.9.

7.2 Minimize the number of states in Table 7.10.

7.3 Design the network that realizes Table 7.11 by using D flip-flops. In this case, which of the two assignments in Table 7.12 produces the simpler network?

Table 7.11 State table.

	$x=0$	$x=1$
Q_0	Q_2, 0	Q_1, 1
Q_1	Q_3, 1	Q_0, 0
Q_2	Q_0, 0	Q_3, 1
Q_3	Q_1, 1	Q_2, 0

Table 7.12 State assignments.

	Assignment 1	Assignment 2
Q_0	0 0	0 0
Q_1	0 1	0 1
Q_2	1 0	1 1
Q_3	1 1	1 0

7.4 Suppose that a sequential network is realized with an AND-OR two-level network and D flip-flips. Let R be the number of states and n be the number of state variables. Then, show that we need only consider at most

$$\frac{(2^n-1)!}{(2^n-R)!n!}$$

different state assignment to obtain the assignment that minimizes the number of gates. We assume that the complements of input variables as well as state variables are available.

7.5 In an *m-out-of-n code*, m bits out of n bits are 1's and other $(n - m)$ bits are 0's. If we realize a sequential network using an m-out-of-n code, inverters are not necessary. Prove this.

8 DELAY AND ASYNCHRONOUS BEHAVIOR

Design methods of sequential networks in Chapters 6 and 7 do not consider the effects of delay of logic elements, nor the influence on circuit operation produced by propagation delay in the interconnections (parasitic delay). However, in practical design, we have to consider the delays of gates and their variance.

In combinational networks, the delays of logic elements and interconnections can produce a spurious pulse called **hazard**.

Synchronous sequential networks use clock pulses, while asynchronous sequential networks do not. By using asynchronous networks, we can design faster networks. However, their design is more complicated than synchronous networks. When we try to change more than one state variables, a **race** will occur. A **critical race** can cause transition of the network into an erroneous state.

8.1 TRANSIENT RESPONSE OF COMBINATIONAL NETWORKS

Up to this chapter, we neglected gate **delays** in combinational logic networks. However, real gates require some time to change the output after input changes. Gate delays depend on the type of gates, the number of fan-in and fan-out, wiring length, load, supply voltage, temperature, and variance of gates. Delays are not constant. Consider the network shown in Fig. 8.1; this network realizes the function $f = xy \vee \bar{y}z$. Suppose that x and z are set to $x = z = 1$, and the value of y is changed from 1 to 0. By assigning $x = z = 1$, we have

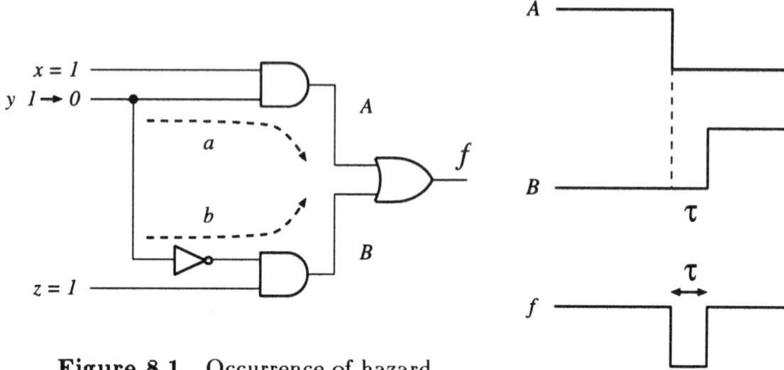

Figure 8.1 Occurrence of hazard.

Figure 8.2 1-hazard.

Figure 8.3 Static hazard.

Figure 8.4 Dynamic hazard.

$F = y \vee \bar{y} = 1$. This means that $f = 1$ independently of the value of y. When $(x, y, z) = (1, 1, 1)$, the path a is activated and $f = 1$. When $(x, y, z) = (1, 0, 1)$, the path b is activation and $f = 1$. Now, assume that the delay of the AND gate and the OR gate are 0, and the delay time for the NOT gate is τ. Fig. 8.2 shows the waveform in points A, B and the output f of Fig. 8.1, when the value y is changed from 1 to 0. Note that, erroneous 0 pulse of width τ occurs at the output. Such a transient error is called a **hazard**. As shown in Fig. 8.3(a), **0-hazard** is the hazard where the outputs before and after the input change are 0. On the other hand, as shown in Fig. 8.3(b), **1-hazard** is the hazard where the outputs before and after the input change are 1. As shown in Fig. 8.3, hazards where the output values before and after the input change are the same are **static hazards**. On the other hand, as shown in Fig. 8.4, the hazard where the output values before and after the input change are different is a **dynamic hazard**.

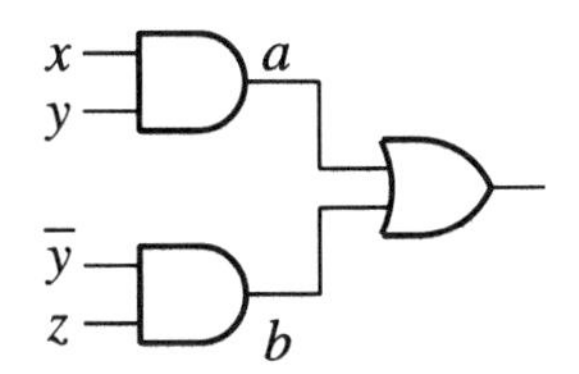

Figure 8.5 A network with 1-hazard.

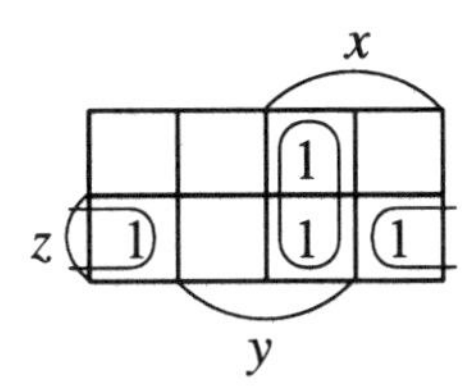

Figure 8.6 A sum-of-products expression with 1-hazard.

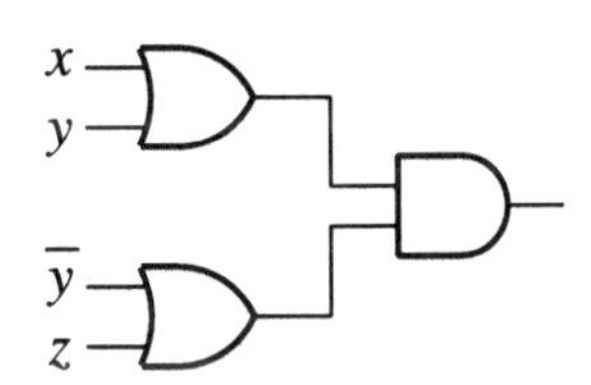

Figure 8.7 A network with 0-hazard.

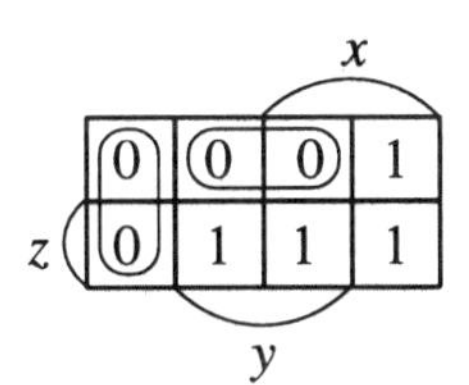

Figure 8.8 A product-of-sums expression with 0-hazard.

Detection of Hazards

In an AND-OR two-level network corresponding to a sum-of-products expression, consider a method to detect hazards produced by the single change of an input variable. In Fig. 8.5, suppose that x and z are set to 1 and the value y is changed from 1 to 0. If the delay of the path a is smaller than that of the path b, then the 1-hazard occurs at the output, as stated before. This can be explained by the map in Fig. 8.6. The input change corresponds to the transition from the product term xy to the product term $\bar{y}z$. The difference of delays of two product terms can produce the hazard. The **occurrence of hazard** means there is an erroneous-pulse in the actual network. On the other hand, the **existence of hazard** means that there is a possibility to produce a hazard. In a two-level network that corresponds to a sum-of-products expression, a single change of an input variable can produce only 1-hazards. It cannot produce 0-hazards nor dynamic hazards. On the other hand, in a two-level network that corresponds to a product-of-sums expression, a single change of an input variable can produce only 0-hazards. For example, the OR-AND two-level network in Fig. 8.7, corresponds to the map in Fig. 8.8. In this case, when x and z are set to $x = z = 0$, which corresponds to the 0-cell loop, a change of y can produce a 0-hazard.

Elimination of Hazard

In the case of AND-OR or OR-AND two-level networks, any hazard produced by a single input change can be eliminated completely. In Fig. 8.5, a hazard occurs when the input (x, y, z) is changed from (1, 1, 1) to (1, 0, 1), since no product term cover two inputs at the same time in Fig. 8.6. Therefore, as shown in Fig. 8.9, adding the AND gate that realizes the product xz eliminates this 1-hazard. In the map, this operation corresponds to adding the loop xz in Fig. 8.10. In an AND-OR two-level network, a 1-hazard for two adjacent inputs $\boldsymbol{a}$ and $\boldsymbol{b}$ can be eliminated by adding the product term that covers the minterms for $\boldsymbol{a}$ and $\boldsymbol{b}$ at the same time. In an OR-AND two-level network, a 0-hazard can be eliminated by the similar manner.

Logic Hazard and Function Hazard

In this part, we consider a hazard produced by two or more inputs change at a time. In Fig. 8.11, suppose that the input (x, y, z, w) changes from (0, 1, 0, 0) to (1, 1, 0, 1). The corresponding map is shown in Fig. 8.12. Two different paths of input changes exist: (x, y, z, w)=(0, 1, 0, 0)→(1, 1, 0, 0)→(1, 1, 0, 1), and (0, 1, 0, 0)→(0, 1, 0, 1)→ (1, 1, 0, 1). In the former case, as shown in Fig. 8.13(a), the 1-hazard changes output values four times. In the latter case, as shown in Fig. 8.13(b), the 1-hazard changes output values twice. However, these hazards can be eliminated by adding the product term $y\bar{z}$. Next, in the network in Fig. 8.14, consider the case where the two inputs change at the same time. As shown in Fig. 8.15, the first path produces the 1-hazard, but the second path does not. Also the first hazard cannot be eliminated by adding a product term. As shown in Fig. 8.12, the **logic hazard** is the static hazard where the function values are the same in all possible paths. The **function hazard** is one where the function values are different in the different paths as shown in Fig. 8.15. A static logic hazard can be eliminated by realizing an AND-OR two-level network consisting of all the prime implicants for f. However, the function hazard cannot be eliminated. Note that a single input change does not produce any function hazard.

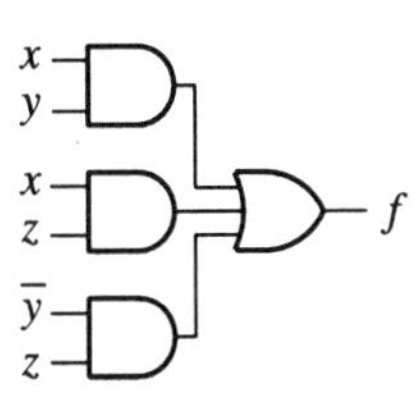

Figure 8.9 Elimination of hazard.

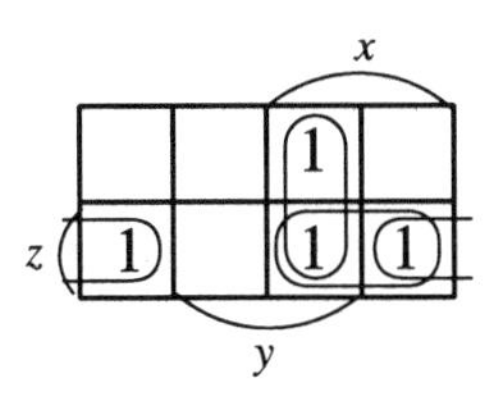

Figure 8.10 A sum-of-products expression without 1-hazard.

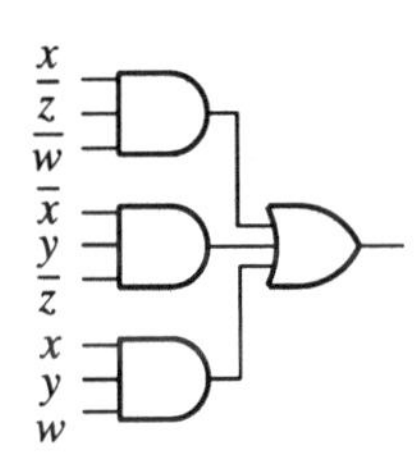

Figure 8.11 A network with logic hazard.

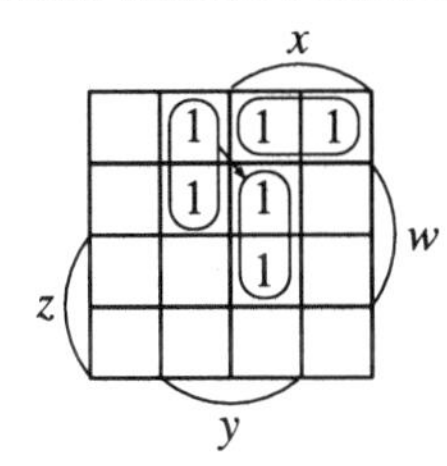

Figure 8.12 A sum-of-products expression with logic hazard.

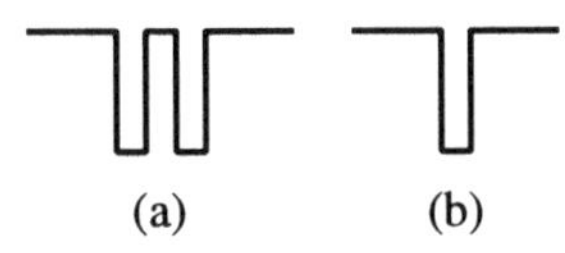

Figure 8.13 1-hazard.

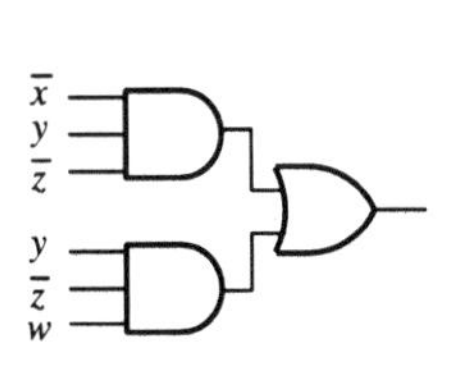

Figure 8.14 A network with function hazard.

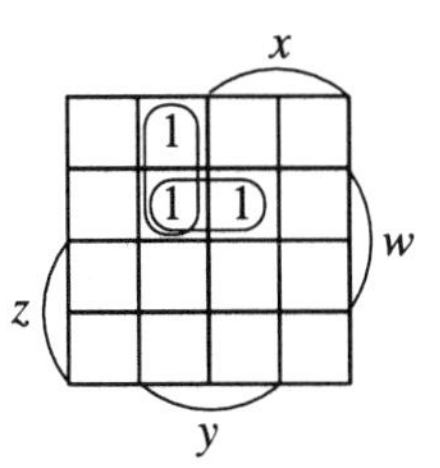

Figure 8.15 Function hazard.

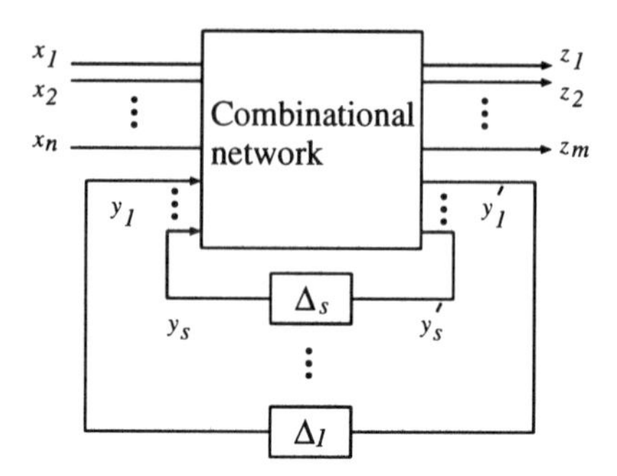

Figure 8.16 A feedback delay model of an asynchronous sequential network.

8.2 ASYNCHRONOUS SEQUENTIAL NETWORKS

Network Models

Fig. 8.16 shows a **feedback delay model (Huffman model)** of an asynchronous sequential network, where $x_1, x_2, \ldots, x_n$ denote input variables, $y_1, y_2, \ldots, y_s$ denote state variables, and $z_1, z_2, \ldots, z_m$ denote output variables. $\Delta_1, \Delta_2, \ldots, \Delta_s$ are delays representing state variables and the delay of feedback loops. The amount of delays can be different. Let y_i' $(i = 1, \ldots, s)$ be the next state value of variables y_i, and let z_j $(j = 1, \ldots, m)$ be the output variables, then these values are represented by

$$\begin{aligned} y_i' &= f_i(x_1, x_2, \ldots, x_n, y_1, y_2, \ldots, y_s) \quad (i = 1, \ldots, s), \\ z_j &= g_j(x_1, x_2, \ldots, x_n, y_1, y_2, \ldots, y_s) \quad (j = 1, \ldots, m). \end{aligned}$$

Example 8.1 (Problem) Represent the network in Fig. 8.17 by using a feedback delay model.

(Solution) The network in Fig. 8.17 seems to have two feedback loops. However, it can be modified to the network with only one feedback loop as shown in Fig. 8.18. Logic functions for y_1', z_1, and z_2 are given by $y_1' = (x_1 \vee y_1)\bar{x}_2$, $z_1 = y_1$, and $z_2 = \overline{x_1 \vee y_1}$. ∎

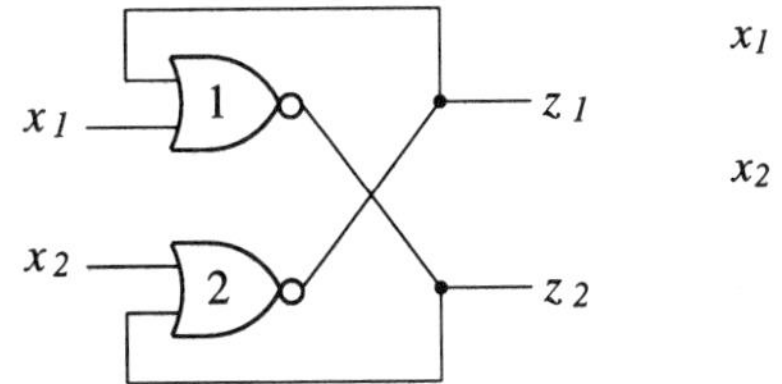

Figure 8.17 Network for Example 8.1.

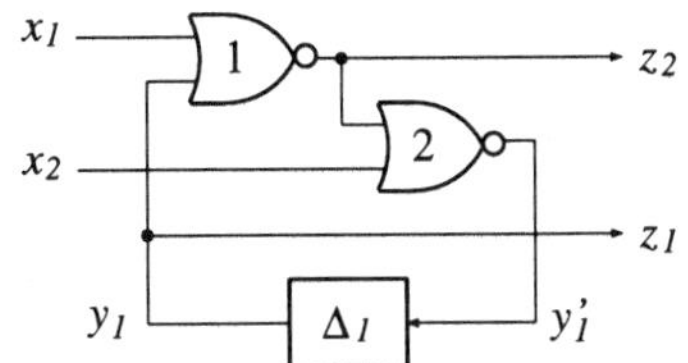

Figure 8.18 A feedback delay model for Fig. 8.17.

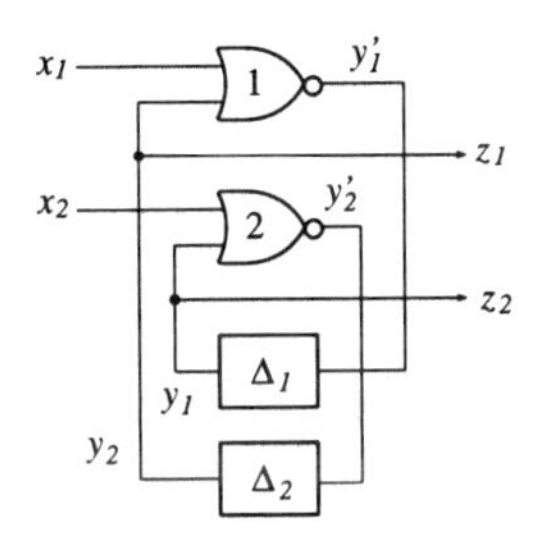

Figure 8.19 A model where each gate has different delay.

Example 8.2 (Problem) Represent the network in Fig. 8.17 by using a model where each gate has independent delay.

(Solution) Fig. 8.19 shows the network model. Logic functions for y_1', y_2', z_1, and z_2 are $y_1' = \overline{x_1 \vee y_2}$, $y_2' = \overline{x_2 \vee y_1}$, $z_1 = y_2$, and $z_2 = y_1$, respectively. ∎

Clock Pulse and Signal

As shown in Chapter 6, in a synchronous sequential network, all the operation are synchronized with the **clock pulse.** All the signals are defined only when the clock pulse is active. Thus, the input signals in Fig. 8.20(a) are interpreted as 01100011 shown in Fig. 8.20(c). The hazards between clock pulses are ignored. On the other hand, in an asynchronous sequential network, there is no clock pulse. Only the changes of the input signals are significant. For example, the signals in Fig. 8.20(a) are interpreted as 01010101.

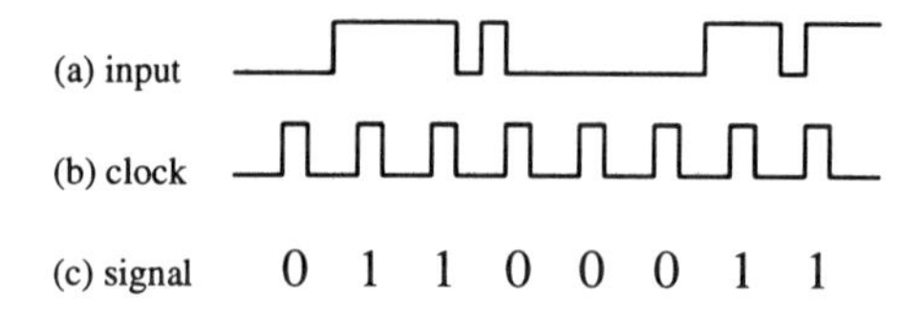

Figure 8.20 Relation of clock pulse and signal.

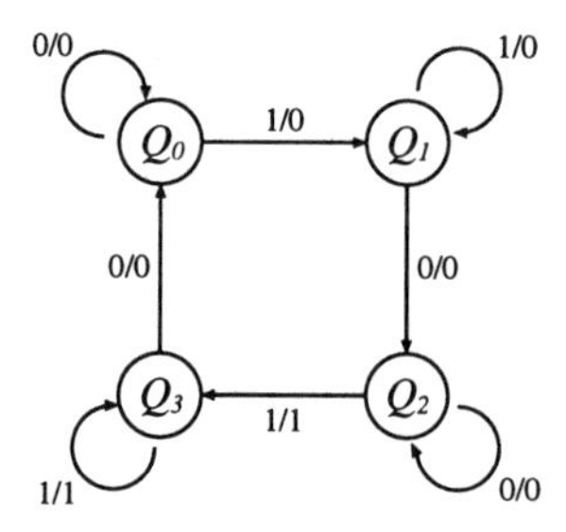

Figure 8.21 State diagram.

State Table

Similar to the case of a synchronous sequential network, an asynchronous sequential network is represented by a state table such as Table 8.1, or by a state diagram such as Fig. 8.21. In the case of an asynchronous network, a state is either **stable** or **unstable**. For an input, if the next state is the same as the present state, then the present state is stable under this input. The stable states are encircled by ○ symbols.

Example 8.3 Consider the state table in Table 8.1. In the initial state Q_0, if the input is changed from 0 to 1, then it moves to the stable state Q_1 through the intermediate state Q_0. Q_1 is stable with the input 1. In the state Q_1, if the input is changed from 1 to 0, then it moves to the stable state Q_2 through an unstable state Q_1. Q_2 is stable with the input 0. In state Q_2, if the input is changed from 0 to 1, then it moves to the stable state Q_3 through an unstable state Q_2. In this case, the output becomes 1. The following figure summarizes these:

Table 8.1 State table.

Present state	Input x = 0	Input x = 1
Q_0	(Q_0) 0	Q_1, 0
Q_1	Q_2, 0	(Q_1) 0
Q_2	(Q_2) 0	Q_3, 1
Q_3	Q_0, 0	(Q_3) 1

Table 8.2 State table for Fig. 8.18.

Present state	Input x_1x_2 = 00	01	11	10
Q_0	(Q_0) 01	(Q_0) 01	(Q_0) 00	Q_1, 00
Q_1	(Q_1) 10	Q_0, 10	Q_0, 10	(Q_1) 10

$$\begin{array}{lccccccccccccccccc}
\text{Input}: & 0 & & 1 & & 1 & & 0 & & 0 & & 1 & & 1 & & 0 & & 0 \\
\text{State}: & (Q_0) & \to & Q_0 & \to & (Q_1) & \to & Q_1 & \to & (Q_2) & \to & Q_2 & \to & (Q_3) & \to & Q_3 & \to & (Q_0) \\
\text{Output}: & 0 & & 0 & & 0 & & 0 & & 0 & & 1 & & 1 & & 0 & & 0
\end{array}$$

This show that Table 8.1 works as the binary counter. ∎

Total State

As shown in Example 8.3, in an asynchronous network, a state can be stable for some inputs, but unstable for other inputs. The **total state** is a pair (x, Q_i) of an input x and a state Q_i. In Table 8.1, the stable total states are $(0, Q_0)$, $(1, Q_1)$, $(0, Q_2)$, and $(1, Q_3)$. Other total states are unstable.

Oscillation

If all the states are unstable for an input value, then the machine undergoes state transitions under this input values. This is **oscillation**. Even if the machine has some stable states for an input value, if the machine does not reach them, then oscillation can occur.

Fundamental Mode

In an asynchronous sequential machine, if the next input change occurs before the completion of the state transition caused by the previous input change, then the machine can make erroneous state transition. Thus, we usually assume that "The input changes only when the machine is in a stable state, i.e., only after

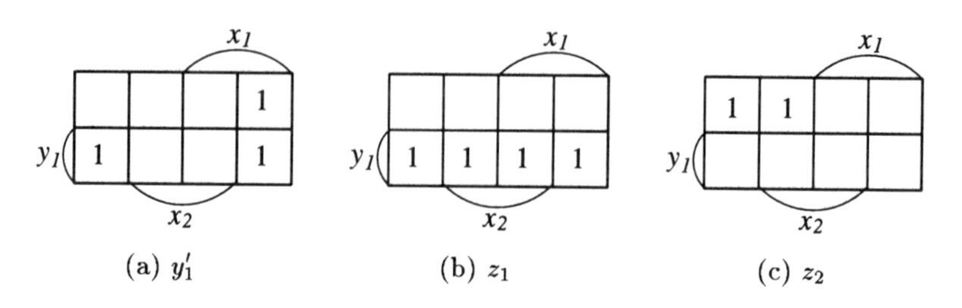

Figure 8.22 Maps for y_1', z_1, z_2.

the state transition." The behavior under this condition is the **fundamental mode**.

Single Input Change

The **single input change** means that no two inputs change at the same time. For example, if we analyze the change with a small time slot, then the input change from 00 to 11 is considered as either 00, 01, 11 or 00, 10, 11. In this case, we cannot predict the order of changes precisely. Since different order of changes can produce different state transitions, we often assume the single input change, when we consider the operation of asynchronous sequential networks.

Normal Fundamental Mode

The operation is **normal fundamental mode** if the operation is single input change, and **single output change** (i.e., for a single input change, at most one output changes).

Example 8.4 (Problem) Derive the state table for Fig. 8.18.

(Solution) From Example 8.1, we obtain the maps for y_1', z_1, and z_2 in Figs. 8.22(a), 8.22(b), and 8.22(c), respectively. Let the state be Q_0 when $y_1 = 0$, and let the state be Q_1 when $y_1 = 1$. Then, we have Table 8.2. ∎

Example 8.5 (Problem) Analyze the network in Fig. 8.19 under the condition of fundamental mode and without the condition of fundamental mode.

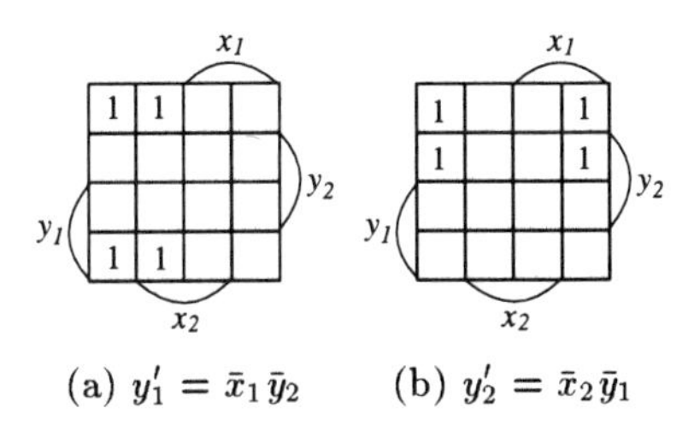

(a) $y_1' = \bar{x}_1\bar{y}_2$ (b) $y_2' = \bar{x}_2\bar{y}_1$

Figure 8.23 Maps for y_1' and y_2'.

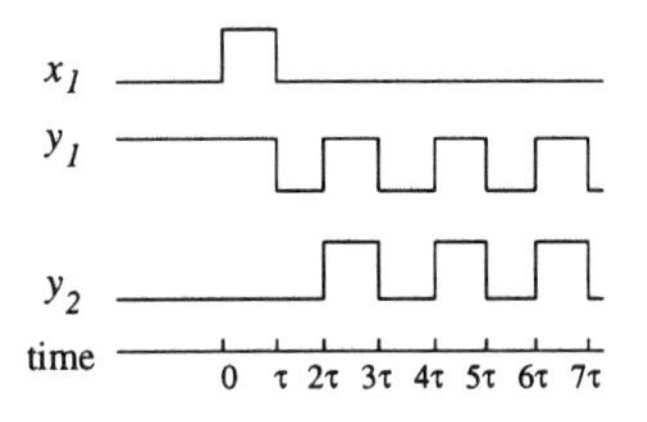

Figure 8.24 Change of signals in Fig. 8.19.

Table 8.3 State assignment.

State	Code
	$y_1 y_2$
Q_0	00
Q_1	01
Q_2	11
Q_3	10

Table 8.4 State table for fundamental mode.

Present	Input $x_1 x_2$			
state	00	01	11	10
Q_0	Q_2	Q_3	(Q_0)	Q_1
Q_1	(Q_1)	Q_0	Q_0	(Q_1)
Q_2	Q_0	Q_0	Q_0	Q_0
Q_3	(Q_3)	(Q_3)	Q_0	Q_0

(Solution)

1. In the case of the fundamental mode:
   From the result of Example 8.2, we have the map in Fig. 8.23. Next, the state assignment in Table 8.3 produces the state table in Table 8.4. For the input (11), the state Q_0 is stable. In Q_0, if the input is changed as (11) $\rightarrow$ (01), then the machine transits to state Q_3. Next, if the input is changed as (01)$\rightarrow$ (00), the machine remains to the state Q_3. Furthermore, if the input changed as (00)$\rightarrow$(10), the machine transitions to state Q_1 through the intermediate state Q_0.
2. In the case of the non-fundamental mode:
   If the input changes while the network is in an unstable state, then we have the following phenomena: Let the delay time of two gates be the same in Fig. 8.19: $\Delta_1 = \Delta_2 = \tau$. Suppose that the input is $(x_1, x_2) = (0, 0)$ in the state $(y_1, y_2) = (1, 0)$. Clearly, the network is in a stable state. Assume that x_1 is set to 1 for the period τ, and at the time τ, x_1 is reset to 0. At time τ, y_1, the output of the gate 1 changes as 1$\rightarrow$0. Next, at the time 2τ, y_2, the output of gate 2 changes as 0$\rightarrow$1. During time τ to 2τ, the both inputs for gate 1 are 0. Thus, at time 2τ, the output y_1 of gate 1 changes

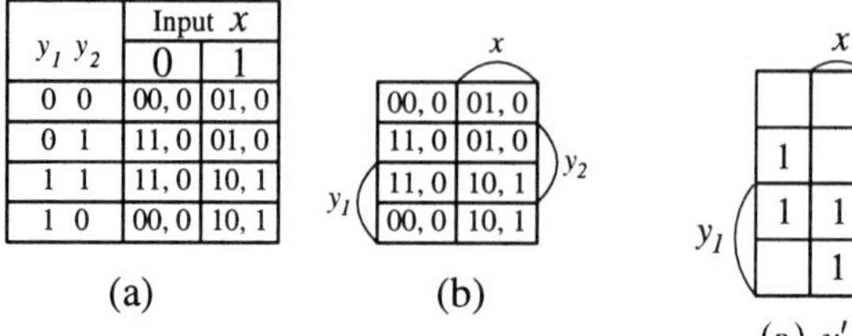

$y_1\ y_2$	Input x = 0	Input x = 1
0 0	00, 0	01, 0
0 1	11, 0	01, 0
1 1	11, 0	10, 1
1 0	00, 0	10, 1

Figure 8.25 Transition table and transition diagram.

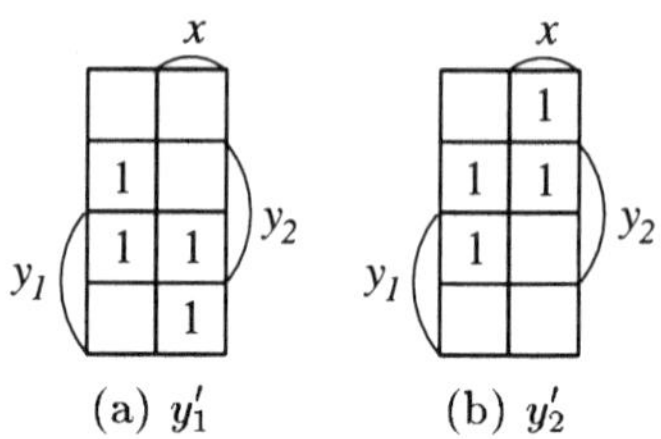

Figure 8.26 Maps for y_1' and y_2'.

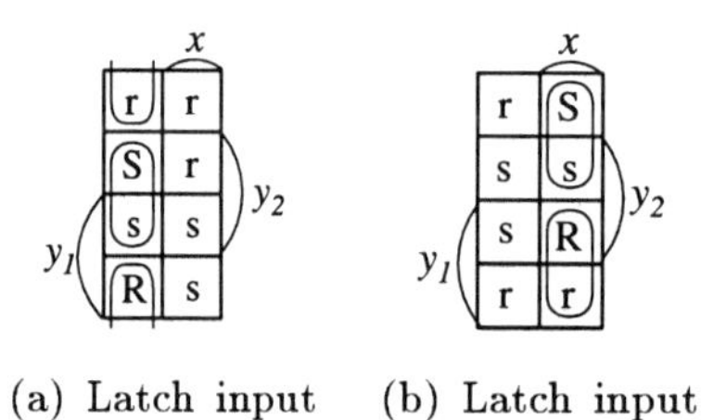

Figure 8.27 Maps for latch inputs.

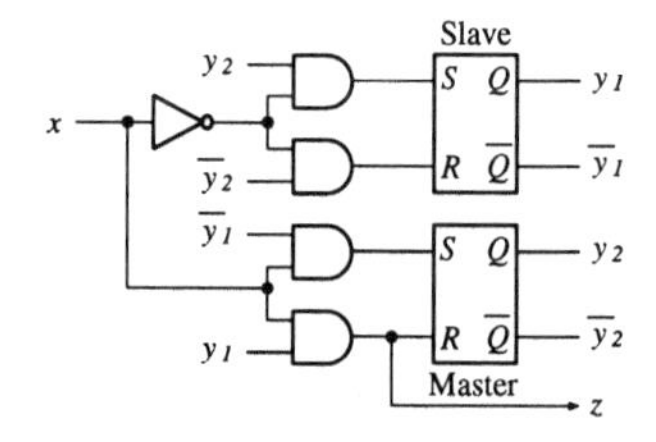

Figure 8.28 Realization of Table 8.1.

as 0→1. During time 2τ to 3τ, 1 is applied to the both gates. Thus, at time 3τ, both gate outputs y_1 and y_2 change as 1→0 at the same time. After this, y_1 and y_2 continue to change between 0 and 1, every τ time as shown in Fig. 8.24.

(Note) The above discussion assumes that the gate operations are ideal. In the practical network, the signals are different from Fig. 8.24, and the waveform are distorted. Thus, y_1 and y_2 cannot change infinitely, but falls into some state. In this case, we cannot predict the stable state. ∎

Design of Asynchronous Sequential Networks

Example 8.6 (Problem) Realize the network for the state table in Table 8.1 (binary counter) by using SR-latches.

(Solution) Let the state assignment be Table 8.3. Then, we have Fig. 8.25(a) as the transition table. Fig. 8.25(b) is the map for the transition table. This is the transition map. From the transition map, we have the map in Fig. 8.26 showing the conditions such that y_1' and y_2' are 1.

Next, we will derive logical expressions for SET inputs and RESET inputs for the latches by using maps. In a SR-latch, when the state variable changes from 0 to 1, the SET input must be 1. Thus, enter S to the corresponding cells in the map in Fig. 8.27. Also, when the state variables changes from 1 to 0, the RESET input must be 1. Thus, enter R to the corresponding cells in the map. Also, when the value of the state variable of the latch is 1, application of 1 to the SET input will not change the state variable. Thus, the SET input is insensitive, and the input is denoted by a lower case letter s. These s are used as *don't cares* in the simplification of the expression for SET inputs. Similarly, when the value of the state variable is 0, application of 1 to the RESET input will not change the state variable. Thus, the RESET input is insensitive, and the input is denoted by a lower case letter r. These r are used as *don't cares* in the simplification of the expression for RESET inputs. Since, S, R, s, and r cannot be 1 at the same time, they can be represented at a time by the map as shown in Fig. 8.27(a). In a similar way, the map for y_2 latch is derived as shown in Fig. 8.27(b). From these maps, we have the inputs for latches:

$$SET(y_1) = \bar{x}y_2, \qquad RESET(y_1) = \bar{x}\bar{y}_2,$$
$$SET(y_2) = x\bar{y}_1, \qquad RESET(y_2) = xy_1.$$

Also, the output function is represented as $z = xy_1$. Fig. 8.28 shows the network for state transition.

Suppose this network is in state Q_0, and let the input be $x = 0$. Then, $(y_1, y_2) = (0, 0)$. When the input x changes from 0 to 1, the SET input for y_2 latch becomes 1, and we have $y_2 = 1$. On the other hand, since $x = 1$, in the y_1 latch, both inputs are 0. Thus, y_1 remains to $y_1 = 0$. Thus, the machine goes to the state Q_1. Next, when the input x is changed to 0, the SET input for y_1 latch becomes 1, and we have $y_1 = 1$. On the other hand, in the y_2 latch, since $x = 0$, both inputs are 0, and y_2 remains 1. Thus, the machine goes to the state Q_2. Furthermore, when the input x is changed to 1, the RESET input of y_2 latch becomes 1, and we have $y_2 = 0$. On the other hand, in the y_1 latch, $x = 1$ and both inputs are 0, and y_1 remains 1. Thus, the machine goes to state Q_3. ∎

The network in Fig. 8.28 consists of two latches, and forms the **master-slave flip-flop** (See Section 6.2). In this case, y_2 corresponds to the master latch, and y_1 corresponds to the slave latch. When $x = 1$, the master latch and the slave latch are separated, and the value of y_1 is transferred to the master. When $x = 0$, the y_1 input is separated from the master latch, and the value of y_2 is transferred from the master to the slave. In this case, the value of master latch and the slave latch are the same.

Table 8.5 State table for Example 8.7.

Present state	Input x_1x_2 00	01	11	10
Q_0	Q_0	Q_1	Q_0	Q_0
Q_1	Q_0	Q_1	Q_2	Q_2
Q_2	Q_3	Q_2	Q_2	Q_2
Q_3	Q_3	Q_2	Q_0	Q_0

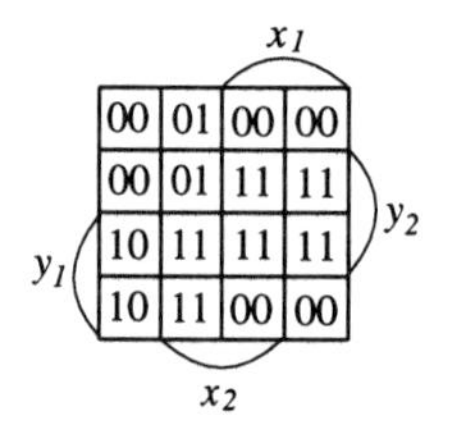

Figure 8.29 Transition map.

(a) y_1'

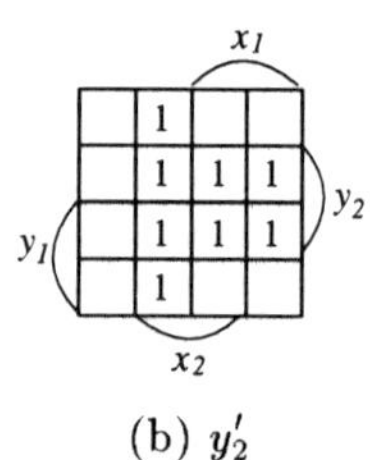

(b) y_2'

Figure 8.30 Maps for y_1' and y_2'.

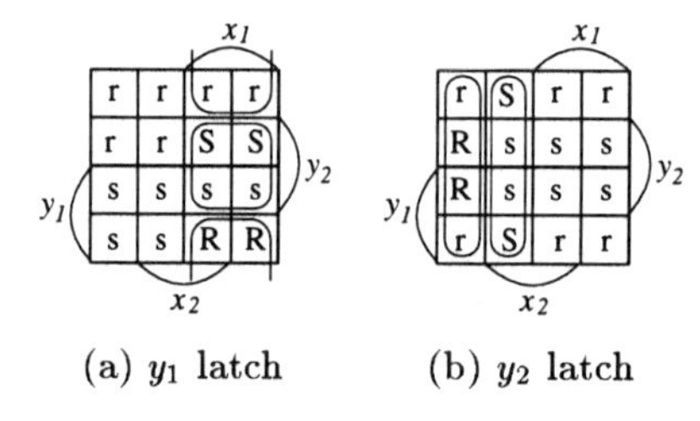

(a) y_1 latch (b) y_2 latch

Figure 8.31 Maps for SR-latch input.

Figure 8.32 MS type D·FF.

Example 8.7 (Problem) Realize Table 8.5 by using SR-latches.

(Solution) Let the state assignment be Table 8.3, then we have the transition map shown in Fig. 8.29. From this, we have Fig. 8.30, the map showing the next state variables y_1' and y_2'. To obtain the latch inputs for y_1 and y_2, derive the map in Fig. 8.31. The expressions for the inputs are obtained as follows:

$$SET(y_1) = x_1y_2, \qquad RESET(y_1) = x_1\bar{y}_2,$$
$$SET(y_2) = \bar{x}_1x_2, \qquad RESET(y_2) = \bar{x}_1\bar{x}_2.$$

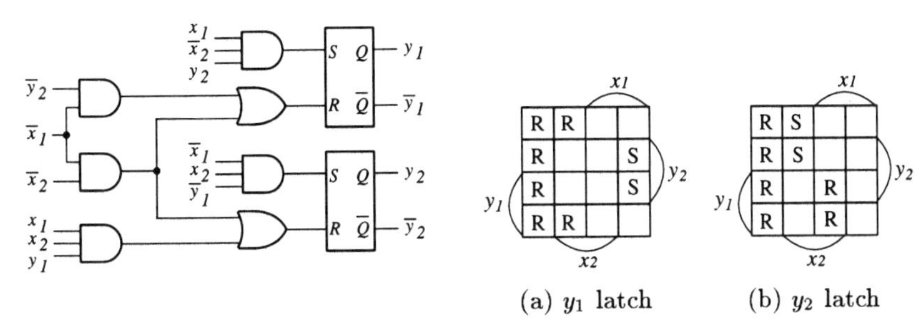

Figure 8.33 Excitation network for $SR\cdot$ latches.

Figure 8.34 Maps for latch inputs.

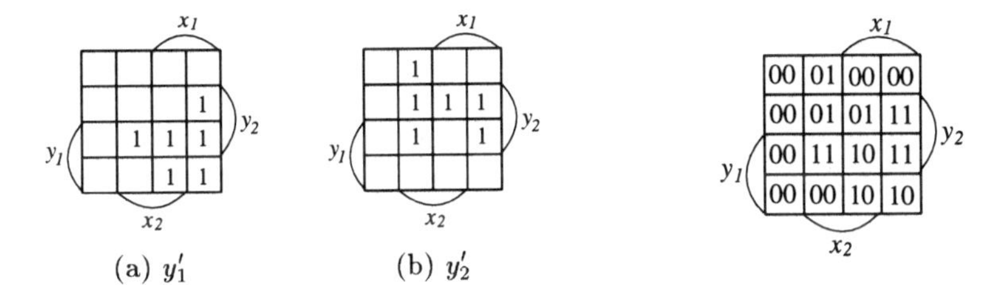

Figure 8.35 Maps for y_1' and y_2'.

Figure 8.36 Transition map.

From these, we have the network in Fig. 8.32. ∎

If we consider $x_1 = c$ as the clock pulse, and $x_2 = D$ as the signal input, then it is the master-slave D flip-flop (See Section 6.2.5), where the latch for y_2 is the master, and the latch for y_1 is the slave. When $x_1 = 0$, the master latch and the slave latch are separated, and the value of x_2 is transferred to the master. When $x_1 = 1$, the x_2 input is separated from the master latch, and the value of the master latch is transferred to the slave. In this case, the value of master latch and the slave latch are the same.

Example 8.8 (Problem) Derive the state table for Fig. 8.33.

(Solution)

(1) Derive the map for each latch input (See Fig. 8.34). These maps should be prepared for the SET input and the RESET input separately. However, since $R \cdot S = 0$, they are represented in the same map. In this case, S denotes the SET input, and R denotes the RESET input. (Note that they have a different meaning from R and S in Fig. 8.31.)

Table 8.6 State table for Fig. 8.33.

y_1y_2	Present state	Input x_1x_2			
		00	01	11	10
00	Q_0	(Q_0)	Q_1	(Q_0)	(Q_0)
01	Q_1	Q_0	(Q_1)	(Q_1)	Q_2
11	Q_2	Q_0	(Q_2)	Q_3	(Q_2)
10	Q_3	Q_0	Q_0	(Q_3)	(Q_3)

(2) Derive the maps for the next state variables y_1 and y_2. First, derive the map for y_1. In the cells where $y_1 = 1$, enter 1's to the all cells except for R cell in Fig. 8.34. Also, in the cells where $y_1 = 0$, enter 1's to the cells corresponding S cells in Fig. 8.34. Thus, we have Fig. 8.35.
(3) Derive the transition map for y_1 and y_2 (See Fig. 8.36).
(4) Let the state assignment be Table 8.3, then we have the state table shown in Table 8.6. ∎

8.3 MALFUNCTIONS OF ASYNCHRONOUS SEQUENTIAL NETWORKS

Asynchronous sequential networks designed by using the method in the previous section may not operate as we expect, resulting in a malfunction of asynchronous sequential networks. Critical races and hazard, which will be explained later, produce malfunctions.

8.3.1 Race

Example 8.9 In the state table in Table 8.7, consider the transition table in Table 8.9, where two state variables y_1 and y_2 are assigned as in Table 8.8. The total state $(10, Q_3)$ is stable. When the input changes to (00), the state should transition to (00, Q_1). However, two state variables must change from $(y_1, y_2) = (1,0)$ to $(y_1, y_2) = (0,1)$ at the same time. If (y_1, y_2) changes as $(1,0)$, $(0,0)$, $(0,1)$, it goes to Q_1. However, if (y_1, y_2) changes as $(1,0)$, $(1,1)$,

Table 8.7 State table.

Present state	Input x_1x_2 00	01	11	10
Q_0	Q_1	Q_0	Q_0	Q_3
Q_1	Q_1	Q_1	Q_1	Q_3
Q_2	Q_2	Q_1	Q_2	Q_3
Q_3	Q_1	Q_1	Q_0	Q_3

Table 8.8 State assignment.

State	Code y_1y_2
Q_0	00
Q_1	01
Q_2	11
Q_3	10

Table 8.9 Transition table with critical race.

	y_1y_2	x_1x_2 00	01	11	10
Q_0	00	01	00	00	10
Q_1	01	01	01	01	10
Q_2	11	11	01	11	10
Q_3	10	01	01	00	10

(0, 1), it goes to Q_2. In other words, the machine should go to the stable state Q_1, but it may go to the erroneous state Q_2. ∎

As shown in Example 8.9, when two or more state variables change at a time, we have a **race**. If the race can produce malfunction, then it is a **critical race**, otherwise it is a **non-critical race**.

Example 8.10 (Problem) Is it possible to avoid a critical race in Example 8.9 by changing the state assignment shown in Table 8.8?

(Solution) Consider various assignments of codes to Q_0, Q_1, Q_2, and Q_3.

(1) When the Hamming distance between the codes for Q_0 and Q_1 is two:
In Q_0, for the input (00), the machine should go to Q_1. However, since the distance between the codes is two, it goes to Q_2 or Q_3 as an intermediate state. In Q_2, for the input (00), it stays in the stable state Q_2. In Q_3, for the input (00), it goes to the stable state Q_1. In other words, when the distance between Q_0 and Q_1 is two, a critical race exists.

(2) When the Hamming distance between the codes for Q_1 and Q_2 is two:
In Q_2, for the input (01), the machine should go to Q_1. However, since the

distance between the codes is two, it goes to Q_0 or Q_3 as an intermediate state. In Q_0, for the input (01), it stays in the stable state Q_0. In Q_3, for the input (01), it goes to the stable state Q_1. In other words, when the distance between Q_0 and Q_3 is two, a critical race exists.

(3) When the Hamming distance between the codes for Q_1 and Q_3 is two: In Q_3, if the input (00) is applied, it should go to Q_1. However, since the distance between the codes is two, it goes to Q_0 or Q_2 as an intermediate state. In Q_2, for the input (00), it stays in the stable state Q_2. In Q_0, for the input (00), it goes to the stable state Q_1. In other words, when the distance between the codes for $Q1$ and Q_3 is two, a critical race exists.

Therefore, the Hamming distances between codes for Q_1 and Q_0, Q_1 and Q_2, and Q_1 and Q_3, must be all 1. However, for the two-bit codes this is impossible. ∎

Example 8.11 (Problem) For the Table 8.10, find the state assignment without a critical race.

(Solution) To find the critical race, it is sufficient to consider a column with two or more different stable states.

(1) When the distance of codes for Q_0 and Q_1 is two: In the state Q_0, when the input (01) is applied, the machine tries to go to the state Q_1. In this case, it visits either Q_2 or Q_3 as an intermediate state. If Q_2 is the intermediate state, then Q_2 will also be the stable state. If Q_3 is the intermediate state, then machine goes to Q_2. Thus, this state assignment produces a critical race.

(2) When the distance of codes for Q_0 and Q_2 is two: In the state Q_0, when the input (10) is applied, it tries to go to the state Q_2. In this case, the machine visits either Q_1 or Q_3 as an intermediate state. If Q_1 is the intermediate state, then Q_1 will also be the stable state. If Q_3 is the intermediate state, then the machine goes to Q_2. Thus, this state assignment also produces a critical race.

(3) Therefore, the only possibility is to assign the codes as $Q_0 = (a, b)$, $Q_1 = (a, \bar{b})$, $Q_2 = (\bar{a}, b)$, and $Q_3 = (\bar{a}, \bar{b})$, where a and b denote either 0 or 1. This assignment will not produce any critical race. ∎

8.3.2 Hazard

Consider the transition table in Table 8.9. In the stable state (01, Q_1), assume that the input change (00) $\rightarrow$ (01) occurs. In this case, the state variables will not change, and the machine should remain in the same state. However,

Table 8.10 State table for Example 8.11.

Present state	Input x_1x_2 00	01	11	10
Q_0	(Q_0)	Q_1	—	Q_2
Q_1	Q_0	(Q_1)	(Q_1)	(Q_1)
Q_2	Q_0	(Q_2)	(Q_2)	(Q_2)
Q_3	Q_0	Q_2	Q_2	Q_2

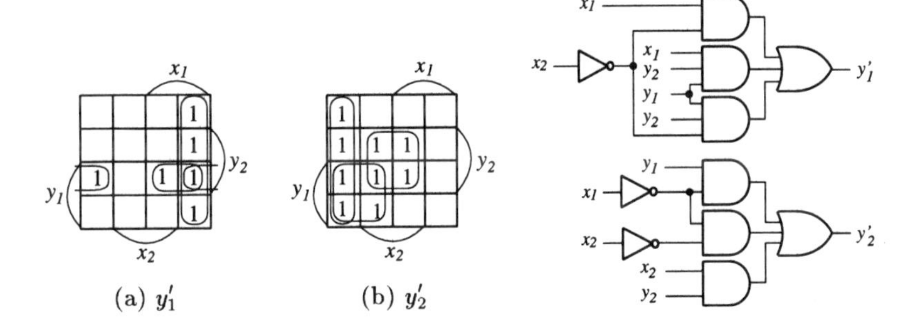

Figure 8.37 Maps for y_1' and y_2'.

Figure 8.38 A network with steady-state hazard.

the design of the network for this transition table requires special care not to produce malfunction.

Example 8.12 (Problem) Realize the network for the transition table in Table 8.9.

(Solution) Let the values of the next state variables be y_1' and y_2'. Then, the maps for y_1' and y_2' are obtained as Fig. 8.37. From these, the expressions for y_1' and y_2' are obtained as

$$\begin{aligned} y_1' &= x_1\bar{x}_2 \vee x_1y_1y_2 \vee \bar{x}_2y_1y_2, \\ y_2' &= \bar{x}_1\bar{x}_2 \vee x_2y_2 \vee \bar{x}_1y_1. \end{aligned}$$

The AND-OR two-level networks for y_1' and y_2' are obtained as Fig. 8.38. In the stable state (00, Q_1), consider the case where the input is changed from (00)

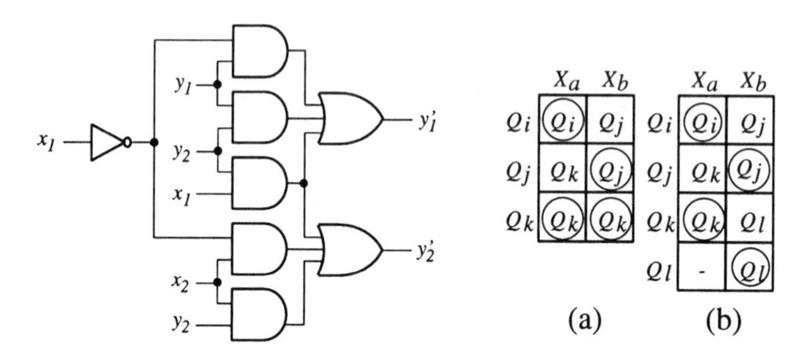

Figure 8.39 Realization of state table in Table 8.5.

Figure 8.40 State table with essential hazard.

to (01). In Fig. 8.37(b), this corresponds to the change of the total state (x_1, x_2, y_1, y_2): (0001) → (0101). Since the network for y_2' has a static 1-hazard, y_2' can change as 1→0→1 due to the delay of gates. If y_2' becomes 0 for a short time, then the total state will be (0100) and the machine goes to the erroneous state (0100), or Q_0. ∎

As shown in Example 8.12, in a sequential network, if it can go to the erroneous state due to gate delay, then network has a **steady-state hazard**. On the other hand, in a combinational network, a hazard can produce spurious pulse only for a short time, so it is a **transient hazard**.

Example 8.13 (Problem) Realize the Table 8.5 by not using SR-latches, but using only gates.

(Solution) In order not to produce steady-state hazards, obtain networks without static hazards for y_1' and y_2'. This can be done by producing SOPs with all the prime implicants:

$$y_1' = \bar{x}_1 y_1 \vee y_1 y_2 \vee y_2 x_1,$$
$$y_2' = \bar{x}_1 x_2 \vee x_2 y_2 \vee y_2 x_1.$$

From this, we have the network in Fig. 8.39. ∎

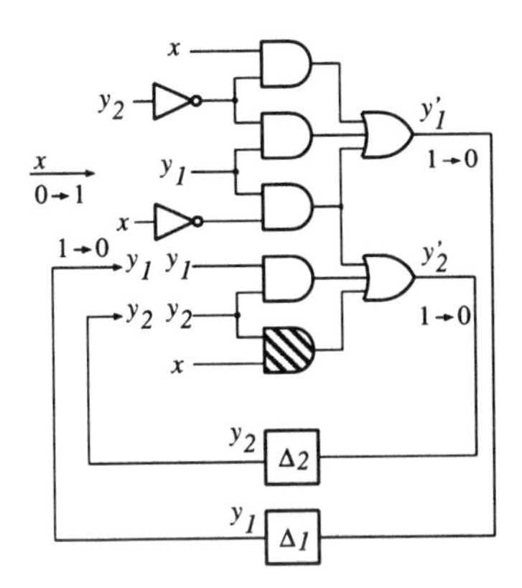

Figure 8.41 Analysis of essential hazard.

8.3.3 Essential Hazard

If the state table contains either Figs. 8.40(a) or 8.40(b) as a subtable, then the table has an **essential hazard**.

Example 8.14 Consider the operation of Fig. 8.41. For the state assignment in Table 8.8, we have the state table in Table 8.11. Assume that this machine is in Q_2 for the input $x = 0$. In this case, $(y_1, y_2) = (1, 1)$. Let the delay of the AND gate (shown by shaded gate in Fig. 8.41) for xy_2 be very large, and the delays for other gates be negligibly small. When x changes as $0 \rightarrow 1$, the signals of the network changes as follows:

1. The output of $\bar{x}y_1$ gate : $1 \rightarrow 0$;
   The value of y_1' : $1 \rightarrow 0$;
2. The output of y_1 : $1 \rightarrow 0$;
   The output of y_1y_2 gate : $1 \rightarrow 0$;
   The value of y_2' : $1 \rightarrow 0$.

During this period, the output for xy_2 remains to be 0 due to the gate delay. In this case, since $x = 1$ and $y_2 = 0$, the output for $x\bar{y}_2$ become 1, and so the value of y_1' will also be 1. In other words, the network goes to the state $(0, 0)$, i.e., Q_0, and then goes from Q_0 to $Q_3 = (1, 0)$. In Table 8.11, this phenomena can be explained as follows: When the input x is changed from 0 to 1 in Q_2, it will not be Q_1, but the erroneous state Q_3. ∎

Table 8.11 State table with essential hazard.

Present state	Input x	
	0	1
Q_0	(Q_0)	Q_3
Q_1	Q_0	(Q_1)
Q_2	(Q_2)	Q_1
Q_3	Q_2	(Q_3)

Essential hazards cannot be eliminated by changing state assignments. In the feedback delay model shown in Fig. 8.16, a malfunction will not occur even if the essential hazard exist. However, in the delay model where delays exist in the inputs of the combinational networks, essential hazard can produce malfunction.

Bibliographical Notes

Recommended textbooks on asynchronous networks are [127, 413]. On asynchronous FSMs, see [292, 373]. Hazard-free two-level network [300].

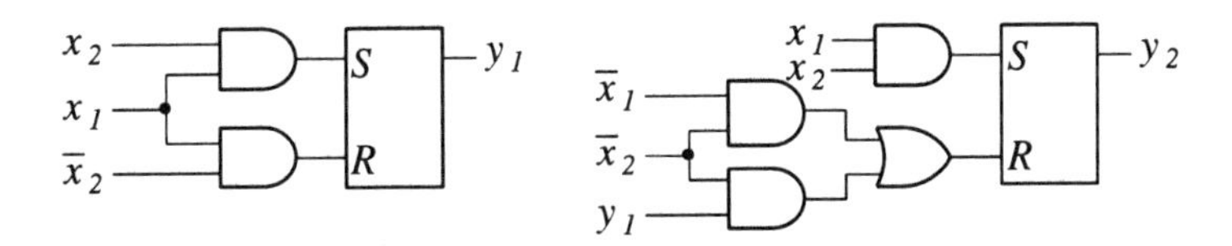

Figure 8.42 Network by using SR-latch.

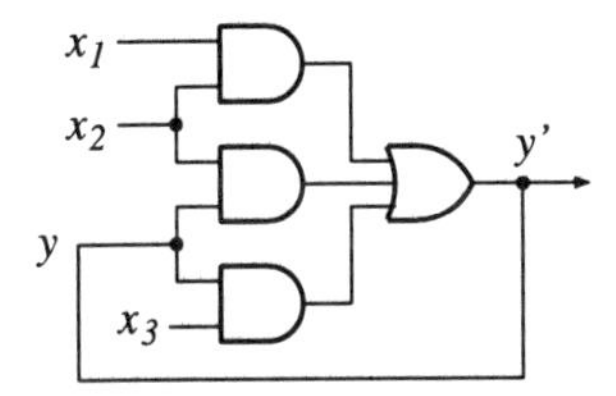

Figure 8.43 Gate network with feedback.

Exercises

8.1 For the following functions, realize hazard free AND-OR two-level networks, and hazard free OR-AND two-level networks:

$$\begin{aligned} f &= \bar{x}_1\bar{x}_2\bar{x}_3 \vee x_2x_3x_4, \\ g &= \bar{x}_1\bar{x}_2 \vee x_2x_3 \vee x_1\bar{x}_2\bar{x}_3. \end{aligned}$$

8.2 Consider the AND-OR two-level network for the SOP: $F = \bar{x}_1\bar{x}_4 \vee \bar{x}_1x_3 \vee x_2x_4 \vee x_2\bar{x}_3$. Is there any logic hazard? To realize the AND-OR two-level network without logic hazard, is it necessary to realize all the prime implicants?

8.3 Make the transition table for the network in Fig. 8.42.

8.4 Describe the behavior of the network in Fig. 8.43.

8.5 Analyze the behavior of the network in Fig. 8.44.

8.6 Suppose that in Table 8.12, the total state is in $(00, Q_0)$. Show an input sequence of (x_1, x_2) that derives the stable state $(01, Q_2)$. Assume that single input change, i.e., x_1 and x_2 cannot change at the same time.

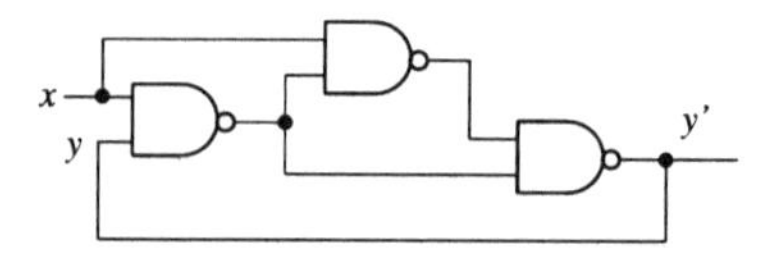

Figure 8.44 Gate network with feedback.

Table 8.12 State table for Problem 8.6.

Present state	Input x_1x_2 00	01	11	10
Q_0	(Q_0)	Q_3	Q_1	(Q_0)
Q_1	Q_0	Q_2	(Q_1)	(Q_1)
Q_2	Q_3	(Q_2)	(Q_2)	Q_1
Q_3	(Q_3)	(Q_3)	(Q_3)	Q_0

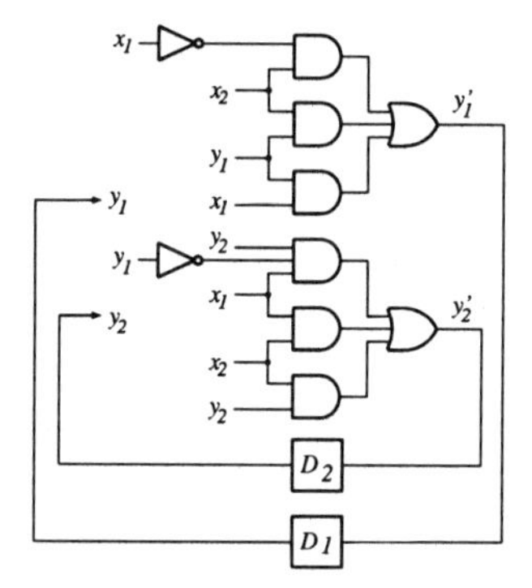

Figure 8.45 Gate network with feedback.

8.7 Derive the transition table for Fig. 8.45.

8.8 Realize the network for Table 8.13 by using SR-latches.

8.9 Analyze the behavior of the network in Fig. 8.46.

Table 8.13 State table for problem 8.8.

Present state	Input x_1x_2			
	00	01	11	10
Q_0	Q_0	Q_0	Q_1	Q_0
Q_1	Q_1	Q_1	Q_1	Q_2
Q_2	Q_3	Q_1	Q_2	Q_2
Q_3	Q_3	Q_0	Q_2	Q_3

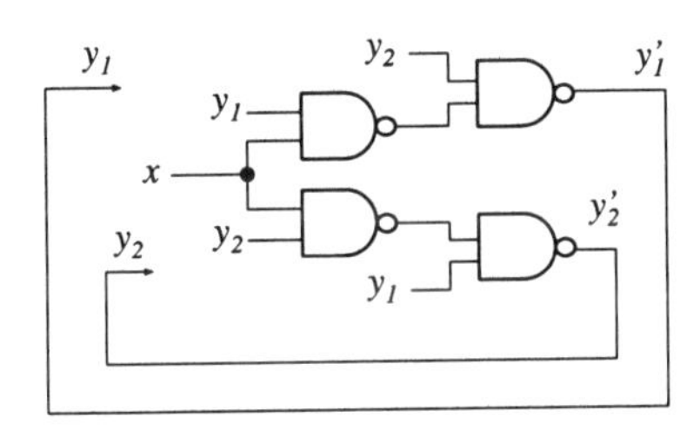

Figure 8.46 Gate network using feedback.

Table 8.14 Transition table with race condition.

y_1y_2	x_1x_2			
	00	01	11	10
00	00	11	00	11
01	11	01	00	11
11	00	11	00	10
10	00	10	10	10

8.10 In the transition table in Table 8.14:
(a) Find all races.
(b) If there is a critical race, show a method to eliminate it.

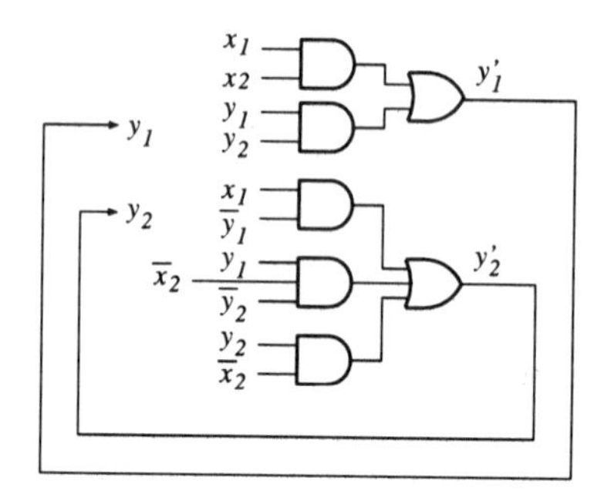

Figure 8.47 Gate network with feedback.

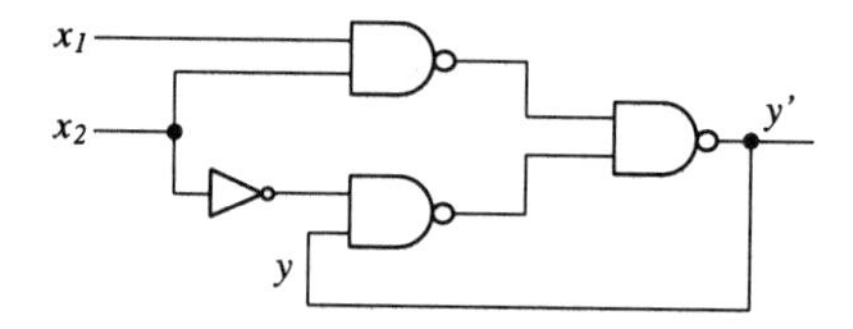

Figure 8.48 Gate network with feedback.

8.11 Derive the transition table for Fig. 8.47, and analyze its behavior. If a critical race exists, then re-design the network so it does not have a critical race.

8.12 Analyze the behavior of the network in Fig. 8.48. If a hazard exists, then re-design the network so it does not have a hazard.

9

MULTI-VALUED INPUT TWO-VALUED OUTPUT FUNCTION

In the preceding chapters, we have considered only two-valued input functions. This chapter considers multi-valued input two-valued output functions. These functions are useful for the simplification of PLAs with input decoders, simplification of multi-output functions, state assignment, code assignment, and design of multi-level logic networks.

9.1 MULTI-VALUED INPUT TWO-VALUED OUTPUT FUNCTION

Before showing the formal definitions of multi-valued input two-valued output functions, let us start with a simple example.

Example 9.1 An automobile dealer has various models of a car. Each model is classified by the features shown in Table 9.1. Among these models, the following 9 models are in the inventory: (manual, 2 doors, white), (manual, 3 doors, blue), (manual, 3 doors, black), (manual, 4 doors, red), (automatic, 2 doors, white), (automatic, 2 doors, black), (automatic, 3 doors, blue), (automatic, 3 doors, black), (automatic, 4 doors, red). We need to design a logic network that shows the inventory of the car models for the customers' request. The **function table** shown in Table 9.2 represents the car models in the inventory. In this table, X_1 shows the transmission type, X_2 shows the number of doors, and X_3 shows the color. If the car model is in the inventory, then $F = 1$, otherwise $F = 0$, where X_1 is two-valued, X_2 is three-valued, and X_3 is four-valued. ∎

Table 9.1 Features of automobiles.

		0	1	2	3
X_1	Transmission	Manual	Automatic		
X_2	Number of doors	2 doors	3 doors	4 doors	
X_3	Color	White	Blue	Red	Black

Table 9.2 Function table.

X_1	X_2	X_3	F	X_1	X_2	X_3	F	X_1	X_2	X_3	F
0	0	0	1	0	2	0	0	1	1	0	0
0	0	1	0	0	2	1	0	1	1	1	1
0	0	2	0	0	2	2	1	1	1	2	0
0	0	3	0	0	2	3	0	1	1	3	1
0	1	0	0	1	0	0	1	1	2	0	0
0	1	1	1	1	0	1	0	1	2	1	0
0	1	2	0	1	0	2	0	1	2	2	1
0	1	3	1	1	0	3	1	1	2	3	0

Definition 9.1 *A mapping $F : P_1 \times P_2 \times \cdots \times P_n \rightarrow B$ is a* **multi-valued input two-valued output function**, *where $P_i = \{0, 1, \ldots, p_i - 1\}$ $(i = 1, 2, \ldots, n)$, and $B = \{0, 1\}$. Let X be a variable that takes one value in $P = \{0, 1, \ldots, p - 1\}$. Let S be a subset $(S \subseteq P)$ of P. Then, X^S is a* **literal** *of X. When $X \in S$, $X^S = 1$, and when $X \notin S$, $X^S = 0$. Let $S_i \subseteq P_i (i = 1, 2, \ldots, n)$, then ${X_1}^{S_1} {X_2}^{S_2} \cdots {X_n}^{S_n}$ is a* **logical product**. *$\bigvee_{(S_1, S_2, \ldots, S_n)} {X_1}^{S_1} {X_2}^{S_2} \cdots {X_n}^{S_n}$ is a* **sum-of-products expression** *(***SOP***). When $S_i = P_i$, ${X_i}^{S_i} = 1$ and the logical product is independent of X_i. In this case, literal ${X_i}^{P_i}$ is redundant and can be deleted. A logical product is also called a* **term**, *or a* **product term**, *and corresponds to a* **cube**. *When $|S_i| = 1$ $(i = 1, 2, \ldots, n)$, a logical product corresponds to an element of the domain. This product is a* **minterm of f**. *When $S_i = P_i$ $(i = 1, 2, \ldots, n)$, the logical product corresponds to the constant 1. This product corresponds to a* **universal cube**. *When $p_i = 2$ $(i = 1, 2, \ldots, n)$, a function is a two-valued logic function. When we consider two-valued logic function only, we often represent the literal $X^{\{0\}}$ by $\overline{X}$, and $X^{\{1\}}$ by X.*

An arbitrary multi-valued input two-valued output function is represented by an SOP. Many SOPs exist that represent the same function. Among them, the one with the minimum number of products is the **minimum SOP** (**MSOP**).

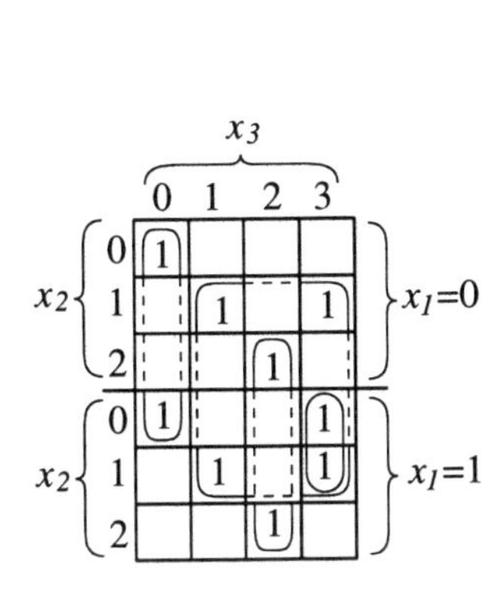

Figure 9.1 Multi-valued input two-valued output function.

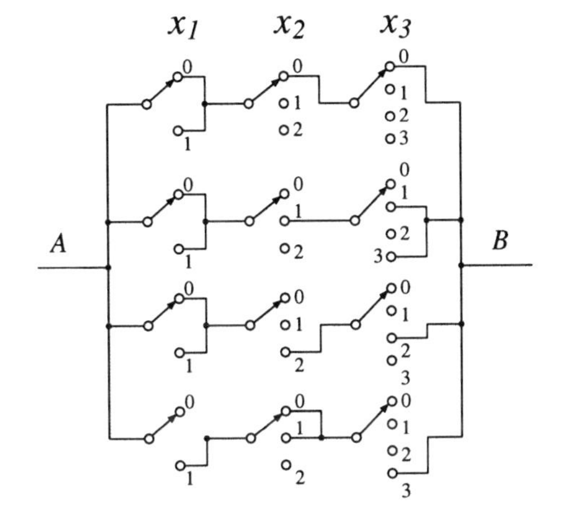

Figure 9.2 Contact network.

Example 9.2 Table 9.2 is an example of a multi-valued input two-valued output function, where $F : P_1 \times P_2 \times P_3 \to B$, such that $P_1 = \{0, 1\}$, $P_2 = \{0, 1, 2\}$, and $P_3 = \{0, 1, 2, 3\}$. X_1 takes two values, X_2 takes three values, and X_3 takes four values. The minterm expansion of F is

$$\begin{aligned} F = \; & {X_1}^{\{0\}}{X_2}^{\{0\}}{X_3}^{\{0\}} \vee {X_1}^{\{0\}}{X_2}^{\{1\}}{X_3}^{\{1\}} \vee {X_1}^{\{0\}}{X_2}^{\{1\}}{X_3}^{\{3\}} \\ & \vee {X_1}^{\{0\}}{X_2}^{\{2\}}{X_3}^{\{2\}} \vee {X_1}^{\{1\}}{X_2}^{\{0\}}{X_3}^{\{0\}} \vee {X_1}^{\{1\}}{X_2}^{\{0\}}{X_3}^{\{3\}} \\ & \vee {X_1}^{\{1\}}{X_2}^{\{1\}}{X_3}^{\{1\}} \vee {X_1}^{\{1\}}{X_2}^{\{1\}}{X_3}^{\{3\}} \vee {X_1}^{\{1\}}{X_2}^{\{2\}}{X_3}^{\{2\}}. \end{aligned}$$

Fig. 9.1 shows the map for this function. The minimum SOP is

$$\begin{aligned} F = \; & {X_1}^{\{0,1\}}{X_2}^{\{0\}}{X_3}^{\{0\}} \vee {X_1}^{\{0,1\}}{X_2}^{\{1\}}{X_3}^{\{1,3\}} \\ & \vee {X_1}^{\{0,1\}}{X_2}^{\{2\}}{X_3}^{\{2\}} \vee {X_1}^{\{1\}}{X_2}^{\{0,1\}}{X_3}^{\{3\}}. \end{aligned}$$

By removing redundant literals, we have

$$F = {X_2}^{\{0\}}{X_3}^{\{0\}} \vee {X_2}^{\{1\}}{X_3}^{\{1,3\}} \vee {X_2}^{\{2\}}{X_3}^{\{2\}} \vee {X_1}^{\{1\}}{X_2}^{\{0,1\}}{X_3}^{\{3\}}.$$

This SOP corresponds to the contact network shown in Fig. 9.2. When $F = 1$, a path exists between terminal A and terminal B. ∎

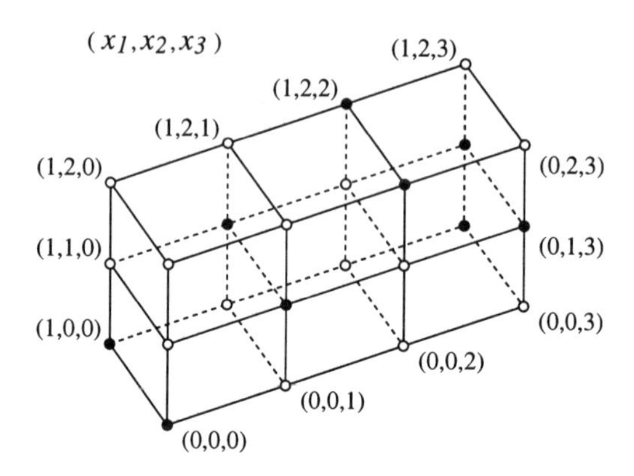

Figure 9.3 Three-dimensional representation.

9.2 BIT REPRESENTATION

Let the domain of a multi-valued input two-valued output function be $P_1 \times P_2 \times \cdots \times P_n$. In this case, the logical product $c = X_1^{S_1} X_2^{S_2} \cdots X_n^{S_n}$, $S_i \subseteq P_i$ corresponds to a **cube** in an n-dimensional hyper-cube. The **bit representation** (positional cube notation) of a cube c is the concatenation of binary numbers showing the cube. $c = \pi_1\text{-}\pi_2\text{-} \cdots \text{-}\pi_n$, where $\pi_i = (\xi_0 \xi_1 \cdots \xi_{p_i-1})$, such that

$$\xi_j = \begin{cases} 1 & (\text{when } j \in S_i), \\ 0 & (\text{when } j \notin S_i). \end{cases}$$

The bit representation of an SOP is an **array**. An array is set of cubes.

Example 9.3 The SOP of Example 9.2 is represented by an array:

$$\begin{array}{c} X_1 \quad X_2 \quad X_3 \\ 01 \text{ - } 012 \text{ - } 0123 \end{array}$$
$$\begin{bmatrix} 11 \text{ - } 100 \text{ - } 1000 \\ 11 \text{ - } 010 \text{ - } 0101 \\ 11 \text{ - } 001 \text{ - } 0010 \\ 01 \text{ - } 110 \text{ - } 0001 \end{bmatrix}.$$

Notice that a variable with all 1's in the bit representation can be omitted in the SOP. Fig. 9.3 shows a three-dimensional representation of the function. ∎

9.3 RESTRICTION

For a multi-valued input two-valued output function F, the function which is obtained by restricting the domain to D is a **restriction** of F, and is denoted by $F(|D)$. For an SOP, the restriction is defined as follows:

Definition 9.2 *Let F be an SOP, and $c = X_1^{S_1} X_2^{S_2} \cdots X_n^{S_n}$ be a product. Then, the restriction $F(|c)$ of F to c is obtained as follows:*

(1) For each product term in F, make a logical product with c. Delete the zero terms.
(2) Let $d = X_1^{T_1} X_2^{T_2} \cdots X_n^{T_n}$ be a product obtained in (1). Replace d with $X_1^{(T_1 \cup \overline{S_1})} X_2^{(T_2 \cup \overline{S_2})} \cdots X_n^{(T_n \cup \overline{S_n})}$.

Example 9.4 Let

$$F = \begin{bmatrix} \text{11-100-1000} \\ \text{11-010-0101} \\ \text{11-001-0010} \\ \text{01-110-0001} \end{bmatrix}.$$

(1) When c =(01-101-1111), $F(|c)$ is obtained as follows:

$$F \cdot c = \begin{bmatrix} \text{01-100-1000} \\ \text{01-001-0010} \\ \text{01-100-0001} \end{bmatrix}.$$

(2) Let $\bar{c}$ be a bit-wise complement of c. Then, $\bar{c}$ =(10-010-0000). For each cube of $F \cdot c$, bit-wise OR with $\bar{c}$, and we have

$$F(|c) = \begin{bmatrix} \text{11-110-1000} \\ \text{11-011-0010} \\ \text{11-110-0001} \end{bmatrix}.$$

∎

Theorem 9.1 $c \cdot F = c \cdot F(|c)$.

The restriction is also called **cofactor**. By Shannon's expansion theory, an arbitrary two-valued logic function is expanded as

$$f(x_1, x_2, \ldots, x_n) = \bar{x}_1 f(0, x_2, \ldots, x_n) \vee x_1 f(1, x_2, \ldots, x_n).$$

In this case, $f(0, x_2, \ldots, x_n)$ and $f(1, x_2, \ldots, x_n)$ represent f with $x_1 = 0$ and $x_1 = 1$, respectively.

Example 9.5 The bit representation of $f = xy \vee yz \vee zx$ is

$$F = \begin{bmatrix} 01 \text{ - } 01 \text{ - } 11 \\ 11 \text{ - } 01 \text{ - } 01 \\ 01 \text{ - } 11 \text{ - } 01 \end{bmatrix}.$$

(columns: x, y, z)

By the Shannon expansion of f with respect to x, we have $f = \bar{x} \cdot f(0, y, z) \vee x \cdot f(1, y, z)$.

(1) Obtain $f(0, y, z)$: Let c_0 =(10-11-11), we have $F \cdot c_0$ =[10-01-01] and $F(|c_0)$ =[11-01-01]. $F(|c_0)$ represents $f(0, y, z)$.

(2) Obtain $f(1, y, z)$: Let c_1 =(01-11-11), we have

$$F \cdot c_1 = \begin{bmatrix} \text{01-01-11} \\ \text{01-01-01} \\ \text{01-11-01} \end{bmatrix} \text{ and } F(|c_1) = \begin{bmatrix} \text{11-01-11} \\ \text{11-01-01} \\ \text{11-11-01} \end{bmatrix}.$$

$F(|c_1)$ represent $f(1, y, z)$. ∎

9.4 TAUTOLOGY

When the logical expression F is equal to logical 1 for all the input combinations, F is a **tautology**. The problem of determining whether a given logical expression is a tautology or not is the **tautology decision problem**.

Example 9.6 Let,

$$F_1 = \begin{bmatrix} \text{01-100-1100} \\ \text{11-111-0010} \end{bmatrix} \text{and } F_2 = \begin{bmatrix} \text{11-110-1110} \\ \text{11-110-0001} \\ \text{11-001-1111} \end{bmatrix}.$$

F_2 is a tautology but F_1 is not. This can be verified by forming the Karnaugh maps for F_1 and F_2. ∎

The tautology decision problem is important in the simplification of SOPs (e.g., expansion of products, detection of essential prime implicants, deletion of redundant products, reduction of redundant literals). It is also important in the equivalence checking of two logical expressions. We consider this in more detail in Section 9.9.

9.5 INCLUSION RELATION

Definition 9.3 *Let F and G be logic functions. For all the minterms c such that $F(c) = 1$, if $G(c) = 1$, then $F \leq G$, and G* **contains** *F. When G does not contain F, we denote it by $F \not\leq G$. If a logic function F contains a product c, then c is an* **implicant** *of F.*

The following theorem shows the determination of c as an implicant of F is transformed to a tautology decision problem.

Theorem 9.2 *Let c be a logical product and F be a logical expression. Then, $c \leq F \Leftrightarrow F(|c) \equiv 1$.*

Example 9.7 Let

$$F = \begin{bmatrix} \text{11-100-1000} \\ \text{11-010-0101} \\ \text{11-001-0010} \\ \text{01-110-0001} \end{bmatrix}.$$

1) When c_1 =(01-100-1001).

$$F(|c_1) = \begin{bmatrix} \text{11-111-1110} \\ \text{11-111-0111} \end{bmatrix}.$$

Since, $F(|c_1) \equiv 1$, we have $c_1 \leq F$.

2) When c_2 =(11-010-1101).

$$F(|c_2) = \begin{bmatrix} \text{11-111-0111} \\ \text{01-111-0011} \end{bmatrix}.$$

Since $F(|c_2) \neq 1$, we have $c_2 \not\leq F$. ∎

9.6 EQUIVALENCE

The following theorem shows that the determination of the equivalence of two SOPs is transformed to the tautology problem:

Theorem 9.3 *Let*

$$F = \bigvee_{i=1}^{p} f_i \text{ and } G = \bigvee_{j=1}^{q} g_j.$$

Then, $F \equiv G \Leftrightarrow F(|g_j) \equiv 1\ (j = 1, \ldots, q)$ *and* $G(|f_i) \equiv 1\ (i = 1, \ldots, p)$.

(Proof) $F \equiv G \Leftrightarrow (F \leq G$ and $F \geq G)$.
From $F \leq G \Leftrightarrow f_i \leq G (i = 1, \ldots, p) \Leftrightarrow G(|f_i) \equiv 1 (i = 1, \ldots, p)$,
and $G \leq F \Leftrightarrow g_j \leq F (j = 1, \ldots, q) \Leftrightarrow F(|g_j) \equiv 1 (j = 1, \ldots, q)$.
Thus, we have the theorem. □

Determining the equivalence of two logical expressions is a fundamental problem in the verification of logic networks. In this case, verification using BDDs (Section 3.7) is often faster than one using SOPs.

Example 9.8 Let $F = x\bar{y} \vee y$ and $G = x \vee \bar{x}y$. Then, $F(|x) \equiv 1$, $F(|\bar{x}y) \equiv 1$, $G(|x\bar{y}) \equiv 1$, and $G(|y) \equiv 1$. Thus, $F \equiv G$. ∎

9.7 DIVIDE AND CONQUER METHOD

The **divide and conquer method** partitions the given problem into smaller ones, and solve them independently. It is quite effective for the manipulation of SOPs. Simplification algorithms for SOPs such as MINI2 and ESPRESSO-II use it.

Theorem 9.4 *Let* F *be an SOP and* c_i $(i = 1, 2, \ldots, k)$ *be the cubes satisfying the following conditions:*

$$\bigvee_{i=1}^{k} c_i \equiv 1 \text{ and } c_i \cdot c_j = 0\ (i \neq j).$$

Then, we have

$$F = \bigvee_{i=1}^{k} c_i \cdot F(|c_i). \tag{9.1}$$

(Proof) (The left-hand side of the equation) $= (\bigvee_{i=1}^{k} c_i) \cdot F = \bigvee_{i=1}^{k} c_i \cdot F = \bigvee_{i=1}^{k} c_i \cdot F(|c_i) =$ (the right-hand side of the equation). □

By using Theorem 9.4, a given SOP is partitioned into k SOPs. In performing some operation on F, first expand F into the form (9.1). Then, for each $F(|c_i)$, apply the operation independently. Finally, by combining the results appropriately, we have the result for the operation on F. Since, the same operation can be applied to $F(|c_i)$ as to F, this method can be computed by a recursive program.

Definition 9.4 *Let $t(F)$ be the number of products in an SOP F.*

Using Theorem 9.4, we try to minimize $F(|c_i)$. That is, we expand it so that

$$\sum_{i=1}^{k} t(F(|c_i))$$

will be minimum. In this case, for simplicity, we consider the partition where $k = 2$, $c_1 = X_j^{S_A}$, $c_2 = X_j^{S_B}$, $S_A \cup S_B = P_j$, and $S_A \cap S_B = \phi$. When we partition the SOP by using Theorem 9.4, the recursive application of restriction operation will attempt to create as many columns of all 0's and all 1's as possible. That is, a column with all 1's or all 0's, can be ignored in the partition. A column that contains both 0 and 1 is **active**.

Selection Method for Variables:

1. Choose all variables with that have the maximum number of active columns 1.
2. Among the variables chosen in step 1, choose variables where the total sum of 0's in the array is maximum.
3. For all variables chosen in step 2, find a column that has the maximum number of 0's. Choose the variable that has the minimum number of 0's in these columns.

After deciding the variables to expand, we have to choose S_A and S_B. We need to find the partition S_A and S_B that minimizes the value of $t(F(|X^{S_A}))+t(F(|X^{S_B}))$. However, this is very time consuming, so we will choose the partition S_A and S_B, such that the active columns of X will be divided into half.

Example 9.9 Let

$$F = \begin{array}{c} \quad X_1 \quad\; X_2 \quad\;\; X_3 \\ \left[\begin{array}{c} 11 \text{ - } 100 \text{ - } 1000 \\ 11 \text{ - } 010 \text{ - } 0100 \\ 11 \text{ - } 001 \text{ - } 0010 \\ 01 \text{ - } 110 \text{ - } 0001 \end{array}\right] \end{array}.$$

In F, X_2 and X_3 have the largest number of active columns. Choose X_3, and we expand the expression by X_3. Let $S_A = \{0,1\}$ and $S_B = \{2,3\}$. Then, both the cubes $F(|X_3^{\{0,1\}})$ and $F(|X_3^{\{2,3\}})$ are partitioned into two. If we expand the array with X_1 instead of X_3, then the number of cubes in $F(|X_1^{\{0\}})$ is three, and the number of cubes in $F(|X_1^{\{1\}})$ is four. This requires more computation time. ∎

In Sections 9.8–9.11, we will show divide and conquer algorithms for complementing the logical expression, tautology decision, prime implicants generation and disjoint sharp operation for multi-valued input two-valued output functions.

9.8 COMPLEMENTATION OF SOPS

Theorem 9.5 *Let*

$$\bigvee_{i=1}^{k} c_i \equiv 1 \text{ and } c_i \cdot c_j = 0 \; (i \neq j).$$

Then, we have

$$\overline{F} = \bigvee_{i=1}^{k} c_i \cdot \overline{F}(|c_i).$$

To select c_i in Theorem 9.5, we have three options:

1. $c_1 = X_i^{S_A}$ and $c_2 = X_i^{S_B}$, where $S_A \cap S_B = \phi$ and $S_A \cup S_B = P_i$.
2. $c_k = X_i^{\{k\}} (k = 0, 1, \ldots, p_i - 1)$.
3. $c_1 = X_1^{\overline{S_1}}, c_2 = X_1^{S_1} \cdot X_2^{\overline{S_2}}, \ldots,$
   $c_n = X_1^{S_1} \cdot X_2^{S_2} \cdot \cdots \cdot X_{n-1}^{S_{n-1}} \cdot X_n^{\overline{S_n}},$
   $c_{n+1} = X_1^{S_1} \cdot X_2^{S_2} \cdot \cdots \cdot X_{n-1}^{S_{n-1}} \cdot X_n^{S_n}.$

Here, we will use 1, the partition into the pair.

Algorithm 9.1 *(Complementation of SOPs)*

1. *When* F *consists of one product* c, *i.e., if* $F = X_1^{S_1} \cdot X_2^{S_2} \cdot \cdots \cdot X_n^{S_n}$, *then we have* $\overline{F} = X_1^{\overline{S_1}} \vee X_1^{S_1} \cdot X_2^{\overline{S_2}} \vee \cdots \vee X_1^{S_1} \cdot X_2^{S_2} \cdot \cdots \cdot X_{n-1}^{S_{n-1}} \cdot X_n^{\overline{S_n}}$.
2. *When* F *consists of more than one product, then, expand* F *into* $F = c_1F(|c_1) \vee c_2F(|c_2)$, *where* $c_1 = X_j^{S_A}, c_2 = X_j^{S_B}, S_A \cup S_B = P_j$, *and* $S_A \cap S_B = \phi$. *In this case, we have* $\overline{F} = c_1\overline{F}(|c_1) \vee c_2\overline{F}(|c_2)$.

When $n = 10$, this method is about 100 times faster than one using De Morgan's theorem.

Example 9.10 When

$$F = \begin{array}{c} \begin{array}{ccc} X_1 & X_2 & X_3 \end{array} \\ \left[\begin{array}{ccccc} 11 & - & 100 & - & 1000 \\ 11 & - & 010 & - & 0100 \\ 11 & - & 001 & - & 0011 \\ 01 & - & 110 & - & 0001 \end{array}\right] \end{array},$$

obtain $\overline{F}$, the complement of F.

1. Since X_3 has four active columns, expand F with respect to X_3. Let $S_A = \{0, 1\}$, $S_B = \{2, 3\}$. We have
   $c_1 = (11\text{-}111\text{-}1100)$,
   $c_2 = (11\text{-}111\text{-}0011)$,
   $F_1 = F(|c_1) = \begin{bmatrix} 11\text{-}100\text{-}1011 \\ 11\text{-}010\text{-}0111 \end{bmatrix}$, and
   $F_2 = F(|c_2) = \begin{bmatrix} 11\text{-}001\text{-}1111 \\ 01\text{-}110\text{-}1101 \end{bmatrix}$.
2. Next, expand F_1 by variable X_2. We have
   $F_1 = c_3F_1(|c_3) \vee c_4F_1(|c_4)$, where
   $c_3 = (11\text{-}100\text{-}1111)$,
   $c_4 = (11\text{-}011\text{-}1111)$,
   $F_3 = F_1(|c_3) = (11\text{-}111\text{-}1011)$, and
   $F_4 = F_1(|c_4) = (11\text{-}110\text{-}0111)$.
3. Similarly, expand F_2 by variable X_2. We have
   $F_2 = c_5F_2(|c_5) \vee c_6F_2(|c_6)$, where
   $c_5 = (11\text{-}110\text{-}1111)$,
   $c_6 = (11\text{-}001\text{-}1111)$,
   $F_5 = F_2(|c_5) = (01\text{-}111\text{-}1101)$, and
   $F_6 = F_2(|c_6) = (11\text{-}111\text{-}1111)$.

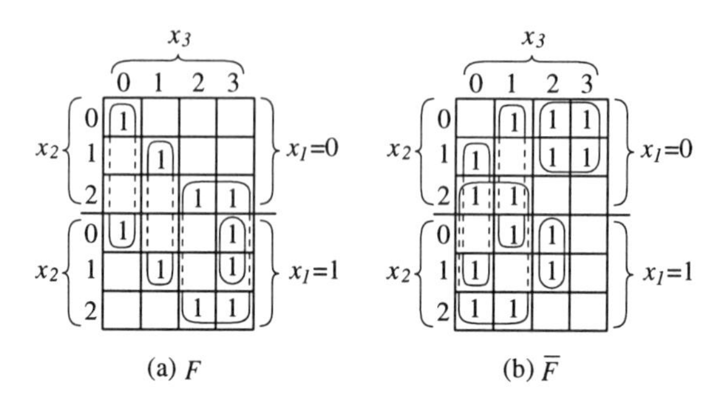

Figure 9.4 Example of complementation.

4. F_3–F_6 consist of single products. We have
   $\overline{F_3}$ =(11-111-0100),
   $\overline{F_4} = \begin{bmatrix} \text{11-001-1111} \\ \text{11-110-1000} \end{bmatrix}$,
   $\overline{F_5} = \begin{bmatrix} \text{10-111-1111} \\ \text{01-111-0010} \end{bmatrix}$, and
   $\overline{F_6} = 0$.
5. By combining all the products, we have:
   $$\begin{aligned} \overline{F} &= c_1\overline{F_1} \vee c_2\overline{F_2} \\ &= c_1(c_3\overline{F_3} \vee c_4\overline{F_4}) \vee c_2(c_5\overline{F_5} \vee c_6\overline{F_6}) \\ &= c_1c_3\overline{F_3} \vee c_1c_4\overline{F_4} \vee c_2c_5\overline{F_5} \vee c_2c_6\overline{F_6} \\ &= \begin{bmatrix} \text{11-100-0100} \\ \text{11-001-1100} \\ \text{11-010-1000} \\ \text{10-110-0011} \\ \text{01-110-0010} \end{bmatrix}. \end{aligned}$$

Fig. 9.4(a), (b) show F and $\bar{F}$, respectively. Note that $\bar{F}$ is disjoint. ∎

9.9 TAUTOLOGY DECISION

Definition 9.5 *When there is a variable X_i and at least one constant $a \in P_i$ satisfying $F(|X_i = a) \leq F(X_1, \ldots, X_i, \ldots, X_n)$, the function F is* **weakly unate** *with respect to the variable X_i.*

Definition 9.6 *In an array F, consider the sub-array consisting of cubes that depend on X_i. In the variable X_i in this array, if all the values in a column are 0, then SOP F is* **weakly unate** *with respect to the variable X_i.*

Example 9.11 Consider the following array:

$$F = \begin{bmatrix} 1111 \text{ - } 1110 \text{ - } 1110 \\ 1111 \text{ - } 1101 \text{ - } 1101 \\ 0110 \text{ - } 0110 \text{ - } 1101 \\ 0101 \text{ - } 0111 \text{ - } 1101 \end{bmatrix} \begin{matrix} c_1 \\ c_2 \\ c_3 \\ c_4 \end{matrix} .$$

(columns: X_1, X_2, X_3)

The cubes that depend on the variable X_1 are c_3 and c_4. Since the first column of c_3 and c_4 are all 0, F is weakly unate with respect to the variable X_1. ∎

Theorem 9.6 *Let an SOP F be weakly unate with respect to the variable X_i. Among the cubes of F, let G be the set of cubes that do not depend on the variable X_i. Then, $G \equiv 1 \Leftrightarrow F \equiv 1$.*

In determining tautology, we can ignore the columns of F with all 1's.

Theorem 9.7 *Let $c_1 = X_j^{S_A}$ and $c_2 = X_j^{S_B}$, where $S_A \cup S_B = P_j$ and $S_A \cap S_B = \phi$. Then, $F \equiv 1 \Leftrightarrow F(|c_1) \equiv 1$ and $F(|c_2) \equiv 1$.*

Algorithm 9.2 *(Tautology Decision)*

1. *If F has a column with all 0's, then F is not a tautology.*
2. *Let $F = \{c_1, c_2, \ldots, c_k\}$, where c_i is a cube. If $\sum_{i=1}^{k} vol(c_i) < \prod_{i=1}^{n} p_i$, then F is not a tautology, where $vol(c_i)$ denotes the number of the minterms in cube c_i.*
3. *If there is a cube with all 1's (i.e., universal cube) in F, then F is a tautology.*
4. *When we consider only the active columns in F, if they are all two-valued, and if the number of the variables is less than 7, then decide the tautology of F by the truth table.*
5. *When there is a weakly unate variable, simplify the problem by using Theorem 9.6.*
6. *When F consists of more than one cube: F is a tautology iff $F(|c_1) \equiv 1$ and $F(|c_2) \equiv 1$, where, $c_1 = X_j^{S_A}$, $c_2 = X_j^{S_B}$, $S_A \cup S_B = P_j$, and $S_A \cap S_B = \phi$.*

Example 9.12

1. $G = \begin{bmatrix} \text{01-100-1100} \\ \text{11-111-0010} \end{bmatrix}$.

   Since the last column of G consist of all 0's, G is not a tautology.

2. $F = \begin{bmatrix} \text{11-110-1110} \\ \text{11-110-0001} \\ \text{11-001-1111} \end{bmatrix}.$

F does not depend on X_1. Let, the partition be c_1=(11-110-1111) and c_2=(11-001-1111).

$F_1 = F(|c_1) = \begin{bmatrix} \text{11-111-1110} \\ \text{11-111-0001} \end{bmatrix} \Rightarrow \text{(11-111-1111)} \equiv 1.$

$F_2 = F(|c_2) = \text{(11-111-1111)} \equiv 1.$

Thus, F is a tautology. ∎

9.10 GENERATION OF PRIME IMPLICANTS

Definition 9.7 *Let c be an implicant of a logic function F. If there is no other implicant c_1 ($c_1 \neq c$) of F that contains c, then c is a* **prime implicant** *of F. In other words, c is a maximal implicant of F. The set of prime implicants for F is denoted by $PI(F)$.*

Definition 9.8 *Let X be a variable that takes a value in $P = \{0, 1, \ldots, p-1\}$. If there is total order ($\preceq$) on the values of variable X in function F, such that $j \preceq k (j, k \in P)$ implies $F(|X = j) \leq F(|X = k)$, then the function F is* **strongly unate** *with respect to X. If F is strongly unate with respect to all the variables, then the function F is strongly unate.*

Next, assume that F is an SOP. If there is total order ($\preceq$) among the values of variable X, and if $j \preceq k$ ($j, k \in P$), then each product term of the SOP $F(|X = j)$ is contained by all the product terms of the SOP $F(|X = k)$. In this case the SOP F is strongly unate with respect to X.

Lemma 9.1 *If F is strongly unate with respect to X_i, then F is weakly unate with respect to X_i.*

Example 9.13 Let

$$F = \begin{bmatrix} \overset{X_1}{1111} & - & \overset{X_2}{1001} \\ 0111 & - & 0111 \\ 0011 & - & 0110 \\ 0001 & - & 0101 \end{bmatrix}.$$

Since $F(|X_1 = 0) < F(|X_1 = 1) = F(|X_1 = 2) = F(|X_1 = 3)$, the function F is strongly unate with respect to X_1. F is also strongly unate with respect to X_2.

$$G = \begin{array}{c} \begin{array}{cc} X_1 & X_2 \end{array} \\ \begin{bmatrix} 1011 \text{ - } 1100 \\ 1111 \text{ - } 0010 \end{bmatrix} \end{array}$$

is strongly unate with respect to X_1. Also, it is strongly unate with respect to X_2. This can be observed as follows: $G(|X_2 = 0)$ =(1011-1111), $G(|X_2 = 1)$ =(1011-1111), $G(|X_2 = 2)$ =(1111-1111), and $G(|X_2 = 3) = \phi$. Thus, we have $G(|X_2 = 3) < G(|X_2 = 0) = G(|X_2 = 1) < G(|X_2 = 2)$. These facts are also clear because G is also represented as $G = \begin{array}{c} \begin{array}{cc} X_1 & X_2 \end{array} \\ \begin{bmatrix} 1011 \text{ - } 1110 \\ 1111 \text{ - } 0010 \end{bmatrix} \end{array}$. ∎

In algorithms for cubes, we use symbol $\cup$. Let A and B be arrays. Then $A \cup B$ denotes the union of cubes in A and B.

Theorem 9.8 *Let* $c_1 = X_j^{S_A}$ *and* $c_2 = X_j^{S_B}$, *where* $S_A \cup S_B = P_j$ *and* $S_A \cap S_B = \phi$. *Let* $F = c_1 F_1 \vee c_2 F_2$, *where* $F_1 = F(|c_1)$ *and* $F_2 = F(|c_2)$. *Then,* $PI(F) \subseteq c_1 PI(F_1) \cup c_2 PI(F_2) \cup PI(F_1)PI(F_2)$.

Algorithm 9.3 *(Generation of all the prime implicants)*

1. *If* F *consists of a single cube* c_1, *then* $PI(F) = c_1$.
2. *When* F *is a one-variable function, i.e., when* $c_1 = X_i^{T_1}$, $c_2 = X_i^{T_2}$, $\dots$, *and* $c_k = X_i^{T_k}$, *then* $PI(F) = X_i^S$, *where* $S = \cup_{i=1}^{k} T_i$.
3. *If* F *is strongly unate with respect to all the variables as an SOP, then* $PI(F)$ *is obtained from* F *by deleting the cubes that are contained by another cube.*
4. *Suppose that* F *is not strongly unate with respect to the variable* X_j. *Expand* F *as* $F = c_1 F_1 \vee c_2 F_2$, *where* $c_1 = X_j^{S_A}$, $c_2 = X_j^{S_B}$, $S_A \cup S_B = P_j$, $S_A \cap S_B = \phi$, $F_1 = F(|c_1)$, *and* $F_2 = F(|c_2)$. *Then,* $PI(F)$ *is obtained from* $c_1 \cdot PI(F_1) \cup c_2 \cdot PI(F_2) \cup PI(F_1) \cdot PI(F_2)$ *by deleting cubes that are contained by another cube.*

Example 9.14 Let us obtain the prime implicants of the following function:

$$F = \begin{bmatrix} \text{1000-1100} \\ \text{1100-0010} \\ \text{0010-1111} \\ \text{1111-0010} \end{bmatrix}.$$

1. Expand F with respect to X_1. Let $S_A = \{0,1\}$ and $S_B = \{2,3\}$. Then, we have $F = c_1F_1 \vee c_2F_2$, where
c_1 =(1100-1111), c_2 =(0011-1111),
$F_1 = F(|c_1) = \begin{bmatrix} \text{1011-1100} \\ \text{1111-0010} \\ \text{1111-0010} \end{bmatrix}$, and $F_2 = F(|c_2) = \begin{bmatrix} \text{1110-1111} \\ \text{1111-0010} \end{bmatrix}$.
2. Expand F_1 with respect to X_2. Let $S_A = \{0,1\}$ and $S_B = \{2,3\}$. Then, we have $F_1 = c_3F_3 \vee c_4F_4$, where
c_3 =(1111-1100), c_4 =(1111-0011),
$F_3 = F_1(|c_3)$ =(1011-1111), and $F_4 = F_1(|c_4) = \begin{bmatrix} \text{1111-1110} \\ \text{1111-1110} \end{bmatrix}$. Thus, $PI(F_3)$ =(1011-1111), and $PI(F_4)$ =(1111-1110).
3. $PI(F_1)$ is obtained from $c_3 \cdot PI(F_3) \cup c_4 \cdot PI(F_4) \cup PI(F_3) \cdot PI(F_4)$ by deleting cubes that are contained by another cube.
$c_3 \cdot PI(F_3)$ =(1011-1100) $*$, $c_4 \cdot PI(F_4)$ =(1111-0010), and $PI(F_3) \cdot PI(F_4)$ =(1011-1110).
The cube with $*$ mark are contained by the last cube, so we have
$PI(F_1) = \begin{bmatrix} \text{1111-0010} \\ \text{1011-1110} \end{bmatrix}$.
4. F_2 is strongly unate. There is no cube that is contained by another cubes. So, we have
$PI(F_2) = \begin{bmatrix} \text{1110-1111} \\ \text{1111-0010} \end{bmatrix}$.
5. $PI(F)$ is obtained from $c_1 \cdot PI(F_1) \cup c_2 \cdot PI(F_2) \cup PI(F_1) \cdot PI(F_2)$ by deleting cubes that are contained by another cube. Note that
$c_1 \cdot PI(F_1) = \begin{bmatrix} \text{1100-0010} \\ \text{1000-1110} \end{bmatrix}\begin{matrix} * \\ * \end{matrix}$, and $c_2 \cdot PI(F_2) = \begin{bmatrix} \text{0010-1111} \\ \text{0011-0010} \end{bmatrix}\begin{matrix} \\ * \end{matrix}$.
Generate $PI(F_1) \cdot PI(F_2)$, and delete the cubes that are contained by another cube. Then, we have
$\begin{bmatrix} \text{1111-0010} \\ \text{1010-1110} \end{bmatrix}$.
Since the cube with $*$ mark is contained by another cube, we have
$PI(F) = \begin{bmatrix} \text{0010-1111} \\ \text{1111-0010} \\ \text{1010-1110} \end{bmatrix}$.

Fig. 9.5(a), (b) show F and $PI(F)$, respectively. ∎

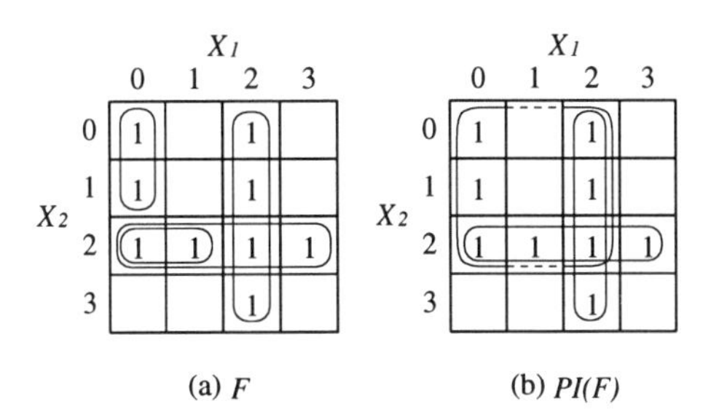

Figure 9.5 Generation of prime implicants.

9.11 SHARP OPERATION

Sharp operation (#) and **disjoint sharp operation** ($\oplus\!\!\!\!\#$) compute $F \cdot \overline{G}$, and both $F\#G$ and $F \oplus\!\!\!\!\# G$ generate arrays for $F \cdot \overline{G}$. Let U be a universal cube (that corresponds to the constant 1 function). Then $U\#G$ is equal to the complement of G obtained by using De Morgan's theorem. $U\#G$ may have the same cubes as well as cubes contained by another cube. By deleting such cubes, we have the set of the prime implicants of G. In the case of functions with two-valued variables input function, an n-variable function has $O(3^n/n)$ prime implicants. Thus, the # operation is quite time consuming. $U \oplus\!\!\!\!\# G$ also produces the complement of G. However, all the cubes are disjoint in the result. Therefore, we need not check the inclusion relation. Also, the number of cubes in the result is at most 2^n in the case of an n-variable logic function.

Definition 9.9 *(# and $\oplus\!\!\!\!\#$ operations)*
Let $a = \pi_1\text{-}\pi_2\text{-}\cdots\text{-}\pi_n$ and $b = \mu_1\text{-}\mu_2\text{-}\cdots\text{-}\mu_n$. Then,

$$a\#b = \begin{bmatrix} \pi_1\bar{\mu}_1 & - & \pi_2 & - & \pi_3 & - \cdots - & \pi_n \\ \pi_1 & - & \pi_2\bar{\mu}_2 & - & \pi_3 & - \cdots - & \pi_n \\ \pi_1 & - & \pi_2 & - & \pi_3\bar{\mu}_3 & - \cdots - & \pi_n \\ & & \cdots & \cdots & \cdots & & \\ \pi_1 & - & \pi_2 & - & \pi_3 & - \cdots - & \pi_n\bar{\mu}_n \end{bmatrix}, \text{ and}$$

$$a \oplus\!\!\!\!\# b = \begin{bmatrix} \pi_1\bar{\mu}_1 & - & \pi_2 & - & \pi_3 & - \cdots - & \pi_n \\ \pi_1\mu_1 & - & \pi_2\bar{\mu}_2 & - & \pi_3 & - \cdots - & \pi_n \\ \pi_1\mu_1 & - & \pi_2\mu_2 & - & \pi_3\bar{\mu}_3 & - \cdots - & \pi_n \\ & & \cdots & \cdots & \cdots & & \\ \pi_1\mu_1 & - & \pi_2\mu_2 & - & \pi_3\mu_3 & - \cdots - & \pi_n\bar{\mu}_n \end{bmatrix}.$$

Let $A = \{a_1, a_2, \ldots, a_r\}$ and $B = \{b_1, b_2, \ldots, b_k\}$. Then,

$$a_i\#B = (\cdots((a_i\#b_1)\#b_2)\cdots)\#b_k,$$
$$a_i \oplus\!\!\!\!\# B = (\cdots((a_i \oplus\!\!\!\!\# b_1) \oplus\!\!\!\!\# b_2)\cdots) \oplus\!\!\!\!\# b_k,$$

$$A \# B = \bigcup_{i=1}^{r} (a_i \# B), \text{ and}$$

$$A \oplus B = \bigcup_{i=1}^{r} (a_i \oplus B).$$

In computing $a\#b$ and $a \oplus b$, the order of the variables may be changed. Also, in computing $a\#B$ and $a \oplus B$, the order of cubes in B may be changed. By changing the order of variables and the order of cubes, we can simplify the results and reduce the computation time. For the computation of $F \cdot \overline{G}$, $F \oplus G$ is much more efficient than $F\#G$.

Example 9.15 Let a =(11-11-11) and b =(01-01-01). We have

$$a\#b = \begin{bmatrix} \text{10-11-11} \\ \text{11-10-11} \\ \text{11-11-10} \end{bmatrix} \text{ and } a \oplus b = \begin{bmatrix} \text{10-11-11} \\ \text{01-10-11} \\ \text{01-01-10} \end{bmatrix}.$$

Also, let

$$A = [\text{11-11-11}] \text{ and } B = \begin{bmatrix} \text{01-01-01} \\ \text{10-10-10} \end{bmatrix} \begin{matrix} (b_1) \\ (b_2) \end{matrix}.$$

Let

$$B = \begin{bmatrix} b_1 \\ b_2 \end{bmatrix},$$

where b_1 =(01-01-01) and b_2 =(10-10-10). Then, we have

$$C = A\#b_1 = \begin{bmatrix} \text{10-11-11} \\ \text{11-10-11} \\ \text{11-11-10} \end{bmatrix}.$$

Furthermore, let $C = \{c_1, c_2, c_3\}$, c_1 =(10-11-11), c_2 =(11-10-11), and c_3 =(11-11-10). We have the following:

$$c_1\#b_2 = \begin{bmatrix} \text{00-11-11} \\ \text{10-01-11} \\ \text{10-11-01} \end{bmatrix} \begin{matrix} \times \\ \\ \\ \end{matrix},$$

$$c_2\#b_2 = \begin{bmatrix} \text{01-10-11} \\ \text{11-00-11} \\ \text{11-10-01} \end{bmatrix} \begin{matrix} \\ \times \\ \\ \end{matrix}, \text{ and}$$

$$c_3\#b_2 = \begin{bmatrix} \text{01-11-10} \\ \text{11-01-10} \\ \text{11-11-00} \end{bmatrix} \begin{matrix} \\ \\ \times \end{matrix},$$

where $\times$ denotes a null cube and will be deleted. Therefore, we have

$$A\#B = \begin{bmatrix} \text{10-01-11} \\ \text{10-11-01} \\ \text{01-10-11} \\ \text{11-10-01} \\ \text{01-11-10} \\ \text{11-01-10} \end{bmatrix}.$$

In the case of $\oplus\!\!\!\!\#$, we have the following:

$$C = A \oplus\!\!\!\!\# b_1 = \begin{bmatrix} \text{10-11-11} \\ \text{01-10-11} \\ \text{01-01-10} \end{bmatrix}.$$

Let $C = \{c_1, c_2, c_3\}$, c_1 =(10-11-11), c_2 =(01-10-11), and c_3 =(01-01-10). We have the following:

$$\begin{aligned} c_1 \oplus\!\!\!\!\# b_2 &= \begin{bmatrix} \text{00-11-11} \\ \text{10-01-11} \\ \text{10-10-01} \end{bmatrix} \begin{matrix} \times \\ \\ \\ \end{matrix}, \\ c_2 \oplus\!\!\!\!\# b_2 &= \begin{bmatrix} \text{01-10-11} \\ \text{00-00-11} \\ \text{00-10-01} \end{bmatrix} \begin{matrix} \\ \times \\ \times \end{matrix}, \text{ and} \\ c_3 \oplus\!\!\!\!\# b_2 &= \begin{bmatrix} \text{01-01-10} \\ \text{00-01-10} \\ \text{00-00-00} \end{bmatrix} \begin{matrix} \\ \times \\ \times \end{matrix}. \end{aligned}$$

From these, we have

$$A \oplus\!\!\!\!\# B = \begin{bmatrix} \text{10-01-11} \\ \text{10-10-01} \\ \text{01-10-11} \\ \text{01-01-10} \end{bmatrix}.$$

Fig. 9.6 shows Karnaugh maps for $A\#B$ and $A \oplus\!\!\!\!\# B$. Note that the number of products in $A\#B$ is 6, while the number of products in $A \oplus\!\!\!\!\# B$ is only 4. Also, note that the product terms in $A \oplus\!\!\!\!\# B$ have no common minterm, i.e., are mutually disjoint. ∎

Disjoint Sharp Using Divide and Conquer

Let c be a cube, and F be an array. By complementing both sides of the equation $cF = cF(|c)$, and then by making a logical product with c, we have $c\overline{F} = c\overline{F}(|c)$. This implies that the computation of $c\overline{F}$ can be done by $c\overline{F}(|c)$. Since $\overline{F}(|c)$ has fewer cubes than $\overline{F}$, in general, the computation of $\overline{F}(|c)$ is, in most cases, faster than that of $\overline{F}$. Thus, the computation of $c\overline{F}$ is done by $c \oplus\!\!\!\!\# F$ using divide and conquer method. This method is faster and produces fewer products than the method shown in Definition 9.9.

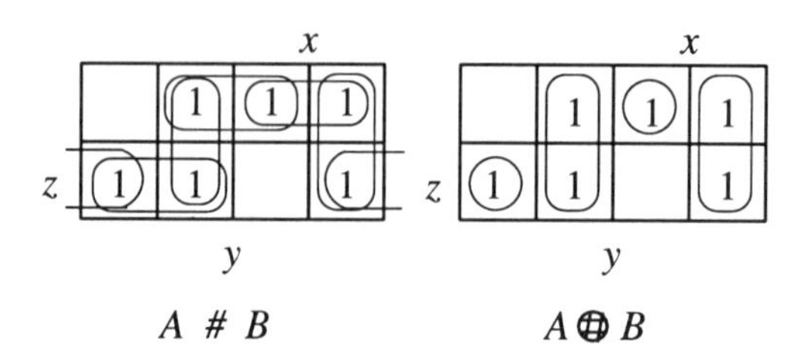

Figure 9.6 # and $\oplus$ operations.

Algorithm 9.4 *(Computation of $c \oplus F$)*

1. *Let $G \leftarrow F(|c)$.*
2. *Obtain $\overline{G}$, the complement of G, by Algorithm 9.1.*
3. *Output $c \cdot \overline{G}$.*

Bibliographical Notes

Many textbooks on logic minimization for CAD are available [42, 103, 105, 171, 252]. Detailed treatments of the topics can be found as follows: Multi-valued input two-valued output functions [349, 394, 395, 406]; multi-valued logic [118, 176, 177, 282, 327, 390]; divide and conquer method [37, 255, 262, 324]; complementation of SOPs [63, 173, 294, 346, 419]; tautology [152, 344, 345]; sharp operations [103, 174].

Table 9.3 Multi-valued input two-valued output function.

x	y	z	f	x	y	z	f
0	0	0	1	1	0	0	0
0	0	1	1	1	0	1	1
0	0	2	1	1	0	2	1
0	0	3	1	1	0	3	0
0	1	0	0	1	1	0	0
0	1	1	1	1	1	1	1
0	1	2	0	1	1	2	0
0	1	3	1	1	1	3	1
0	2	0	1	1	2	0	0
0	2	1	0	1	2	1	0
0	2	2	0	1	2	2	0
0	2	3	1	1	2	3	0

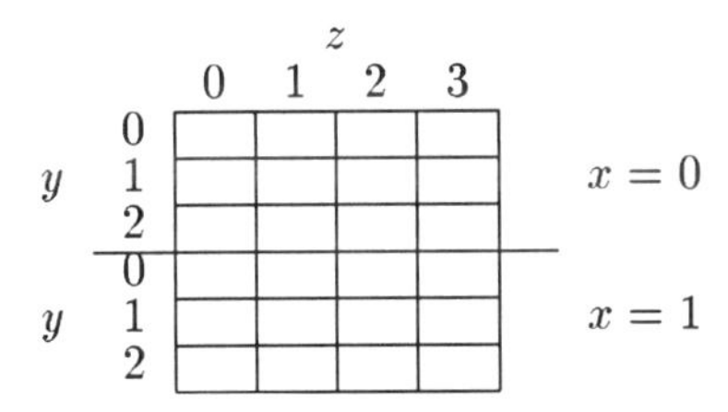

Figure 9.7 Multi-valued Karnaugh map.

Exercises

9.1 Show a practical example which can be represented by a multi-valued input two-valued output function (e.g., Example 9.1).

9.2 Draw the Karnaugh map for the function in Table 9.3 in Fig. 9.7. Obtain the MSOP for this function.

9.3 Is the following function a tautology or not?

$$f = xyzw \vee \bar{x} \vee x\bar{y} \vee xy\bar{z} \vee xyz\bar{w}.$$

9.4 (M) Prove the following:

$$x_1 \vee x_2 \vee x_3 \vee x_4 \geq (x_1 \vee x_2 \vee y_1) \cdot (x_3 \vee \bar{y}_1 \vee y_2) \cdot (x_4 \vee \bar{y}_2 \vee y_3) \cdot (x_2 \vee \bar{y}_3 \vee y_1) \cdot (x_3 \vee x_4 \vee \bar{y}_1).$$

9.5 Do the following two SOPs represent the same function?

$$F_1 = x_1\bar{x}_2 \vee x_2\bar{x}_3 \vee \bar{x}_1 x_3 \vee x_4\bar{x}_5 \vee x_5\bar{x}_6 \vee \bar{x}_4 x_6,$$
$$F_2 = \bar{x}_1 x_2 \vee \bar{x}_2 x_3 \vee x_1\bar{x}_3 \vee \bar{x}_4 x_5 \vee \bar{x}_5 x_6 \vee x_4\bar{x}_6.$$

9.6 Let $F = x_1x_2x_3 \vee \bar{x}_1\bar{x}_2\bar{x}_3 \vee x_4x_5x_6 \vee \bar{x}_4\bar{x}_5\bar{x}_6$. Then,

(a) Obtain the SOP for the complement of F by the De Morgan's theorems.

(b) Obtain the SOP for the complement of F by the divide and conquer method.

9.7

(a) Let $F = g_1h_1 \vee g_2h_2$. Prove that if $h_1 \vee h_2 = 1$ and $h_1h_2 = 0$, then $\overline{F} = \bar{g}_1h_1 \vee \bar{g}_2h_2$.

(b) Let $F = g_1h_1 \vee g_2h_2 \vee \cdots \vee g_ph_p$. Prove that if $h_1 \vee h_2 \vee \cdots \vee h_p = 1$ and $h_ih_j = 0$ $(i \neq j)$, then $\overline{F} = \bar{g}_1h_1 \vee \bar{g}_2h_2 \vee \cdots \vee \bar{g}_ph_p$.

9.8 Obtain the complements of each of the following functions as SOPs by using the De Morgan's theorem as well as the divide and conquer method. Then, compare the number of products obtained by these two methods. Also, show the MSOPs for the complement functions in the Karnaugh maps.

(a) $f = x_1x_2x_3x_4 \vee \bar{x}_1\bar{x}_2\bar{x}_3\bar{x}_4 \vee x_1x_2x_3x_5 \vee \bar{x}_1\bar{x}_2\bar{x}_3\bar{x}_5 \vee x_1x_2x_4x_5 \vee \bar{x}_1\bar{x}_2\bar{x}_4\bar{x}_5$ $\vee x_1x_3x_4x_5 \vee \bar{x}_1\bar{x}_3\bar{x}_4\bar{x}_5 \vee x_2x_3x_4x_5 \vee \bar{x}_2\bar{x}_3\bar{x}_4\bar{x}_5$.

(b) $g = x_1x_2x_3x_4x_5 \vee \bar{x}_1\bar{x}_2\bar{x}_3\bar{x}_4\bar{x}_5$.

9.9 Let $f = x_1x_2\cdots x_n \vee \bar{x}_1\bar{x}_2\cdots\bar{x}_n$.

(a) Obtain the complement of f by De Morgan's theorem, and show the number of products in the SOP.

(b) Prove that $f = (x_1 \vee \bar{x}_2)(x_2 \vee \bar{x}_3)\cdots(x_{n-1} \vee \bar{x}_n)(x_n \vee \bar{x}_1)$ holds.

(c) Show that the complement of f can be represented by n products.

9.10 By using the divide and conquer method, obtain an SOP for the complement of $F = x_1x_2x_3 \vee x_1\bar{x}_2x_4 \vee \bar{x}_1x_2x_5 \vee \bar{x}_1\bar{x}_2x_6$.

9.11 Is F a tautology or not, where $F = \begin{bmatrix} \text{110-110-101} \\ \text{111-101-111} \\ \text{011-110-111} \\ \text{110-110-110} \end{bmatrix}$?

9.12 Are the following functions, weakly unate, or strongly unate?

(a) $X_1^{\{0,1,2\}} \cdot X_2^{\{1\}} \vee X_1^{\{1,2\}} \cdot X_2^{\{1,2\}} \vee X_1^{\{1\}} \cdot X_2^{\{0,1,2\}}$.
(b) $X_1^{\{0,1,2\}} \cdot X_2^{\{0\}} \vee X_1^{\{1,2\}} \cdot X_2^{\{1,2\}} \vee X_1^{\{2\}} \cdot X_2^{\{0,1,2\}}$.

9.13 (M) Obtain all the prime implicants for $F = \begin{bmatrix} \text{100-010-110} \\ \text{101-110-001} \\ \text{110-101-011} \\ \text{011-110-101} \end{bmatrix}$.

9.14 Let a =(0110-1010-1100) and b =(1101-1011-1011). Then, obtain $a\#b$ and $a \oplus b$.

9.15 Let

$$A = \begin{bmatrix} \text{10-11-10} \\ \text{10-01-11} \end{bmatrix} \text{ and } B = [\text{ 10-01-11 }].$$

Obtain $A \oplus B$ and $A - B$.

9.16 Prove Lemma 9.1.

9.17 Prove Theorem 9.5.

9.18 Prove Theorem 9.6.

10

HEURISTIC OPTIMIZATION OF TWO-LEVEL NETWORKS

Simplification of sum-of-products expressions is one of the most important problem in logic synthesis. This chapter introduces heuristic minimization programs such as PRESTO, MINI, and ESPRESSO. It also introduces encoding problems and assignment problems. The reader is recommended to study Chapter 9 first.

10.1 SIMPLIFICATION OF SOPS WITH MANY INPUTS

Simplification of SOPs is one of the key technology in logic synthesis. In total computation time for logic synthesis, the ratio of the time spent for simplifications for SOPs is considerably large. Simplification of SOPs is directly related to the simplification of PLAs (described in Chapter 12). This section presents simplification methods for SOPs suitable for computer manipulation.

Quine-McCluskey Method

The **Quine-McCluskey method** (**QM method**) described in Chapter 4 can be extended for multi-valued input two-valued output functions. However, the straightforward implementation of this algorithm often fails to simplify the medium size SOPs, e.g., a 16-input 16-output SOP that corresponds to a commercially available FPLA (field programmable logic array). In practical applications, the problems are, in most cases, multi-output, and so the number of prime implicants is very large. Therefore, the QM method requires excessive amount of memory. Even for the cases where all the prime implicants can be generated, if the number of input variables is large, then the number of minterms is very large, and so the covering table becomes too large to store. Thus, the QM method is unsuitable for large problems. However, if the number of input variables is small, the QM method is faster than heuristic methods such as MINI2 and ESPRESSO. Also, the QM method always obtains the exact minimum solutions. So, QM method is useful for theoretical research. The QM method utilizing BDD data structure can obtain exact minimum SOPs with many inputs. However, for general problems, this method also fails to minimize SOPs due to the limitation of memory storage and computation time.

Heuristic Methods

Since the QM method is unsuitable for the simplification of SOPs with many inputs, several heuristic algorithms have been developed. These heuristic methods do not necessarily produce optimal solutions, but they obtain fairly good solutions in a short time. In practical applications, we need not to obtain an exact minimum solution, but a near minimum solution may be sufficient. So, these heuristic methods are extensively used for the SOPs with many inputs. Among these heuristic programs, PRESTO developed at UCLA, MINI developed at IBM, and ESPRESSO developed at IBM and University of California at Berkeley are well known. In Sections 10.2–10.4, we will describe the basic operations used in these algorithms.

10.2 MERGE, EXPANSION, AND DELETE

Definition 10.1 *Let two products be*

$$c_1 = X_1^{S_1} X_2^{S_2} \cdots X_n^{S_n} \text{ and } c_2 = X_1^{T_1} X_2^{T_2} \cdots X_n^{T_n}.$$

Then, **L-distance** $L\text{-}dist(c_1, c_2)$ *of the two products is defined as follows:*

$$L\text{-}dist(c_1, c_2) = [\text{ The number of } i\text{'s such that } (S_i \neq T_i)\ ].$$

That is, the number of different literals in two products.

A basic simplification method for SOPs is shown below:

Algorithm 10.1 *(Simplification of SOPs)*

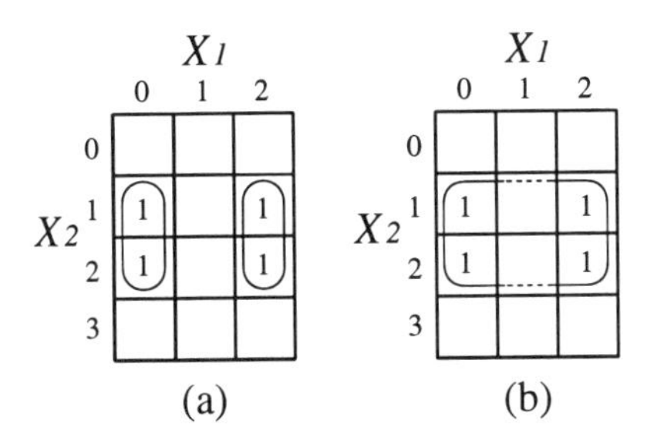

Figure 10.1 Merge operation.

1. *Merge products whose L-distances are 1 (pre-processing).*
2. *Expand each product into a prime implicant.*
3. *Delete redundant products.*

Merge is a simplification of SOPs by using the relation:

$$X^A \cdot c \vee X^B \cdot c = X^{(A \cup B)} \cdot c, \ (A, B \subseteq P).$$

Example 10.1 Consider the SOP of two-variables: $F_1 = X_1^{\{0\}} X_2^{\{1,2\}} \vee X_1^{\{2\}} X_2^{\{1,2\}}$ (Fig. 10.1(a)). In this case, only the literals of X_1 are different in two products. So we can merge them to have Fig. 10.1(b): $(X_1^{\{0\}} \vee X_1^{\{2\}}) X_2^{\{1,2\}} = X_1^{\{0,2\}} X_2^{\{1,2\}}$. ∎

Expand is an operation of increasing the volume of cubes without changing the functions. The expanded cube is larger than the original. Let the given SOP be represented as $F = c \vee G$, where c is a cube having the form $X^S \cdot e$, G is an SOP consisting of cubes in F except for c. Then, since $X^{\{a\}} \cdot e \leq F$ implies $F = X^{\{a\}} \cdot e \vee F$, we have the following relations:

$$\begin{aligned} F &= X^{\{a\}} \cdot e \vee (X^S \cdot e \vee G) = (X^{\{a\}} \vee X^S) e \vee G \\ &= X^{(\{a\} \cup S)} \cdot e \vee G. \end{aligned}$$

This means that S of the cube $X^S \cdot e$ is expanded to $(S \cup \{a\})$. For each variable for each product, check whether the literal can be expanded. Expand it whenever possible. This result in the SOP where no cubes can be expanded. The cubes obtained in this way represent a prime implicant as shown below:

Theorem 10.1 *(Expansion)*
Let the given SOP be represented as $F = c \vee G$, *where* $c = X_i^{S_i} \cdot e$. *Let* $c(a) = X_i^{\{a\}} \cdot e$, *where* $a \in P_i - S_i$.

1. *If* $c(a) \leq F$, *then* $F = X_i^{(S_i \cup \{a\})} \cdot e \vee G$.
2. *If* $c(a) \not\leq F$, *for all* i $(i = 1, \ldots, n)$ *and for all* a $(a \in P_i - S_i)$, *then* c *is a prime implicant of* F. *Note that* $\leq$ *denotes the inclusion relation of functions.*

Example 10.2 Consider the SOP in Fig. 10.2(a):

$$F_2 = X_1^{\{1\}} X_2^{\{2\}} \vee X_1^{\{1,2\}} X_2^{\{0\}} \vee X_1^{\{0\}} X_2^{\{2\}} \vee X_1^{\{2\}} X_2^{\{2,3\}}.$$

Note that this SOP can be represented as:

$$F_2 = c_1 \vee G,$$

where

$$\begin{aligned} c_1 &= X_1^{\{1\}} X_2^{\{2\}}, \text{ and} \\ G &= X_1^{\{1,2\}} X_2^{\{0\}} \vee X_1^{\{0\}} X_2^{\{2\}} \vee X_1^{\{2\}} X_2^{\{2,3\}}. \end{aligned}$$

First, check whether the literal of X_1 in c_1 can be expanded to include logic value 2. Let $c_1 = X_1^{\{1\}} \cdot e_1$ and $e_1 = X_2^{\{2\}}$. Consider $c_1(2) = X_1^{\{2\}} \cdot e_1$. Since $c_1(2) \leq F_2$, the literal of X_1 can be expanded to include logic value 2. In other words, c_1 is expanded into $c_2 = X_1^{\{1,2\}} X_2^{\{2\}}$ as shown in Fig. 10.2(b).
Next, in c_2 of Fig. 10.2(b), check whether the literal of X_2 can be expanded to include logic value 0. Let $c_2 = X_2^{\{2\}} \cdot e_2$ and $e_2 = X_1^{\{1,2\}}$. Consider $c_2(0) = X_2^{\{0\}} \cdot e_2$. Since $c_2(0) \leq F_2$, the literal of X_2 can be expanded to include 0. In other words, c_2 is expanded into $c_3 = X_1^{\{1,2\}} X_2^{\{0,2\}}$ as shown in Fig. 10.2(c). Since c_3 cannot be expanded into further, it is a prime implicant.
In Fig. 10.2(a), if we expand c_1 to include the 0 of variable X_1, then we have c_4 shown in Fig. 10.2(d). c_4 can also expanded to include the 2 of variable X_1, and we have c_5 in Fig. 10.2(e). Since c_5 cannot be further expanded, it is a prime implicant. As shown in the example, different expansions will produce different prime implicants. Fig. 10.2(f) shows all the prime implicants of F_2. ∎

The **delete** operation removes redundant products. For example, in Fig. 10.3(a), removing prime implicant c_1 does not change the function. In general, we can show this operation as follows: Let the given function be represented as $F = c \vee G$, where c is a product, and G denotes the logical sum of products

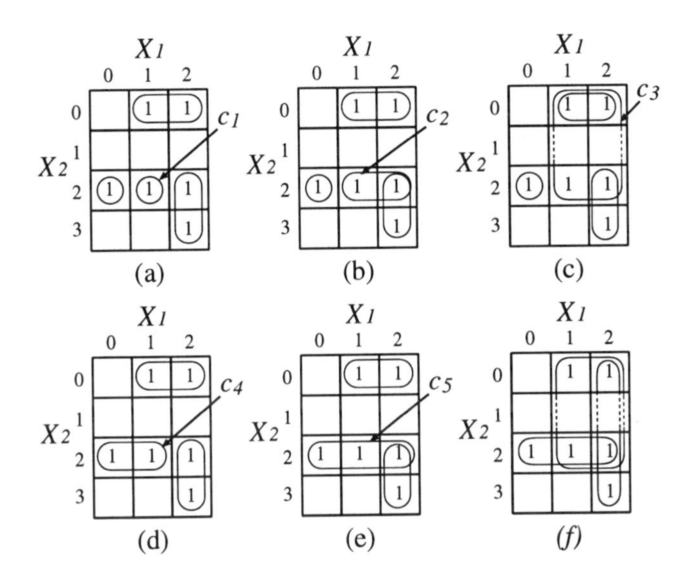

Figure 10.2 Expand operation.

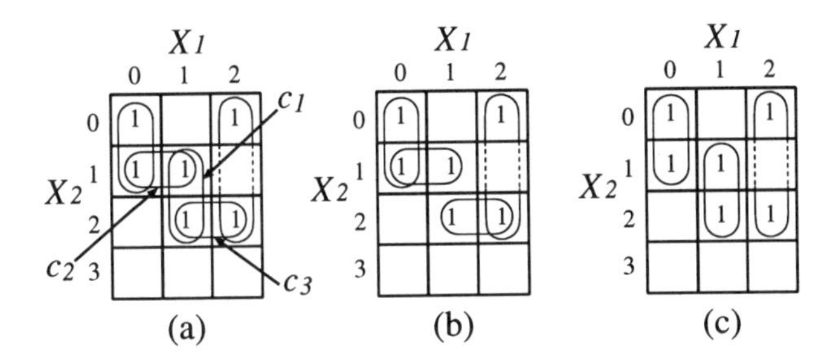

Figure 10.3 Delete operation.

other than c. If $c \leq G$, then we can remove product c from F without changing the function. Consider an SOP F that consists of prime implicants only. For each product, check whether it can be deleted. Delete products while it is possible. After this operation, we have an SOP, where we can delete no product. This SOP is an **irredundant sum-of-products expression (ISOP)** as shown below:

Theorem 10.2 *(Generation of an ISOP)*

1. *Let an SOP be represented as $F = c \vee G$. If $c \leq G$, then the product c can be deleted from F, i.e., $F = G$.*
2. *Let the SOP is represented as a sum of prime implicants: $F = c_1 \vee c_2 \vee \cdots \vee c_p$. Let G_i be the sum of products in F other than c_i. If $c_i \not\leq G_i$ for all i $(i = 1, \ldots, p)$, then F is an ISOP.*

Example 10.3 Fig. 10.3(a) shows the SOP F_3, which consists of prime implicants only. F_3 is represented as

$$\begin{aligned} F_3 &= c_1 \vee G_1, \text{ where } c_1 = X_1^{\{1\}} X_2^{\{1,2\}}, \text{ and} \\ G_1 &= X_1^{\{0\}} X_2^{\{0,1\}} \vee X_1^{\{0,1\}} X_2^{\{1\}} \vee X_1^{\{1,2\}} X_2^{\{2\}} \vee X_1^{\{2\}} X_2^{\{0,2\}}. \end{aligned}$$

Since $c_1 \leq G_1$, we can delete c_1 from F_3, and have the SOP in Fig. 10.3(b). We cannot delete any products in this SOP, so it is an ISOP.
However, we delete both c_2 and c_3 in Fig. 10.3(a). Thus, we have a simpler SOP in Fig. 10.3(c). In this SOP, we cannot delete any product, so it is an ISOP. This example shows that the order in which products are deleted determines the final ISOPs. ∎

10.3 REDUCE AND RESHAPE

Algorithm 10.1 produces an irredundant SOP, where no product can be deleted. For example, in Fig. 10.3(b), deletion of any product produces different functions, and so no product can be deleted. This SOP does not contain any redundant product, so it is an **irredundant sum-of-products expression** (ISOP). On the other hand, Fig. 10.3(c) is a **minimum solution**, which corresponds to a minimum sum-of-products expression (MSOP), since no other SOP require fewer products. An MSOP is an ISOP. However, the converse is not always true. The QM method produces MSOPs, while Algorithm 10.1 produces ISOPs. Usually, ISOPs requires 10 to 15 percent more products than MSOPs. However, for some functions, the ISOPs have many more products than MSOPs. For example, the MSOP in Fig. 10.3(c) has three products, while the ISOP in Fig. 10.3(b) has four products. In this case the ISOP requires 33.3% more products than the MSOP.

Since Algorithm 10.1 cannot improve the solution in Fig. 10.3(b), we will introduce the reduce and reshape operations.

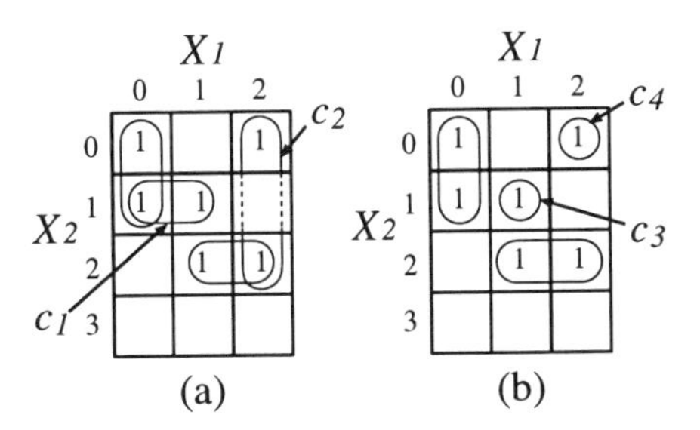

Figure 10.4 Reduce operation.

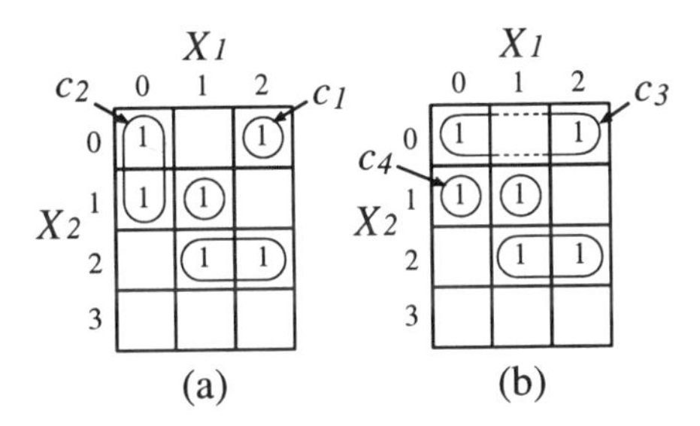

Figure 10.5 Reshape operation.

The **reduce** operation reduces the size of cubes without changing the function. Especially, if the cube is redundant, then delete the cube. For example, the reduce operation changes c_1 and c_2 in Fig. 10.4(a), into c_3 and c_4, in Fig. 10.4(b), respectively. The reduce operation is the converse to the expand operation.

The **reshape** operation replaces a pair of adjacent cubes with an another pair without changing the function. For example, the pair (c_1, c_2) in Fig. 10.5(a) is reshaped into the pair (c_3, c_4) in Fig. 10.5(b).

The reshape operation is formally defined as follows:

Definition 10.2 *Let the L-distance of the two products c_1 and c_2 be two. Let*

$$\begin{aligned} c_1 &= X_1^{S_1} \cdot X_2^{S_2} \cdot \cdots \cdot X_i^{S_i} \cdot \cdots \cdot X_j^{S_j} \cdot \cdots \cdot X_n^{S_n}, \text{ and} \\ c_2 &= X_1^{S_1} \cdot X_2^{S_2} \cdot \cdots \cdot X_i^{T_i} \cdot \cdots \cdot X_j^{T_j} \cdot \cdots \cdot X_n^{S_n}, \end{aligned}$$

where $S_i \cap T_i = \phi$ *and* $S_j \subseteq T_j$ *and* $S_k = T_k$ $(k = 1, 2, \ldots, n,\ k \neq i,\ k \neq j)$. *Then, the reshape operation replaces the pair* (c_1, c_2) *with* (c_3, c_4)*, where*

$$\begin{aligned} c_3 &= X_1^{S_1} \cdot X_2^{S_2} \cdot \cdots \cdot X_i^{(S_i \cup T_i)} \cdot \cdots \cdot X_j^{S_j} \cdot \cdots \cdot X_n^{S_n}, \text{ and} \\ c_4 &= X_1^{S_1} \cdot X_2^{S_2} \cdot \cdots \cdot X_i^{T_i} \cdot \cdots \cdot X_j^{(T_j - S_j)} \cdot \cdots \cdot X_n^{S_n}. \end{aligned}$$

Example 10.4 In Fig. 10.5(a), since the following products satisfy the reshaping condition $c_1 = X_1^{\{2\}} X_2^{\{0\}}$, $c_2 = X_1^{\{0\}} X_2^{\{0,1\}}$, we have the reshaped products: $c_3 = X_1^{\{0,2\}} X_2^{\{0\}}$, $c_4 = X_1^{\{0\}} X_2^{\{1\}}$. ∎

10.4 DETECTION OF ESSENTIAL PRIME IMPLICANTS

By applying the expand, the reduce, and the reshape operations repeatedly, we have an ISOP. An ISOP has no redundancy, but it may not be an minimum (MSOP). However, by using the essential prime implicants, we can often prove the minimality of an ISOP.

Theorem 10.3 *If a logic function* F *has a distinguished minterm* m*, then an MSOP of* F *has an essential prime implicant that contains* m*.*

Example 10.5 The SOP shown in Fig. 10.6 is an MSOP. This can be proved as follows: First, consider the prime implicant c_1. c_1 corresponds to $X_1^{\{0,1,2\}} X_2^{\{2\}}$, and covers three minterms. Among them, only one prime implicant c_1 covers the minterm $v_1 = X_1^{\{0\}} X_2^{\{2\}}$. Thus, v_1 is a distinguished minterm. On the other hand, as for the minterm $v_2 = X_1^{\{1\}} X_2^{\{2\}}$, both c_1 and $c_2 = X_1^{\{1,2\}} X_2^{\{0,2\}}$ cover v_2. Thus, v_2 is not a distinguished minterm. In c_2, $X_1^{\{1\}} X_2^{\{0\}}$ is a distinguished minterm. In c_3, $X_1^{\{2\}} X_2^{\{3\}}$ is a distinguished minterm. Because each of prime implicants c_1, c_2, and c_3 covers a distinguished minterm, they are essential prime implicants. Since the prime implicants in the map are all essential, they form an MSOP. ∎

As illustrated in Example 10.5, if all the prime implicants in an ISOP are essential, then the ISOP is the unique MSOP. Therefore, in this case, we need not continue the expansion, the reduce, and the reshape operations any more. Even if a subset of the products in ISOP are essential, detect all the essential prime implicants, remove them from the ISOP, change the covered minterms

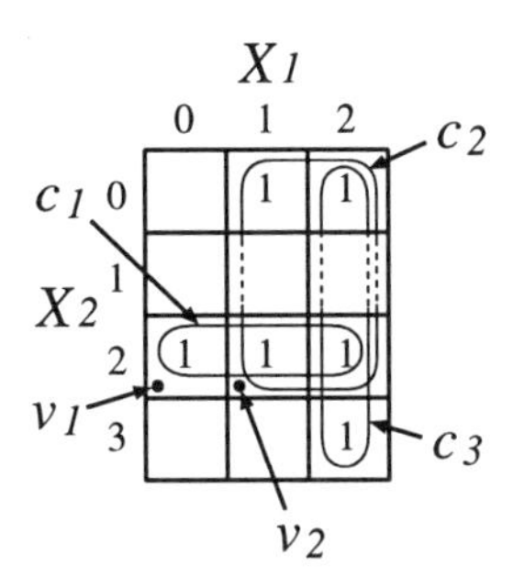

Figure 10.6 Detection of essential prime implicants.

into *don't care*, and continue the simplification operations. In the case of control circuits for microprocessors, about a half of the prime implicants in the MSOP are essential. So, the detection of the essential prime implicants at first will reduce the computation time as well as improve the quality of the solution. Heuristic minimization algorithms programs such as MINI2 and ESPRESSO (Section 10.7) detect all the essential prime implicants.

Definition 10.3 *Let two products be*

$$c_1 = X_1^{S_1} \cdot X_2^{S_2} \cdot \cdots \cdot X_n^{S_n} \text{ and } c_2 = X_1^{T_1} \cdot X_2^{T_2} \cdot \cdots \cdot X_n^{T_n}.$$

The **consensus with respect to variable** x_i *of* c_1 *and* c_2 *is defined as*

$$cons(i : c_1, c_2) = X_1^{(S_1 \cap T_1)} \cdot X_2^{(S_2 \cap T_2)} \cdot \cdots \cdot X_i^{(S_i \cup T_i)} \cdot \cdots \cdot X_n^{(S_n \cap T_n)}.$$

The **consensus** *of* c_1 *and* c_2 *is defined as*

$$cons(c_1, c_2) = \bigcup_{i=1}^{n} cons(i : c_1, c_2).$$

Let c *be a product and* G *be a set of products. Then,*

$$cons(c, G) = \bigcup_{c_k \in G} cons(c, c_k).$$

Algorithm 10.2 *(Decision of essential prime implicant for multi-valued input two-valued output functions)*

1) *Let F be the set of prime implicants of a function. Let c be the prime implicant which we want to determine is essential or not. Let G be the set of prime implicants other than c.*
2) *Partition the products in G into three classes:*
   G_1 : *Set of products that have common elements with c.*
   G_2 : *Set of products that are adjacent to c.*
   G_3 : *Set of other products.*
3) $T = (G_1 \circledast c) \cup G_2$.
   $H = cons(c, T)$.
4) $c \not\leq H$ *iff c is an essential prime implicant. This decision can be done by the tautology decision of $H(|\,c)$.*

Example 10.6 Find the essential prime implicants in Fig. 10.6. The set of prime implicants is

$$\begin{array}{c} \\ c_1 \\ c_2 \\ c_3 \end{array} \begin{array}{c} \begin{array}{cc} X_1 & \quad X_2 \end{array} \\ \left[\begin{array}{c} 111\text{ - }0010 \\ 011\text{ - }1010 \\ 001\text{ - }1011 \end{array}\right] \end{array}.$$

First, decide if c_1 is an essential prime implicant or not. $G = \{c_2, c_3\}$, $G_1 = \{c_2, c_3\}$, $G_2 = \phi$, $G_3 = \phi$, $T = (G_1 \circledast c_1) \cup G_2 = \begin{bmatrix} \text{011-1000} \\ \text{001-1001} \end{bmatrix}$, $H = cons(c_1, T)$, $c_1 =$(111-0010), and $H = \begin{bmatrix} \text{011-1010} \\ \text{001-1011} \end{bmatrix}$. Clearly, $c_1 \not\leq H$. Thus, c_1 is an essential prime implicant. Similarly, we can show that both c_2 and c_3 are also essential. ∎

10.5 MULTI-OUTPUT FUNCTION

As described in Section 4.9, for multi-output functions, the independent minimization of each output does not always produce the exact minimum solutions. To minimize (or simplify) SOPs for multi-output functions, we have to consider all the outputs at the same time. In the past, **multi-output prime implicants** were used to solve this problem. However, that method requires a complicated algorithm. This section presents a simplification method of SOPs for multi-output functions by using characteristic functions.

Let $f_j(X_1, X_2, \ldots, X_n)$ $(j = 0, 1, \ldots, m-1)$ be m functions. Then, the two-valued output function $F(X_1, X_2, \ldots, X_n, X_{n+1})$, where $F(X_1, X_2, \ldots, X_n, j) = f_j(X_1, X_2, \ldots, X_n)$, is the **characteristic function for multi-output func-**

Table 10.2 Characteristic function of a multi-output function.

	X_1	X_2	X_3	X_4	F
1	0	0	0	0	0
2	0	0	0	1	1
3	0	0	0	2	1
4	0	0	1	0	0
5	0	0	1	1	0
6	0	0	1	2	1
7	0	1	0	0	0
8	0	1	0	1	0
9	0	1	0	2	0
10	0	1	1	0	0
11	0	1	1	1	0
12	0	1	1	2	1
13	1	0	0	0	0
14	1	0	0	1	0
15	1	0	0	2	0
16	1	0	1	0	1
17	1	0	1	1	1
18	1	0	1	2	0
19	1	1	0	0	1
20	1	1	0	1	0
21	1	1	0	2	0
22	1	1	1	0	1
23	1	1	1	1	0
24	1	1	1	2	0

Table 10.1 Multi-output function.

Input			Output		
x_1	x_2	x_3	f_0	f_1	f_2
0	0	0	0	1	1
0	0	1	0	0	1
0	1	0	0	0	0
0	1	1	0	0	1
1	0	0	0	0	0
1	0	1	1	1	0
1	1	0	1	0	0
1	1	1	1	0	0

tion. Here, X_{n+1} is the m valued variable representing outputs. In other words, $F(a_1, a_2, \ldots, a_n, j) = 1 \Leftrightarrow (x_1 = a_1, x_2 = a_2, \ldots, x_n = a_n, f_j = 1)$.

Example 10.7 Table 10.1 shows a multi-output function. Table 10.2 shows its characteristic function F, where X_4 is the variable showing the outputs, and takes a value in {0,1,2}. $X_4 = j$ represents $f_j = 1$ $(j = 0, 1, 2)$. If $X_1 = X_2 = X_3 = 0$, then $f_0 = 0$, $f_1 = f_2 = 1$. Thus, the value of F in the first line of Table 10.2 is 0; in the second and the third lines, the values are 1. Also, if $X_1 = X_2 = 0$, $X_3 = 1$, then $f_0 = f_1 = 0$, $f_2 = 1$. Thus, in Table 10.2, the values are 0 in the 4th and 5th line, and 1 in the 6th line. Similarly, other part can be obtained. ∎

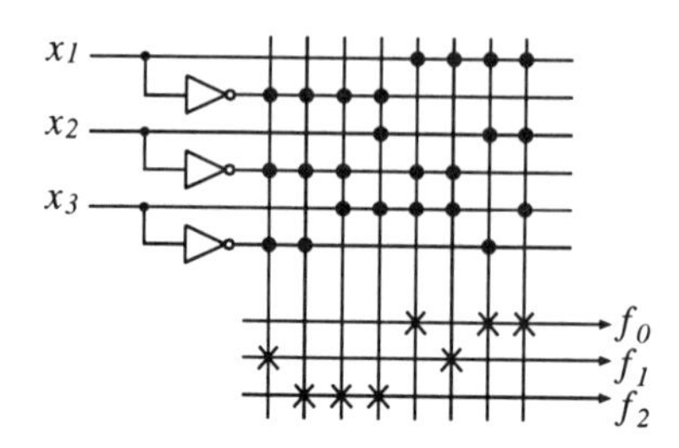

Figure 10.7 PLA for (10.1).

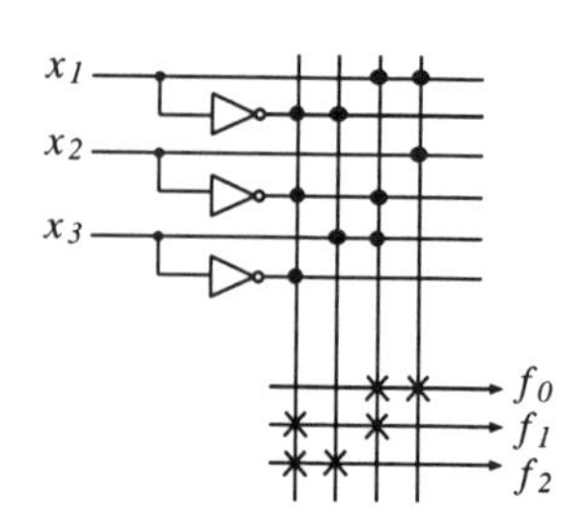

Figure 10.8 PLA for (10.2).

Each product in the SOP of the characteristic function F for the multi-output function $(f_0, f_1, \ldots, f_{m-1})$ corresponds to a product line of the PLA (Section 12.1.1) for the multi-output function.

Example 10.8 The SOP for Table 10.2 is

$$\begin{aligned} F \;=\; & X_1^{\{0\}}X_2^{\{0\}}X_3^{\{0\}}X_4^{\{1\}} \vee X_1^{\{0\}}X_2^{\{0\}}X_3^{\{0\}}X_4^{\{2\}} \\ & \vee X_1^{\{0\}}X_2^{\{0\}}X_3^{\{1\}}X_4^{\{2\}} \vee X_1^{\{0\}}X_2^{\{1\}}X_3^{\{1\}}X_4^{\{2\}} \\ & \vee X_1^{\{1\}}X_2^{\{0\}}X_3^{\{1\}}X_4^{\{0\}} \vee X_1^{\{1\}}X_2^{\{0\}}X_3^{\{1\}}X_4^{\{1\}} \\ & \vee X_1^{\{1\}}X_2^{\{1\}}X_3^{\{0\}}X_4^{\{0\}} \vee X_1^{\{1\}}X_2^{\{1\}}X_3^{\{1\}}X_4^{\{0\}}. \end{aligned} \tag{10.1}$$

Fig. 10.7 shows the PLA for this SOP. The first product shows the relation "if $X_1 = X_2 = X_3 = 0$, then $f_1 = 1$." ∎

An MSOP for the characteristic function F corresponds to a minimum PLA realizing the multi-output function $(f_0, f_1, \ldots, f_{m-1})$.

Example 10.9 The MSOP for the function F in Table 10.1 can be obtained as follows:
By merging the 1st and 2nd products, we have $X_1^{\{0\}}X_2^{\{0\}}X_3^{\{0\}}X_4^{\{1,2\}}$.
By merging the 3rd and 4th products, we have $X_1^{\{0\}}X_2^{\{0,1\}}X_3^{\{1\}}X_4^{\{2\}}$.
By merging the 5th and 6th products, we have $X_1^{\{1\}}X_2^{\{0\}}X_3^{\{1\}}X_4^{\{0,1\}}$.
By merging the 7th and 8th products, we have $X_1^{\{1\}}X_2^{\{1\}}X_3^{\{0,1\}}X_4^{\{0\}}$.
Thus, we have the SOP:

$$\begin{aligned} F \;=\; & X_1^{\{0\}}X_2^{\{0\}}X_3^{\{0\}}X_4^{\{1,2\}} \vee X_1^{\{0\}}X_2^{\{0,1\}}X_3^{\{1\}}X_4^{\{2\}} \\ & \vee X_1^{\{1\}}X_2^{\{0\}}X_3^{\{1\}}X_4^{\{0,1\}} \vee X_1^{\{1\}}X_2^{\{1\}}X_3^{\{0,1\}}X_4^{\{0\}}. \end{aligned} \tag{10.2}$$

Table 10.3 Multi-output function.

Input			Output	
x_1	x_2	x_3	f_0	f_1
0	0	0	1	0
0	0	1	0	0
0	1	0	1	1
0	1	1	1	1
1	0	0	0	0
1	0	1	0	1
1	1	0	0	0
1	1	1	1	0

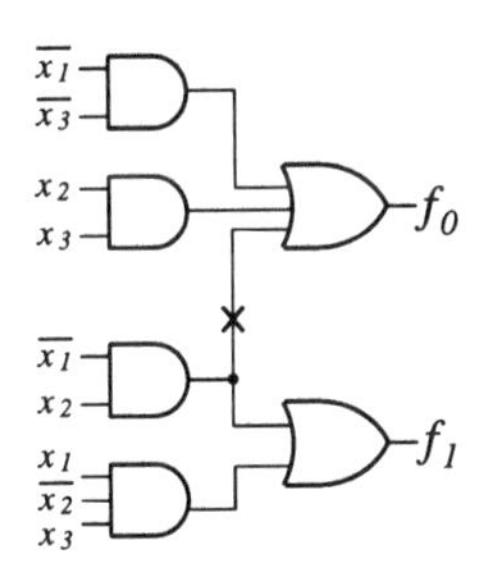

Figure 10.9 Network with redundant OR connection.

We can prove that it is an MSOP. Fig. 10.8 shows the PLA corresponding to this MSOP. ∎

In the two-level logic network that is obtained from the MSOP for multi-output function $(f_0, f_1, \ldots, f_{m-1})$, the number of AND gates is minimum, while the number of connections is not always minimum.

Example 10.10 The MSOP for the characteristic function of the two-output function in Table 10.3 is

$$\begin{aligned} F &= X_1^{\{0\}} X_2^{\{0,1\}} X_3^{\{0\}} X_4^{\{0\}} \vee X_1^{\{0\}} X_2^{\{1\}} X_3^{\{0,1\}} X_4^{\{0,1\}} \\ &\vee X_1^{\{0,1\}} X_2^{\{1\}} X_3^{\{1\}} X_4^{\{0\}} \vee X_1^{\{1\}} X_2^{\{0\}} X_3^{\{1\}} X_4^{\{1\}}. \end{aligned}$$

Fig. 10.9 shows the corresponding AND-OR two-level network. In this case, the connection from the AND gate realizing $\overline{X}_1 X_2$ to the OR gate for f_0, is redundant. Fig. 10.10(a) shows the Karnaugh map with the redundant OR

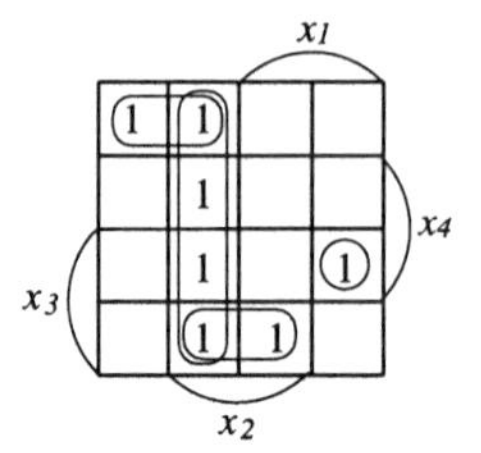

(a) SOP with redundant OR connection.

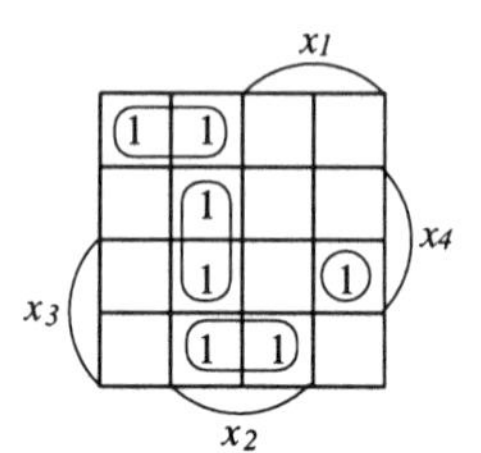

(b) SOP without redundant OR connection.

Figure 10.10 Karnaugh maps for the characteristic function for two-output function.

connection. Fig. 10.10(b) shows the Karnaugh map without redundant OR connection. As shown in this example, to reduce the redundant OR connections, we try not to expand the output part (X_4) of the loops. ∎

The **input-fat** operation reduces the AND connections, while the **output-slim** operation reduces the OR connections. These two operations are used in PRESTO, MINI2, and ESPRESSO algorithms in the next sections.

10.6 PRESTO

This section introduces PRESTO, one of the simplest heuristic algorithm for SOPs with many inputs. The performance of PRESTO is not so good as MINI and ESPRESSO (shown in the next section). However, the algorithm is simple and easy to understand. Also, we can easily predict the size of memory and computation time. Thus, it is easy to use for novice users. However, its demerits are:

1) Quality of the solutions are worse than MINI and ESPRESSO.
2) It often requires more computation time than MINI or ESPRESSO.

The algorithm of PRESTO is as follows:

Algorithm 10.3 *(PRESTO)*
In the set of cubes that represents the characteristic function of the multi-output function.

Table 10.4 Multi-output function.

Input			Output		
x_1	x_2	x_3	f_0	f_1	f_2
0	0	0	1	1	0
0	0	1	0	1	1
0	1	0	1	0	1
0	1	1	0	0	1
1	0	0	0	1	0
1	0	1	0	1	1
1	1	0	1	1	0
1	1	1	0	0	0

1) *Input-fat: For each literal of an input variable in each cube, check whether it can be expanded. If possible, expand it.*
2) *Output-slim: For each bit of the output part, check whether the bit can be converted from 1 to 0. If so, do it. If all the bits in the output of a cube becomes 0, then delete the cube.*

Repeat the above steps until no cube changes.

If we explain the above operations in a PLA, 1) corresponds to removing a connection in the AND array, and 2) corresponds to removing a connection in the OR array. In the OR array, the product lines that are not connected to any output lines are deleted.

Example 10.11 (Problem) Using Algorithm 10.3 (PRESTO), simplify the SOP for the multi-output function in Table 10.4.

(Solution) The bit representation of Table 10.4 is as follows:

Input part Output part
$(X_1X_2X_3)$ $(f_0f_1f_2)$

$$
\begin{array}{r} \\ 1\\2\\3\\4\\5\\6\\7\end{array}
\begin{array}{c}\text{01-01-01-012}\\ \left[\begin{array}{c}\text{10-10-10-110}\\\text{10-10-01-011}\\\text{10-01-10-101}\\\text{10-01-01-001}\\\text{01-10-10-010}\\\text{01-10-01-011}\\\text{01-01-10-110}\end{array}\right]\end{array}
\overset{\text{(i)}}{\Rightarrow}
\begin{array}{r}1\\2\\3\\8\\9\\6\\7\end{array}
\left[\begin{array}{c}\text{10-10-10-110}\\\text{10-10-01-011}\\\text{10-01-10-101}\\\text{10-}\mathbf{11}\text{-01-001}\\\text{01-}\mathbf{11}\text{-10-010}\\\text{01-10-01-011}\\\text{01-01-10-110}\end{array}\right]
\overset{\text{(ii)}}{\Rightarrow}
\begin{array}{r}1\\10\\3\\8\\9\\6\\11\end{array}
\left[\begin{array}{c}\text{10-10-10-110}\\\text{10-10-01-01}\mathbf{0}\\\text{10-01-10-101}\\\text{10-11-01-001}\\\text{01-11-10-010}\\\text{01-10-01-011}\\\text{01-01-10-1}\mathbf{00}\end{array}\right]
\overset{\text{(iii)}}{\Rightarrow}
$$

$$\begin{array}{r}1\\12\\3\\8\\9\\6\\13\end{array}\begin{bmatrix}\text{10-10-10-110}\\\textbf{1}\text{1-10-01-010}\\\text{10-01-10-101}\\\text{10-11-01-001}\\\text{01-11-10-010}\\\text{01-10-01-011}\\\textbf{1}\text{1-01-10-100}\end{bmatrix}\underset{\Rightarrow}{(iv)}\begin{array}{r}1\\12\\14\\8\\9\\15\\13\end{array}\begin{bmatrix}\text{10-10-10-110}\\\text{11-10-01-010}\\\text{10-01-10-}\textbf{0}\text{01}\\\text{10-11-01-001}\\\text{01-11-10-010}\\\text{01-10-01-0}\textbf{0}\text{1}\\\text{11-01-10-100}\end{bmatrix}\underset{\Rightarrow}{(v)}\begin{array}{r}1\\12\\16\\8\\9\\17\\13\end{array}\begin{bmatrix}\text{10-10-10-110}\\\text{11-10-01-010}\\\text{10-01-1}\textbf{1}\text{-001}\\\text{10-11-01-001}\\\text{01-11-10-010}\\\textbf{1}\text{1-10-01-001}\\\text{11-01-10-100}\end{bmatrix}\underset{\Rightarrow}{(vi)}$$

$$\begin{array}{r}1\\12\\16\\9\\17\\13\end{array}\begin{bmatrix}\text{10-10-10-110}\\\text{11-10-01-010}\\\text{10-01-11-001}\\\text{01-11-10-010}\\\text{11-10-01-001}\\\text{11-01-10-100}\end{bmatrix}\Rightarrow \text{Stop}$$

(i) Input-fat

1. In the 4th product, expand X_2 to include 0, and obtain the 8th product.
2. In the 5th product, expand X_2 to include 1, and obtain the 9th product.

(ii) Output-slim

3. In the 2nd product, reduce the bit of f_2 to 0, and obtain the 10th product.
4. In the 7th product, reduce the bit of f_1 to 0, and obtain the 11th product.

(iii) Input-fat

5. In the 10th product, expand X_1 to include 1, and obtain the 12th product.
6. In the 11th product, expand X_1 to include 0, and obtain the 13th product.

(iv) Output-slim

7. In the 3rd product, reduce the bit of f_0 to 0, and obtain the 14th product.
8. In the 6th product, reduce the bit of f_1 to 0, and obtain the 15th product.

(v) Input-fat

9. In the 14th product, expand X_3 to include 1, and obtain the 16th product.
10. In the 15th product, expand X_1 to include 0, and obtain the 17th product.

(vi) Output-slim

11. The 8th product is contained by the 16th product and the 17th product, so reduce the bit of f_2 to 0. And, delete the 8th product.

The procedure tries to expand the input part. However, the products cannot be expanded anymore. The algorithm halts. ∎

In the PRESTO algorithm, most of the computation time is spent in deciding the inclusion relations. It has the following demerits:

1. The algorithm for deciding inclusion relations is inefficient and time consuming.
2. The heuristic for the ordering of expansion is not good.
3. It does not expand the output part.
4. It does not use the reshape operation.

10.7 MINI AND ESPRESSO

This section introduces MINI and ESPRESSO algorithms as PLA simplification methods.

MINI

Algorithm 10.4 *(MINI)*

1. *Merge the cube of the given SOP F. (This operation is not necessary, but reduces the computation time for other steps.)*
2. *Obtain $\overline{F}$, the complement of F. (This operation is necessary for the fast expansion in the next step.)*
3. *Using $\overline{F}$, expand each product of F. If another product in F is covered by the expanded products, then delete the covered products. (In the expansion of MINI, the products are not expanded into the full prime. But, the cubes are expanded into the moderate size. If a product is expanded into the prime, then the possibility that the product is contained by others will be smaller.)*
4. *Reduce each product of F. (This operation reduces the size of cubes as small as possible. Especially, when $F = c \vee G$, if $c \leq G$ then $F = G$, and c can be deleted, where c is a product, and G is the SOP consisting the products other than c. This will reduce the number of products by one.)*
5. *Reshape each product of F. (This operation does not reduce the number of products, but modifies the shape of cubes. Then, it increases the possibility of reduction of cubes in the next expand step.)*

Apply Step 3, 4, and 5 repeatedly until there is no reduction in the number of products.

A feature of MINI is generation of $\overline{F}$, the complement of F, in Step 2. $\overline{F}$ is used in the expansion. Expansion can be done without using $\overline{F}$: It can be done by the decision of $c(i) \leq F$, as shown in Theorem 10.1. However, $c(i) \leq F$ is equivalent to the relation $c(i)\overline{F} = 0$. Also, the decision of $c(i)\overline{F} = 0$ can be done much faster than the decision of $c(i) \leq F$. The reason why MINI is efficient is that it uses $\overline{F}$ to expand products. The method to expand the products efficiently by using $\overline{F}$, is an important invention in the history of SOP minimization.

MINI usually obtains near minimum solutions in a reasonable time. However, MINI has the following demerits:

1. MINI requires the complement of F, and the size of the complement can be very large. Thus, it sometimes requires an excessive amount of memory. This makes the estimation of memory size and computation time difficult.
2. MINI is complicated.

Example 10.12 Let us simplify the SOP for the 4-variable function shown in Fig. 10.11(a) by MINI, where d denotes *don't care.*

1) First, by merging the products of distance 1, we have the map in Fig. 10.11(b).
2) The complement of F is obtained as shown in Fig. 10.11(c). This will be used in the expansion step later.
3) In Fig. 10.11(b), by expanding the products ① and ②, we have Fig. 10.11(d). We use *don't care* in the expansion to reduce the number of literals.
4) In Fig. 10.11(d), by reducing the product ③, we have Fig. 10.11(e).
5) In Fig. 10.11(e), by reshaping the products ④ and ⑤, we have Fig. 10.11(f).
6) In Fig. 10.11(f), by reshaping the products ⑥ and ⑦, we have Fig. 10.11(g).
7) In Fig. 10.11(g), by expanding the products ⑧ and ⑨, we have Fig. 10.11(h).
8) The iterative application of the operations does not decrease the number of products. Thus, stop the operation. If we compare Fig. 10.11(g) with Fig. 10.11(h), Fig. 10.11(h) requires fewer literals than Fig. 10.11(g), although the numbers of products are the same. ∎

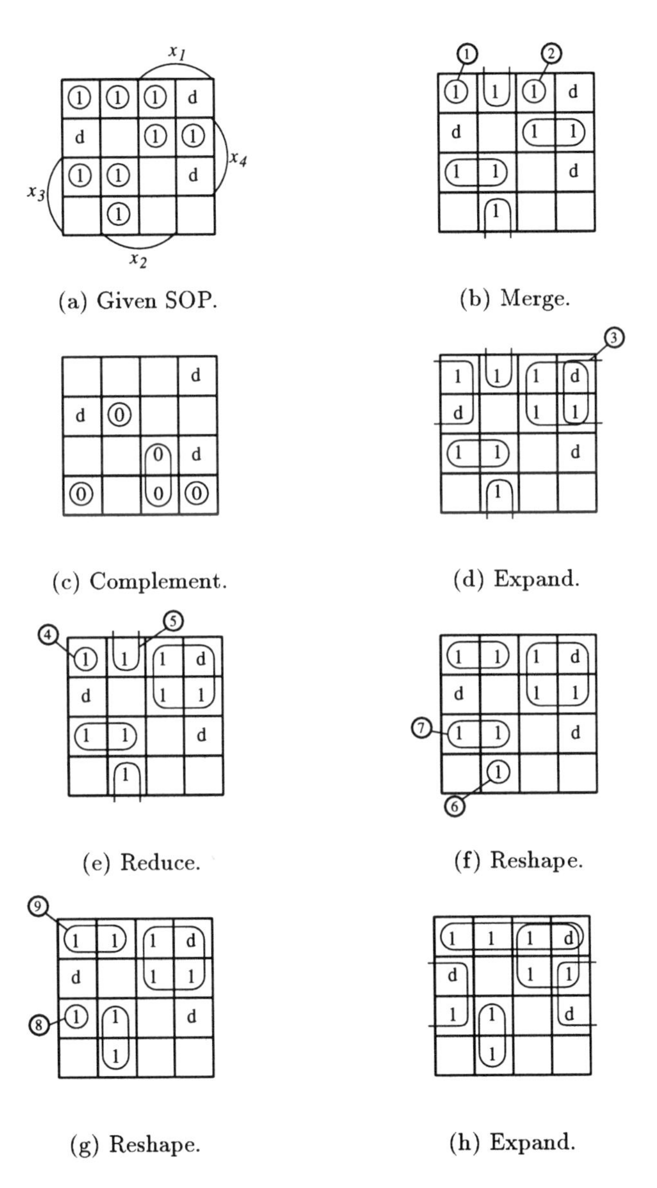

(a) Given SOP. (b) Merge.

(c) Complement. (d) Expand.

(e) Reduce. (f) Reshape.

(g) Reshape. (h) Expand.

Figure 10.11 Simplification of SOP by MINI.

The above example illustrates only the concept of MINI. The real program uses various heuristics to find good directions for expansion and reduction, as well as the ordering of products to improve the performance.

MINI2, developed by the author, is an improvement of MINI, and has the following features:

1. It uses a divide and conquer algorithm to obtain the complement of F. So, it is faster and requires smaller amount of memory than the original MINI which uses the disjoint sharp operation. It can simplify practical PLAs in a computer with small amount of memory.
2. It detects essential prime implicants of F efficiently, and obtains near minimum SOPs in reasonable computation time.

Algorithm 10.5 *(MINI2)*

1. *Merge the products of F.*
2. *In the case of a multi-output function, reshape the cubes so that the input parts are expanded, and the output part is reduced.*
3. *Obtain $\overline{F}$, the complement of F.*
4. *Expand $\overline{F}$ by using F. Delete the products that are covered by other products.*
5. *Expand F by using $\overline{F}$. Delete the products that are covered by other products.*
6. *Detect the set ESS of essential prime implicants of F.*
7. *Let $F \leftarrow F - ESS$.*
8. *Reduce F.*
9. *Reshape F.*
10. *Expand F. Delete the products that are covered by other products.*
11. *During the operations 8–10, if the number of products is reduced, then go to the step 8.*
12. *Let $F \leftarrow F \cup ESS$.*
13. *Expand the input parts. (Reduce the connections in the AND array.)*
14. *In the case of a multi-output function, reduce the output part (Reduce the connections in the OR array).*
15. *In the case of multi-output function, expand the input part. (Reduce the connections in the AND array.)*

The features of MINI and MINI2 are to obtain the complement of the function. However, there is an n-variable function, whose size of the complement is exponentially larger than the original SOP (see Section 12.1.2).

ESPRESSO

ESPRESSO-II is faster than MINI, and can simplify SOPs with more inputs and outputs. The reasons are:

1. It uses divide and conquer methods extensively.
2. It detects unate functions, and utilizes their properties.
3. It adopts better heuristics for the expand operation.
4. The delete operation is formulated as a minimum covering problem to produce better SOPs.
5. It detects essential prime implicants. This reduces the total computation time, and produces better SOPs.
6. It has various improvements in details of the algorithms. They are mainly obtained by analyzing original MINI and MINI2.

Since ESPRESSO-II and MINI2 were developed at the same time and at the same place, they have many common ideas. The C language version of ESPRESSO-II was developed at Berkeley. It is available as a public domain software with detailed experimental results.

10.8 ENCODING METHOD FOR COMBINATIONAL NETWORKS

Encoding for Inputs and Outputs

Up to this section, we assume that inputs and outputs are all represented by two-valued variables in given specifications. However, in the practical specifications, inputs and output are often represented by multi-valued symbols which the circuit is two-valued. In these cases, we have to represent these symbols by two-valued codes. Note that the complexity of the network depends on the assignment of the codes.

Example 10.13 (Problem) Let $I = \{I_0, I_1, I_2, I_3, I_4, I_5\}$ be the set of input symbols, and let $U = \{U_0, U_1, U_2, U_3\}$ be the set of output symbols. Let the relation of the inputs and outputs be shown in Table 10.5. Assume that this table is realized by a 3-input 2-output PLA as shown in Fig. 10.12. Compare the case where the natural binary codes are assigned to the symbols, and the case where the assignments is intended to reduce the number of products.

Table 10.5 Specification represented by multi-valued symbols.

Input	Output
I_0	U_2
I_1	U_1
I_2	U_0
I_3	U_2
I_4	U_2
I_5	U_3

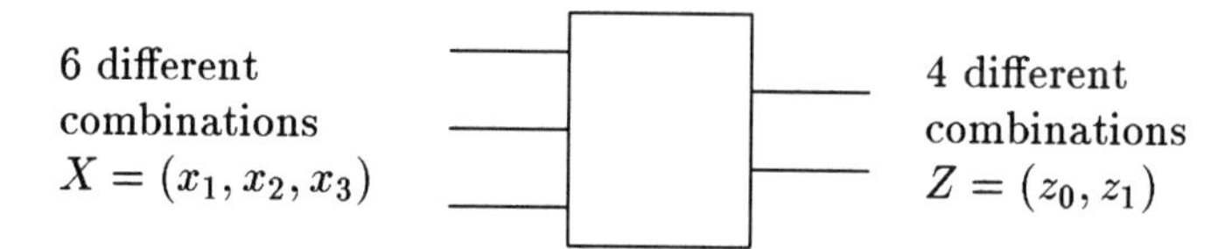

Figure 10.12 Network after the assignment of two-valued codes.

Table 10.6 Natural binary code assignment.

Input $x_1x_2x_3$	Output z_0z_1
000	10
001	01
010	00
011	10
100	10
101	11

Table 10.7 Output encoding method.

Input	Output	
I_0	U_2	$\rightarrow$ 00
I_1	U_1	$\rightarrow$ 01
I_2	U_0	$\rightarrow$ 10
I_3	U_2	$\rightarrow$ 00
I_4	U_2	$\rightarrow$ 00
I_5	U_3	$\rightarrow$ 11

(Solution) By assigning the natural binary codes to the input and the output symbols, we have the logic function shown in Table 10.6. Thus, we have SOPs for z_0 and z_1 that require four products:

$$z_0 = x_1\bar{x}_2 \vee \bar{x}_2\bar{x}_3 \vee \bar{x}_1x_2x_3,$$
$$z_1 = \bar{x}_2x_3.$$

Next, consider an output encoding method that makes the number of the products as small as possible.

Table 10.8 Encoding of the outputs.

Input	Output $z_0 z_1$
I_0	00
I_1	01
I_2	10
I_3	00
I_4	00
I_5	11

Table 10.9 Truth table after code assignment.

Input $x_1 x_2 x_3$	Output $z_0 z_1$
011	00
001	01
010	10
100	00
101	00
000	11

Encoding of outputs
In the case of a PLA realization, for the output $(0, 0, \ldots, 0)$, no product line is necessary. So, assigning the most frequently occurring pattern to the code $(0, 0, \ldots, 0)$, tends to result in a smaller PLA. For example, in Table 10.5, since the output U_2 appears the most frequently, assign the code (0,0) to it. By assigning codes as shown in Table 10.7, we can convert Table 10.5 into Table 10.8. This table requires at most 3 products in a PLA.
Next, we will consider an input encoding method to reduce the number of products.
Encoding of input
Consider the input as a multi-valued variable, and simplify the SOP for a multi-valued input two-valued output function. For example, consider the truth table in Table 10.7. This function is represented by the Karnaugh map in Fig. 10.13. To cover all the true minterms in this Karnaugh map, we need at least two products. In input encoding, in order not to increase the product terms, the Hamming distance between codes I_1 and I_5, as well as codes I_2 and I_5 should be one. The assignment of codes $I_5 = (0, 0, 0)$, $I_1 = (0, 0, 1)$, and $I_2 = (0, 1, 0)$, satisfies this condition. For other inputs, if we assign the codes as in Fig. 10.13, then we have the truth table in Table 10.9. Thus, the SOPs for z_0 and z_1 require only two products:

$$z_0 = \bar{x}_1 \bar{x}_3,$$
$$z_1 = \bar{x}_1 \bar{x}_2.$$

∎

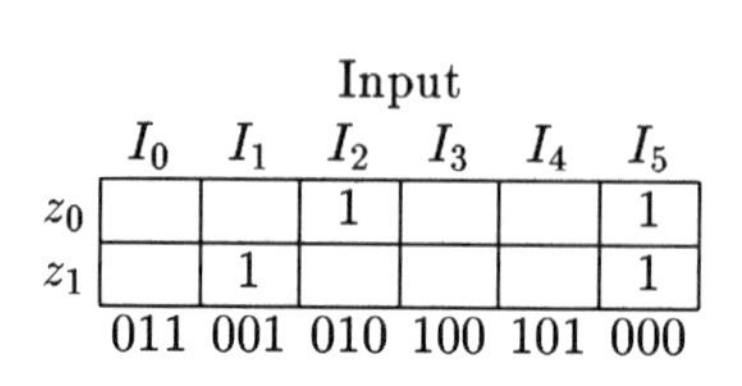

Input	I_0	I_1	I_2	I_3	I_4	I_5
z_0			1			1
z_1		1				1
	011	001	010	100	101	000

Figure 10.13 Assignment of input codes.

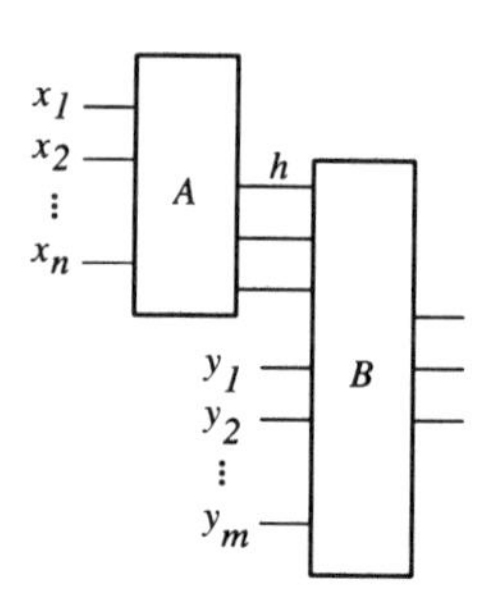

Figure 10.14 Serial decomposition of an AND-OR network.

Serial Decomposition of AND-OR Networks

When the given logic function is realized by an AND-OR two-level network, the number of products often becomes too large to implement. In this case, we can partition the input variables into the two sets, and decompose the network into a serial network as shown in Fig. 10.14. This is a functional decomposition (details are shown in Section 11.5). In this case, we have to partition the input variables so that the number of the output patterns in the network A becomes as small as possible. Let μ be the number of the output patterns. If we use the 1-hot code (see Section 7.3) where 0 and 1 are interchanged, then the number of the products in the network B is equal to the number of products in the MSOP for a μ-valued input function.

10.9 STATE ASSIGNMENT FOR SEQUENTIAL NETWORKS

As stated in Section 7.3, the complexity of a sequential network depends on the state assignment. This section shows a method for state assignment using multi-valued input two-valued output functions.

Example 10.14 (Problem) Realize the PLA for the state table in Table 10.10. In this case, assign the codes to the inputs and states so that the PLA has as few products as possible.

(Solution) First, do the state assignment by considering only the outputs:

Table 10.10 State table.

State	Next state, output		
	I_0	I_1	I_2
S_0	S_1, U_1	S_1, U_1	S_0, U_1
S_1	S_2, U_0	S_1, U_0	S_1, U_1
S_2	S_1, U_0	S_2, U_1	S_2, U_0

State $= \{S_0, S_1, S_2\}$, Input $= \{I_0, I_1, I_2\}$, Output $= \{U_0, U_1\}$

Table 10.11 Output table.

state	input		
	I_0	I_1	I_2
S_0	0	0	0
S_1	1	1	0
S_2	1	0	1

Table 10.12 Output function.

State X	Input Y	Output z
0	0	0
0	1	0
0	2	0
1	0	1
1	1	1
1	2	0
2	0	1
2	1	0
2	2	1

1) Since U_1 appears more frequently than U_0, we assign 0 to the output U_1, and 1 to the output U_0. Then, we have Table 10.11.
2) By representing the output table of Table 10.11 by a multi-valued input two-valued output function, we have Table 10.12.
3) For the simplified SOP for the multi-valued input two-valued output logic function in Table 10.12, we have the SOP with only two products as shown in Fig. 10.15.
4) Assign the codes to inputs and states so that the increase in the number of products is as small as possible. The assignment $S_0 = (0,0)$, $S_1 = (0,1)$, $S_2 = (1,0)$, $I_0 = (0,0)$, $I_1 = (0,1)$, and $I_2 = (1,0)$, will not increase the number of products.
5) The SOPs for the output z is obtained from Fig. 10.16 as follows: $z = \bar{x}_1 y_2 \vee \bar{x}_2 y_1$. ∎

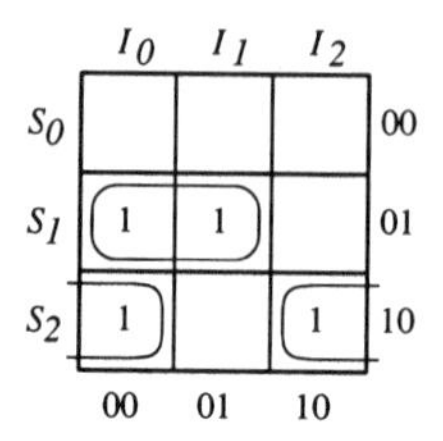

Figure 10.15 Multi-valued Karnaugh map for output table.

Input x_1 x_2

Present state y_1 y_2	00	01	11	10
00			X	
01	1	1	X	
11	X	X	X	X
10	1		X	1

Figure 10.16 Two-valued Karnaugh map for output table.

Bibliographical Notes

Detailed treatments of each topic in this chapter are available as follows: MINI [174]; MINI2 [343]; PRESTO [45]; detection of essential prime implicants using the set of all the PIs [103]; detection of essential prime implicants without using the set of all the PIs [215, 343] (this method is used in ESPERSSO); ESPRESSO-II [38]; ESPRESSO-MV [333]; WSOP (ISOP that has the maximal number of PIs) [372]; optimization of SOPs for multi-output functions using MVL [339]; encoding [96, 100, 433, 434]; state assignment [95, 98, 418]; special hardware for logic minimization [345, 396, 429]; other logic minimization algorithms [4, 12, 24, 29, 59, 82, 255, 326, 425, 435]; implicit PI generation algorithm [77, 226, 366]; verification of combinational networks [152]; benchmark functions [433]. De Micheli's book [97] contains an excellent chapter on two-level optimization.

Exercises

10.1 A **distance** d satisfies the following three axioms:
A1. $d(a,b) \geq 0$,
A2. $d(a,b) = d(b,a)$,
A3. $d(a,b) + d(b,c) \geq d(a,c)$.
Let two products be

$c_1 = X_1^{S_1} X_2^{S_2} \cdots X_n^{S_n}$ and $c_2 = X_1^{T_1} X_2^{T_2} \cdots X_n^{T_n}$.

And let

L-dist(c_1,c_2)=[Number of i's such that $(S_i \neq T_i)$], and
D-dist(c_1,c_2)=[Number of i's such that $(S_i \cap T_i = \phi)$].

Show that L-dist satisfies the axioms of the distance, but D-dist does not. Also, show the following relation is true:

L-dist(c_1,c_2)$\geq$D-dist(c_1,c_2).

10.2 Let A, B, C, $D \subseteq P$, $A \cap C = \phi$, and $D \subseteq B$. Then, prove the following equation (reshape operation):

$$X^A \cdot Y^B \vee X^C \cdot Y^D = X^{A \cup C} \cdot Y^D \vee X^A \cdot Y^{B \cap \overline{D}}.$$

10.3 By using Algorithm 10.2, obtain the essential prime implicants for the function $f = \bar{x}_1 \bar{x}_2 \bar{x}_3 \vee x_1 x_2 \vee x_3 x_4$.

10.4 Let the inputs of a function f be (x_1, x_2, x_3, x_4, x_5). f is 1 iff the number of 1's in the inputs is equal to 2. Design a PLA for f.

10.5 By using the PRESTO algorithm, simplify the following SOP:

x_1	x_2	x_3	x_4	f
10 -	10 -	10 -	10 -	1
10 -	10 -	10 -	01 -	1
10 -	10 -	01 -	01 -	1
10 -	01 -	10 -	10 -	1
10 -	01 -	01 -	01 -	1
10 -	01 -	01 -	10 -	1
01 -	01 -	10 -	10 -	1
01 -	01 -	10 -	01 -	1
01 -	01 -	01 -	01 -	1
01 -	10 -	10 -	01 -	1

10.6 By using the PRESTO algorithm, simplify the SOP for the BCD to 7-segment display converting network (Fig. 10.17). Design a 4-input 7-output PLA for it.

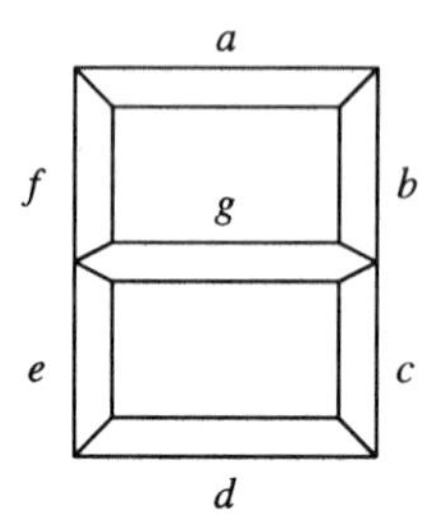

Figure 10.17 7-segment display.

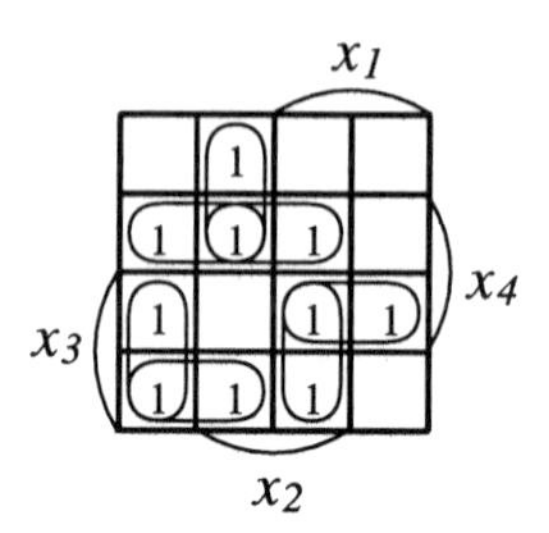

Figure 10.18 Simplification using MINI.

Table 10.13 Code assignment problem.

Input	Output
I_0	U_0
I_1	U_1
I_2	U_2
I_3	U_3
I_4	U_1
I_5	U_2
I_6	U_1

10.7 By using the MINI algorithm, simplify the SOP shown in Fig. 10.18.

10.8 We want to design an AND-OR two-level logic network for Table 10.13. Assign the 2-bit codes to inputs and outputs, so that the network becomes as simple as possible.

10.9 In Example 10.13, an encoding exists that makes the network simpler. Find such an encoding.

11

MULTI-LEVEL LOGIC SYNTHESIS

Multi-level logic networks often require fewer gates and fewer connections than two-level logic networks. On the other hand, designs of multi-level logic networks are far more complex than those of two-level logic networks. In the past, automatic design of multi-level logic networks was considered to be very difficult. However, the recent research have made it possible to synthesize multi-level network in a short time. The qualities of the designs are often comparable to manual designs. However, multi-level logic synthesis has no established algorithm in contrast to the two-level minimization. Design is done by using combinations of ad hoc methods.

11.1 LOGIC SYNTHESIS SYSTEM

Fig. 11.1 shows an example of LSI design system. Such a system is often called a **silicon compiler**, and usually consists of two parts. The first part is the **logic synthesis part**, and transforms functional descriptions into logic networks. Functional description are written by design languages such as **VHDL** (The VHSIC Hardware Description Language), logical expressions, truth tables, transition tables, etc. Logic networks are described by the list of interconnections of logic elements. Such a description is called a **netlist**. The first part is independent of the semiconductor technology. The second part is a **layout/mask pattern generation part**, and transforms netlists into masks for LSIs. This part depends on the semiconductor technology.

Next, a typical design method for multi-level logic networks will be shown. First, a specification described by a high-level language is transformed into

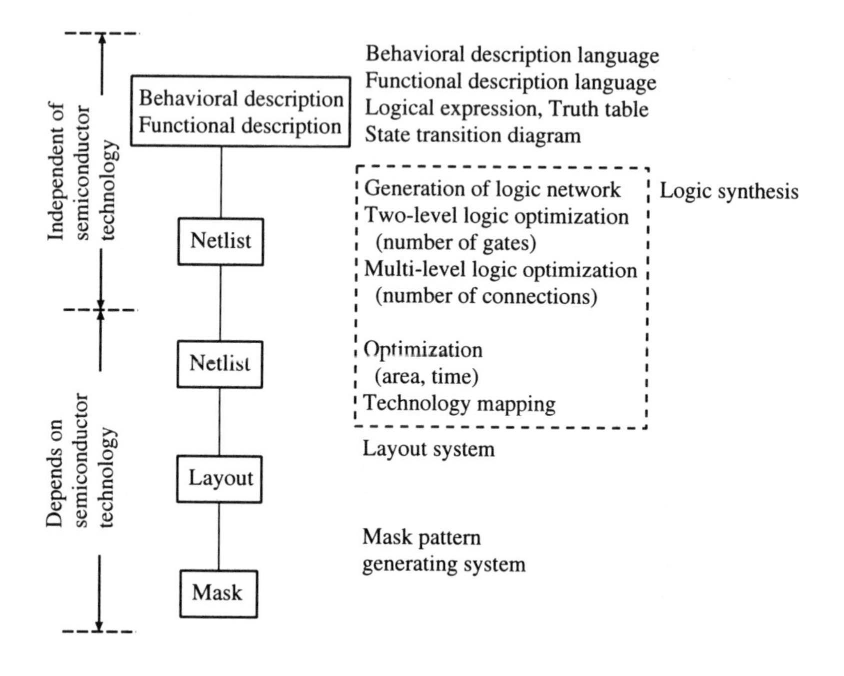

Figure 11.1 Structure of LSI design system.

two-level logic networks. In this case, if the whole network is transformed into a two-level network, then the network size becomes very large. So, in many cases, the network is partitioned into blocks of suitable sizes. Then, each block is transformed into two-level networks, and is simplified (Section 10.1). To transform two-level networks into multi-level networks, factoring (Section 11.2) and division of logical expressions (Section 11.4) are used. These multi-level networks are simplified by using *don't cares* (Section 11.7). The delay time of a multi-level logic network is roughly proportional to the number of levels of the network (Section 11.9). When **timing optimization** (Section 11.9) is necessary, networks are reconstructed to reduce the delay time. The logic synthesis part generates networks that are independent of semiconductor technology, and the networks usually consist of ANDs, ORs, NOTs, etc.

Next, assume that these networks are implemented with gate arrays. In gate arrays, logic elements that can be used are decided in advance. A set of logical elements that can be used by the gate array is called a **library**. In a library, there are complex gates such as AND-OR-NOT in addition to NANDs, NORs, and inverters. The total number of different elements in a library is often

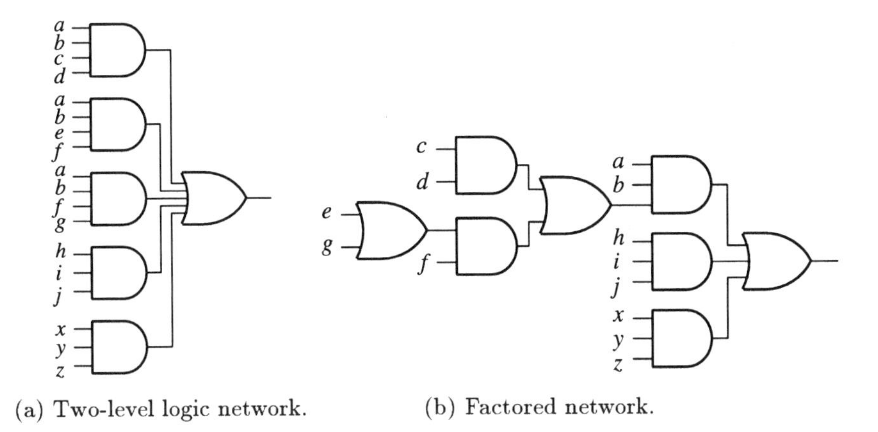

(a) Two-level logic network. (b) Factored network.

Figure 11.2 Factoring.

more than 20, since gates with various numbers of inputs and delay times are available.

Next, the system maps multi-level networks into gate arrays. This operation is called **technology mapping**. In this case, circuits are realized with as few cells as possible by effectively using cells in the library. For this purpose, the procedure allocates cells from the leafs (near inputs) by using the tree structure. Because the propagation delay time is large for a cell with large load, buffers are inserted in gates where the fan-outs are large. Finally, the network is further improved by using local transformations (Section 11.6).

11.2 FACTORING USING PRODUCT TERMS

In a sum-of-products expression (SOP), by finding common products and by factoring with parenthesis, we have a multi-level logical expression. The network structure corresponding to the factored expression is **fan-out free**, when all the literals in the expression is distinct.

Example 11.1 Consider the logical expression:

$$F = abcd \vee abef \vee abfg \vee hij \vee xyz. \tag{11.1}$$

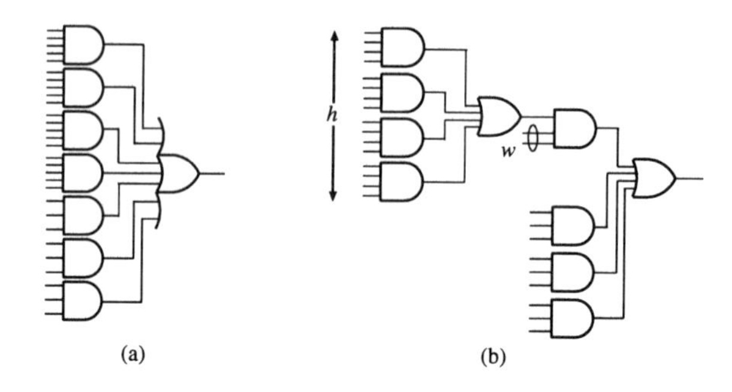

Figure 11.3 Factoring of a two-level network.

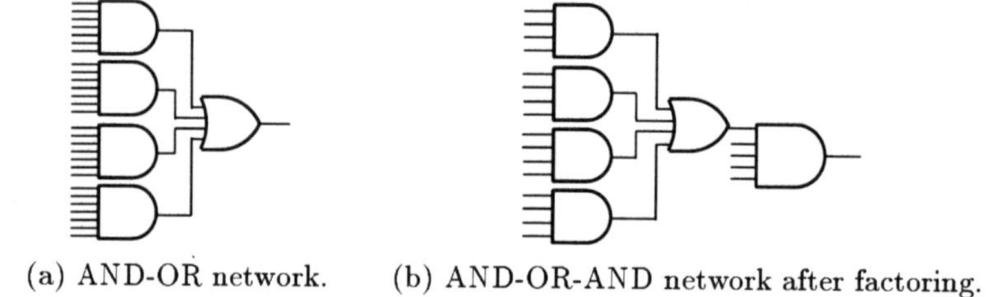

(a) AND-OR network. (b) AND-OR-AND network after factoring.

Figure 11.4 Selection of the optimal network.

The direct realization of this expression is shown in Fig. 11.2(a). However, by finding common logical products for (11.1), we have the factored expression:

$$\begin{aligned} F &= ab(cd \vee ef \vee fg) \vee hij \vee xyz \\ &= ab(cd \vee (e \vee g)f) \vee hij \vee xyz. \end{aligned}$$

This corresponds to the network in Fig. 11.2(b). ∎

Although **factoring** increases the number of gates, it decreases fan-in, since it reduces the number of literals in the expression. Therefore, when the gates have fan-in restrictions, factoring is effective to decreasing the total number of literals. A method to find common products is as follows:

Algorithm 11.1

1. *Enumerate common products.*
2. *Select the product term that maximally reduces the number of literals.*
3. *Reconstruct the expression, and repeat steps 1 and 2.*

Optimal Factoring

Assume that the AND-OR network shown in Fig. 11.3(a) is given. In this case, if the network contains h AND gates that have w common literals, then the network can be transformed into the network shown in Fig. 11.3(b). Note that, this network contains an additional AND and OR gates. However, the number of input lines in the AND gates in the input stage is decreased by hw. Because AND gates have fan-in restrictions, AND gates with many inputs are synthesized with several AND gates. Thus, factoring often reduces the total number of gates. Note that this factoring reduces the number of literals by $hw - w = w(h-1)$. To obtain factoring that maximally reduces the number of literals, we use $w(h-1)$ as the figure-of-merit function, and try to find the factor that maximizes this value. Optimal factoring can be obtained with a **branch and bound method.**

Selection of the Optimal Network

After obtaining a factoring with the maximum figure-of-merit, we realize two networks shown in Fig. 11.4. Fig. 11.4(a) shows an AND-OR network, and Fig. 11.4(b) shows an AND-OR-AND network with factoring. We select the network with smaller cost.

11.3 TWO-VARIABLE FUNCTION GENERATOR

An **OVFG** (One-Variable Function Generator) is a **macro** element that generates all the functions of one variable, i.e., x and $\bar{x}$. On the other hand, a **TVFG** (Two-Variable Function Generator: Fig. 11.5) is a macro element that generates all the functions of up to two variables, except for constants 0 and 1. An AND-OR two-level logic network with OVFGs realizes an arbitrary logic function. Similarly, an AND-OR two-level logic network with TVFGs realizes an arbitrary logic function. In the design of two-level networks with TVFGs, input variables are partitioned into pairs. In this case, each pair $X = (A, B)$ of variables is considered as a four-valued variable that takes 00, 01, 10, or 11.

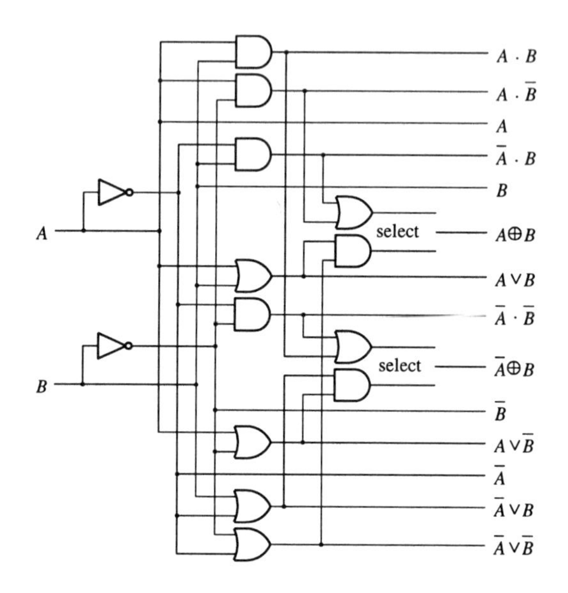

Figure 11.5 Two-variable function generator (TVFG).

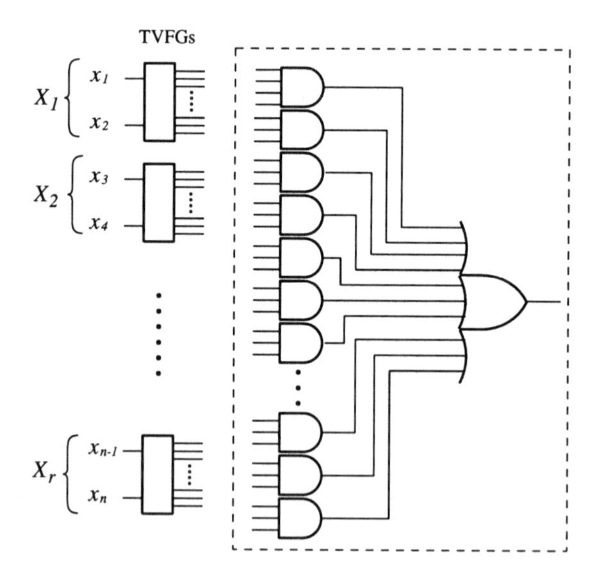

Figure 11.6 AND-OR two-level network with TVFGs.

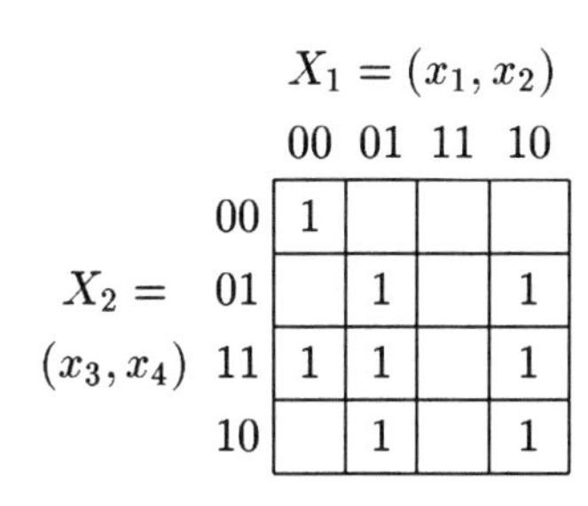

$X_2 = (x_3, x_4)$ \ $X_1 = (x_1, x_2)$	00	01	11	10
00	1			
01		1		1
11	1	1		1
10		1		1

Figure 11.7 Example function.

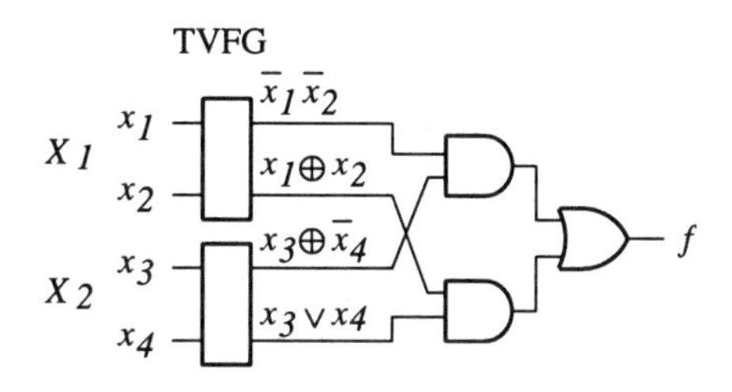

Figure 11.8 Realization of function in Fig. 11.7 using TVFGs.

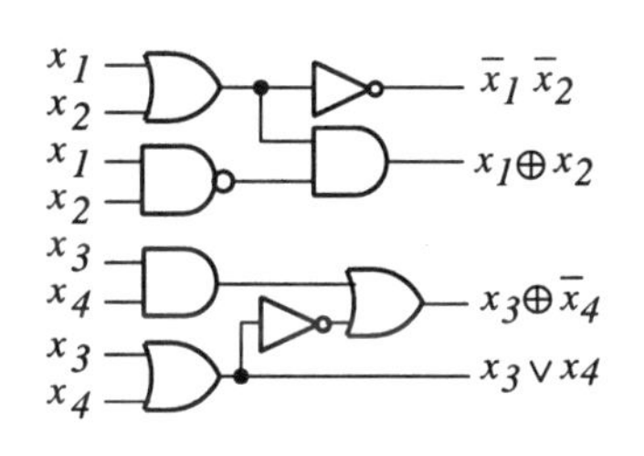

Figure 11.9 Macro expansion of a TVFG.

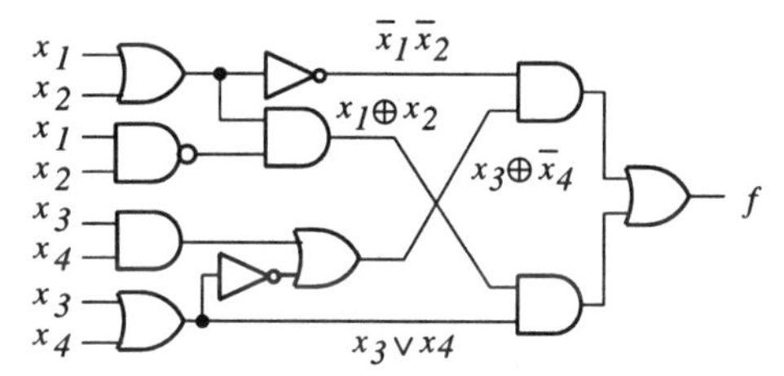

Figure 11.10 Final network.

Theorem 11.1 *An arbitrary logic function $f(x_1, x_2, \ldots, x_n)$ $(n = 2r)$ is represented by an SOP of 4-valued variables.*

$$f(x_1, x_2, \ldots, x_n) = \bigvee_{(S_1, S_2, \ldots, S_r)} X_1^{S_1} X_2^{S_2} \cdots X_r^{S_r}, \tag{11.2}$$

where $X_1 = (x_1, x_2)$, $X_2 = (x_3, x_4)$,... Also, an AND-OR two-level logic network (Fig. 11.6) with TVFGs realizes the above-mentioned SOP. Since each term of the expression (11.2) corresponds to an AND gate, an MSOP of (11.2) corresponds to a network with the minimum number of AND gates.

Example 11.2 In the function of Fig. 11.7, let $X_1 = (x_1, x_2)$ and $X_2 = (x_3, x_4)$. Then, we have the expression:

$$\begin{aligned} f &= X_1^{\{00\}} X_2^{\{00\}} \vee X_1^{\{00\}} X_2^{\{11\}} \vee X_1^{\{01\}} X_2^{\{01\}} \vee X_1^{\{01\}} X_2^{\{10\}} \\ &\quad \vee X_1^{\{01\}} X_2^{\{11\}} \vee X_1^{\{10\}} X_2^{\{01\}} \vee X_1^{\{10\}} X_2^{\{10\}} \vee X_1^{\{10\}} X_2^{\{11\}}. \end{aligned}$$

By simplifying this expression, we have $f = X_1^{\{00\}} X_2^{\{00,11\}} \vee X_1^{\{01,10\}} X_2^{\{01,10,11\}}$. Note that

$$X_1^{\{00\}} = \bar{x}_1 \bar{x}_2, \qquad X_2^{\{00,11\}} = x_3 \oplus \bar{x}_4,$$
$$X_1^{\{01,10\}} = x_1 \oplus x_2, \qquad X_2^{\{01,10,11\}} = x_3 \vee x_4.$$

If we represent the original function by using conventional two-valued logic variables, we have

$$f = (\bar{x}_1 \bar{x}_2)(x_3 \oplus \bar{x}_4) \vee (x_1 \oplus x_2)(x_3 \vee x_4).$$

The network corresponding to this expression is shown in Fig. 11.8. The **macro expansion** of the TVFGs produces Fig. 11.9. Thus, we have a multi-level logic network in Fig. 11.10. ∎

The above example illustrates the following: In an AND-OR two-level logic network using TVFGs, the macro expansion of TVFGs produces a multi-level logic network. Different assignments of the input variables to TVFGs produce networks with different complexities. Therefore, the first thing to do is to find the best assignment of the input variables the TVFGs. In other words, this method decomposes the design problem into the following three sub-problems, and solves each sub-problem independently.

1. Assignment problem of the input variables to the TVFGs.
2. Simplification problem of the AND-OR two-level logic network.
3. The macro expansion problem of TVFGs.

11.4 ALGEBRAIC DIVISION OF LOGICAL EXPRESSIONS

Let $R[x]$ be the set of polynomials that have real-numbered coefficients. Then, we have the following:

Theorem 11.2 *Let $P(x) \neq 0$ be a polynomial with degree p, and let $S(x)$ be a polynomial with degree s. If $1 \leq p \leq s$, then there uniquely exists the polynomial $Q(x)$ with degree q, and the polynomial $R(x)$ with degree r satisfying the following conditions:*

$$S(x) = P(x)Q(x) + R(x), \; q = s - p, \; \text{and} \; 0 \leq r < p.$$

$Q(x)$ is the **quotient**, *and $R(x)$ is the* **reminder**.

Example 11.3 Let $S(x) = x^3 - x + 1$ and $P(x) = x + 2$. Then, we have $Q(x) = x^2 - 2x + 3$ and $R(x) = -5$. Note that $Q(x)$ and $R(x)$ are unique. ∎

Next, when two SOPs F and $P(\neq 0)$ are given, consider the problem whether there exist two SOPs Q and R that satisfies $F = P \cdot Q \vee R$. If we can find Q and R that are simpler than F, then the given SOP F can be represented as the combination of simpler SOPs P, Q, and R. This means that the multi-level logic network can also be designed. However, as the following example illustrates, the problem is not so easy.

Example 11.4 Let $F = xy \vee z \vee w$, and $P = x \vee z$. Since $F = (x \vee z)(y \vee z) \vee w$, we have the solution $Q = y \vee z$ and $R = w$. However, F can be also represented as

$$F = (x \vee z)(y \vee z \vee w) \vee (z \vee w). \tag{11.3}$$

There exist another solution $Q = y \vee z \vee w$ and $R = z \vee w$. ∎

As shown in the above example, the result of division is not unique, in contrast to the polynomial with real-numbered coefficients. Thus, for logical expression, we cannot define the quotient. The reason for this is as follows: In a Boolean algebra, the special relations that does not hold in a ring or field holds: $x \cdot \bar{x} = 0$, $x \cdot x = x$, $x \vee \bar{x} = 1$, $x \vee x = x$, etc. However, if such a relation does not happen, we can uniquely define division, and the solution of division of logical expressions.

In this case, since the handling of logical expression is different from a usual Boolean algebra, such logical expression will be called an **algebraic SOP**. In an algebraic SOP, the quotient and reminder are uniquely defined, and the division of logical expression can be done efficiently. However, the multi-level logic networks obtained by this method are not necessarily optimum.

Here, we introduce the theory of division for algebraic SOPs.

Definition 11.1 *Let two SOPs be*

$$F = \bigvee_{i=1}^{p} f_i \text{ and } G = \bigvee_{j=1}^{q} g_j.$$

When F and G have no common variables, the **algebraic product** *of F and G is defined as*

$$F \cdot G = \bigvee_{(i,j)} (f_i \cdot g_j). \tag{11.4}$$

Since, F and G have no common variables, the above operation does not produce any relation which is specific to the Boolean algebra.

Definition 11.2 *Let F and P be two algebraic SOPs. F/P is the* **quotient** *of algebraic SOP when F is divided by P. $Q = F/P$ is the maximum SOP that satisfies $F = Q \cdot P \vee R$. The SOP R is the* **reminder**. *Here, the product of Q and P is the* **algebraic product**. *The expression which is obtained by expanding the right-hand side of the equation of the above expression is equal to F. In other words, the same product terms appear on both sides. When P is neither a constant 0 nor 1, P is a* **divisor** *of F. Especially, when the number of products in R is the minimum, this division is a* **weak division**.

For given two SOPs F and P, weak division produces the quotient Q and the reminder R uniquely.

Algorithm 11.2 *(Weak Division) In this algorithm, a set of products is considered as an SOP.*
WEAK_DIV(F,P)

$$\textit{Let, } F = \{f_1, f_2, \ldots, f_t\}, \textit{ and } P = \{p_1, p_2, \ldots, p_s\}.$$

Partition the literals of all products as follows:

$$U = (u_1, u_2, \ldots, u_t), \textit{ and } V = (v_1, v_2, \ldots, v_t),$$

where $f_j = u_j \cdot v_j$, and u_j denotes a product of literals which are in p_i among the literals in f_j. Also, v_j denotes the product of literals which are not in p_i among the literals in f_j. Also, assume that if all the literals are deleted from u_j or v_j, then the result is 1. Let,

$$\begin{aligned} V(p_i) &= \bigcup_{u_j = p_i} v_j, \\ Q &= \bigcap_{i=1}^{s} V(p_i), \\ R &= F - P \cdot Q. \\ \textit{Return} \quad & (Q, R) \end{aligned}$$

Example 11.5 Let

$$F = ac \vee ad \vee ae \vee bc \vee bd \vee be \vee \bar{a}b, \; P = a \vee b.$$

In this case, we have:

$$\begin{aligned} U &= (a, a, a, b, b, b, b), \\ V &= (c, d, e, c, d, e, \bar{a}), \\ V(a) &= \{c, d, e\}, \\ V(b) &= \{c, d, e, \bar{a}\}, \\ Q &= \{c, d, e\} = F/P, \text{ and} \\ R &= \bar{a}b. \end{aligned}$$

Therefore, F is modified to $F = (a \vee b)(c \vee d \vee e) \vee \bar{a}b$. ∎

Example 11.6 Let,

$$F = ab \vee ac \vee ad \vee bc \vee bd, \tag{11.5}$$

and

$$P = a \vee b. \tag{11.6}$$

Note that

$$Q = F/P = \{c, d\}, \ R = ab. \tag{11.7}$$

Therefore, we have

$$F = (a \vee b)(c \vee d) \vee ab. \tag{11.8}$$

Note that P and Q do not contain the same variables. By expanding (11.8), we have (11.5). ∎

When a multi-level logic network is realized by complex CMOS gates, the total of number of literals that appears in logical expressions is a measure that estimates the number of transistors. For example, the number of literals in (11.5) is ten. Also, the number of literals in (11.8) is six. Therefore, for division of logical expressions, the divisor P that minimizes the number of the literals in the resultant expression must be found. In the factoring method using product terms described in Section 11.2, product terms were used as divisors. By using kernels as divisors P, we can derive multi-level logical expressions. This is described below.

Definition 11.3 *In an SOP with more than one product, if all the products have no common literal, then the SOP is* **cube-free**. *An SOP with only one product is not cube-free.*

Example 11.7 Since $abc \vee abd$ is represented as $ab(c \vee d)$, it is not cube-free. abc is not cube-free. $c \vee d$ is cube-free. ∎

The **kernel** gives sufficient set of common divisors that are easy to manipulate.

Definition 11.4 *Let F be an SOP, and let c be a product term. The cube-free quotient F/c is a* **kernel** *of F. The set of all the kernels of F is denoted by $K(F)$. If the original expression is cube-free, then the original expression is also a kernel (divided by constant $c = 1$).*

Kernels are derived from SOPs by factoring the common products. For example, consider $F = ab \vee ac \vee ad \vee bc \vee bd$. Since, $ab \vee ac \vee ad = a(b \vee c \vee d)$, we have a kernel $b \vee c \vee d$. Also, since $ac \vee bc = c(a \vee b)$, we have a kernel $a \vee b$.

As shown in the above example, different choices of the common products produce different kernels from the expression. To decrease the number of literals in the expression as much as possible, divide the expression by all possible kernels, and choose the one that produces the expression with the fewest literals.

Example 11.8 Let

$$\begin{aligned} F &= ae \vee be \vee cde, \\ G &= ad \vee ae \vee bd \vee be \vee bf, \text{ and} \\ H &= abc. \end{aligned}$$

Then, we have

$$\begin{aligned} K(F) &= \{a \vee b \vee cd\}, \\ K(G) &= \{a \vee b,\ d \vee e,\ d \vee e \vee f,\ ad \vee ae \vee bd \vee be \vee bf\}, \\ K(H) &= \phi. \end{aligned}$$

∎

When more than two expressions have a common expression A, first introduce the intermediate variable a that represent the expression A, and then replace A in an original expression with a. Such an operation is the **extraction of common divisor**.

Theorem 11.3 *Let F and G be SOPs. If F and G have a cube-free common divisor, then the cube-free common divisor for two SOPs, $H \in K(F)$ and $M \in K(G)$, consists of sum-of-products that are common to H and M.*

Therefore, common divisors for two or more logical expressions are derived from the common product terms of the kernels.

Example 11.9 Let

$$F = ae \vee be \vee cde, \text{ and } G = ad \vee ae \vee bd \vee be \vee bf.$$

The kernels of F and G are

$$\begin{aligned} K(F) &= \{a \vee b \vee cd\}, \text{ and} \\ K(G) &= \{a \vee b,\ d \vee e,\ d \vee e \vee f,\ ad \vee ae \vee bd \vee be \vee bf\}, \end{aligned}$$

respectively. Let $H = a \vee b \vee cd$, and $M = a \vee b$, then the divisor that is common to F and G is $a \vee b$. ∎

As shown in the next example, the kernels can find the solutions that cannot be found by the factoring using product terms (Section 11.2).

Example 11.10 Let us design the multi-level logic network for three functions F, G, and H:

$$\begin{aligned} F &= ade \vee bde \vee cde \vee f, \\ G &= bg \vee cg \vee dg \vee aef, \text{ and} \\ H &= aeg \vee bc. \end{aligned}$$

(1) Factoring using product terms:
The product term that is common to F, G, and H is ae. Let $X = ae$. Then, F, G, and H are modified as follows:

$$\begin{aligned} F &= Xd \vee bde \vee cde \vee f, \\ G &= bg \vee cg \vee dg \vee Xf, \\ H &= Xg \vee bc, \text{ and} \\ X &= ae \text{ (common product term).} \end{aligned}$$

(2) Kernel extraction:
Kernel of F is $a \vee b \vee c$, and the kernel of G is $b \vee c \vee d$. By extracting these kernels, we have the followings:

$$\begin{aligned} F &= (a \vee b \vee c)de \vee f, \\ G &= (b \vee c \vee d)g \vee aef, \text{ and} \\ H &= aeg \vee bc. \end{aligned}$$

The common kernel of F and G is $b \vee c$. Let $X = b \vee c$. Then, F, G, and H can be modified as follows:

$$F = (a \vee X)de \vee f,$$

$$
\begin{aligned}
G &= (X \vee d)g \vee aef, \\
H &= aeg \vee bc, \text{ and} \\
X &= b \vee c.
\end{aligned}
$$

In the case of factoring using product terms, the number of literals is 23, while in the case of kernel extraction, the number of literals is 18. So, kernel extraction gives simpler network. ∎

11.5 FUNCTIONAL DECOMPOSITION

This section describes functional decomposition theory. In general, an n-variable function f requires about $2^n/n$ gates. Suppose that the function f can be decomposed into two networks as shown in Fig. 11.11. Let the numbers of inputs for the network H and G be n_1 and n_2+1, respectively, where $n_1+n_2=n$. Then, H and G can be realized by the networks with at most $2^{n_1}/n_1$ and $2^{n_2}/n_2$ gates, respectively. When n is large, $2^n/n \gg 2^{n_1}/n_1 + 2^{n_2}/n_2$. This implies that the decomposed realization requires many fewer gates than the non-decomposed one. Such a design method is a **functional decomposition.**

Definition 11.5 *Let the input variables be $X = (x_1, x_2, \ldots, x_n)$. Let $\{X\}$ denote the set of variables in X. If $\{X_1\} \cup \{X_2\} = \{X\}$ and $\{X_1\} \cap \{X_2\} = \phi$, then $X = (X_1, X_2)$ is a* **bipartition** *of X. Let the number of variables in X be denoted by $d(X)$.*

Definition 11.6 *Let $f(X)$ a completely specified function, and let X be partitioned as $X = (X_1, X_2)$. The* **decomposition chart** *of f is a table with 2^{n_1} columns and 2^{n_2} rows, and each row and column has a label with a binary number, and the corresponding element denotes the value of f, where $n_1 = d(X_1)$ and $n_2 = d(X_2)$, and each row and column has all the patterns of n_1-bit and n_2-bit, respectively.*

Example 11.11 Table 11.1 shows an example of a decomposition chart for a five-variable function, where $X = (X_1, X_2)$ is a partition of X, $X_1 = (x_1, x_2, x_3)$, and $X_2 = (x_4, x_5)$. ∎

Definition 11.7 *The number of different column patterns of the decomposition chart is the* **column multiplicity** *and it is denoted by μ. $\gamma = \min(2^{n_1}, 2^{2^{n_2}})/\mu$ is the* **gain of the decomposition**.

Table 11.1 Decomposition chart.

		X_1							
		0	0	0	0	1	1	1	1 x_1
		0	0	1	1	0	0	1	1 x_2
	x_4x_5	0	1	0	1	0	1	0	1 x_3
X_2	00	0	1	0	0	0	1	1	0
	01	1	1	1	1	1	1	1	1
	10	1	0	1	1	1	0	0	1
	11	0	1	0	0	0	1	1	0

Table 11.2 Decomposition chart.

		X_1			
		0	0	1	1 x_1
	$x_3x_4x_5$	0	1	0	1 x_2
X_2	000	0	0	0	1
	001	1	1	1	1
	010	1	1	1	0
	011	0	0	0	1
	100	1	0	1	0
	101	1	1	1	1
	110	0	1	0	1
	111	1	0	1	0
		Φ_0	Φ_1	Φ_0	Φ_2

Example 11.12 In the decomposition chart in Table 11.1, $n_1 = 3$, $n_2 = 2$, $\mu = 2$, and $\gamma = \min\{2^3,\ 2^4\}/2 = 8/2 = 4$. ∎

The column multiplicity is defined for a decomposition chart, and depends on the partition $X = (X_1, X_2)$ of the input variables. For a completely specified function, the column multiplicity is unique.

Definition 11.8 *In a decomposition chart, let the logic functions represented by distinct column patterns be $\Phi_i(X_2)$ $(i = 0, 1, \ldots, \mu - 1)$. Let $\Psi_i(X_1)$ be the function showing set of the input variables X_1 that produces $\Phi_i(X_2)$.*

By the definition of the decomposition chart, Ψ_i have the following property: $\Psi_i \cdot \Psi_j = 0$ $(i \neq j)$, $\bigvee_{i=0}^{\mu-1} \Psi_i = 1$. Also, Φ_i is represented as $\Phi_i = f(| X_1 = \Psi_i)$.

Example 11.13 In the decomposition chart in Table 11.2, $X_1 = (x_1, x_2)$ and $X_2 = (x_3, x_4, x_5)$. Also, $n_1 = 2$, $n_2 = 3$, $\mu = 3$, and $\gamma = \min\{2^2, 2^8\}/3 = 4/3 = 1.33$.

$$
\begin{aligned}
\Phi_0(X_2) &= \bar{x}_3\bar{x}_4x_5 \vee \bar{x}_3x_4\bar{x}_5 \vee x_3\bar{x}_4\bar{x}_5 \vee x_3\bar{x}_4x_5 \vee x_3x_4x_5, \\
\Phi_1(X_2) &= \bar{x}_3\bar{x}_4x_5 \vee \bar{x}_3x_4\bar{x}_5 \vee x_3\bar{x}_4x_5 \vee x_3x_4\bar{x}_5, \\
\Phi_2(X_2) &= \bar{x}_3\bar{x}_4\bar{x}_5 \vee \bar{x}_3\bar{x}_4x_5 \vee \bar{x}_3x_4x_5 \vee x_3\bar{x}_4x_5 \vee x_3x_4\bar{x}_5, \\
\Psi_0(X_1) &= \bar{x}_1\bar{x}_2 \vee x_1\bar{x}_2, \\
\Psi_1(X_1) &= \bar{x}_1x_2, \text{ and} \\
\Psi_2(X_1) &= x_1x_2.
\end{aligned}
$$

∎

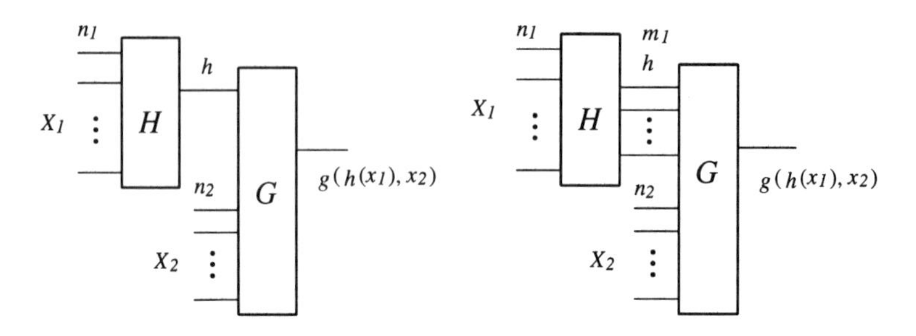

Figure 11.11 Simple disjoint decomposition.

Figure 11.12 Generalized functional decomposition.

Definition 11.9 *Let a p-valued output function* $h : B^{n_1} \rightarrow P$, *where* $B = \{0,1\}$, *and* $P = \{0,1,\ldots,p-1\}$, *be defined as follows:* $h(X_1) = i \Leftrightarrow \Psi_i(X_1) = 1$ $(i = 0,1,\ldots,p-1)$.

Definition 11.10 *Let a function* $g(Y, X_2) : P \times B^{n_2} \rightarrow B$, $B = \{0,1\}$, $P = \{0,1,\ldots,p-1\}$ *be defined as follows:* $h(X_1) = i \Leftrightarrow g(i, X_2) = f(X_1, X_2)$.

Definition 11.11 *For the functions* f, g, *and* h *in Definitions 11.8–11.10,* $f(X) = g(h(X_1), X_2)$ *is a decomposition of the function* f. *When* $\gamma > 1$, *this decomposition is* **non-trivial**. *When a function* f *has a non-trivial decomposition,* f *is* **decomposable**.

The **simple disjoint decomposition** only considers the realization of Fig. 11.11, and the function is decomposable when $\mu = 2$. In this section, we consider the decomposition shown in Fig. 11.12, and the function is decomposable when $\gamma > 1$. Note that the number of the output lines of the network H is, in general, greater than one. When $\gamma \geq 2$, the number of the output lines of H is smaller than the number of input lines. When $1 < \gamma < 2$, we cannot reduce the number of output lines. However, we have $(2^{n_1} - \mu)$ input combinations that do not occur in the inputs of G. In this case, such combination can be used as *don't cares* in optimizing the network G. Thus, to find the partition that maximizes the gain is important.

Theorem 11.4 *When the column multiplicity of the decomposition* $f(X) = g(h(X_1), X_2)$ *is* μ, f *is realizable by the network in Fig. 11.12, where* m_1 *is the minimum integer such that* $\mu \leq 2^{m_1}$.
(Proof) From Definitions 11.8–11.10, we have functions $g : M \times B^{n_2} \rightarrow B$,

Table 11.3 Function for H.

x_1	x_2	h	h_1	h_2
0	0	0	0	0
0	1	1	0	1
1	0	0	0	0
1	1	2	1	0

$x_3x_4x_5$ \ h_1h_2	00	01	11	10
000	0	0	×	1
001	1	1	×	1
011	0	0	×	1
010	1	1	×	0
110	0	1	×	1
111	1	0	×	0
101	1	1	×	1
100	1	0	×	0

Figure 11.13 Map for G.

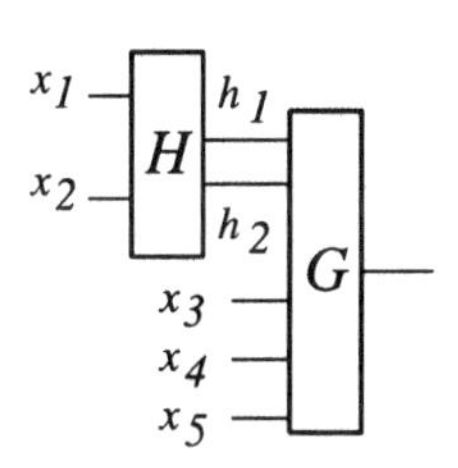

Figure 11.14 Realization of Table 11.2.

and $h : B^{n_1} \rightarrow M$, *such that* $f(X) = g(h(X_1), X_2)$, *where* $B = \{0,1\}$, $M = \{0,1,\ldots,\mu-1\}$, $n_1 = d(X_1)$, $n_2 = d(X_2)$, *and* μ *denotes the column multiplicity. The network* H *realizes* m_1 *functions* $h_1, h_2, \ldots, h_{m_1}$, *where, binary vector* $(h_1, h_2, \ldots, h_{m_1})$ *represents the value of* M. *Also the network* G *realizes the function* g. *Therefore, the network in Fig. 11.12 realizes the function* f. □

Example 11.14 Consider the decomposition chart shown in Table 11.2. Table 11.3 and Fig. 11.13 show the functions for H and G, respectively. Because the column multiplicity is 3, the number of outputs for H is 2. H does not produce the (1,1) output, and this pattern can be used as *don't care* when we design G. In Fig. 11.13, the cells with × denote *don't cares*, and Fig. 11.14 shows the corresponding network. ∎

Definition 11.12 *Let* $\{X_1\}$ *be a subset of the input variables of a function* f. *If* f *is invariant under any permutation of the variables in* X_1, *then* f *is* **partially symmetric** *with respect to* $\{X_1\}$.

Lemma 11.1 *If a function f is partially symmetric with respect to $\{X_1\}$, then the column multiplicity of the decomposition $f(X) = g(h(X_1), X_2)$ is at most $n_1 + 1$, where $n_1 = d(X_1)$.*

Many important arithmetic functions are partially symmetric, and we can apply the generalized decomposition theory.

11.6 TRANSFORMATION OF NETWORKS

A **local transformation** simplifies multi-level logic networks by applying several rules iteratively. The rules are applied locally. For simplicity, we assume that the networks consist of ANDs, ORs, and inverters only.

- **Reduction of Constants**
  Remove a constant 1 that is connected to an AND gate. Remove a constant 0 that is connected to an OR gate. Replace an AND gate that has a constant 0 input, with a constant 0. Replace an OR gate that has a constant 1 input, with a constant 1 (Fig. 11.15).
- **Deletion of Unused Gates**
  Remove gates whose outputs are not connected to other gates or terminals.
- **Reduction of 1-input AND Gates and 1-input OR Gates**
- **Inverter reduction**
  Remove two inverters that are connected in series (Fig. 11.16).
- **Reduction of Duplicated Gates**
  If there are two gates whose inputs and outputs are the same, remove one, and create a fan-out (Fig. 11.17).
- **Gate Merging**
  If two AND gates (OR gates) are connected in series, then merge them into one (Fig. 11.18).
- **Sharing of a Factor**
  The input lines that are common to the same kind of gates form a factor. Add a gate realizing the factor, and reduce the fan-in of the original gates (Fig. 11.19).
- **Simplification Using Inverters**
  If two gates realize complementary functions, then realize the simpler one, and remove the other one. Realize the complement by adding inverter (Fig. 11.20).
- **Reduction of Redundant Connections**
  Let two NAND gates be connected in series as shown in Fig. 11.21. If the

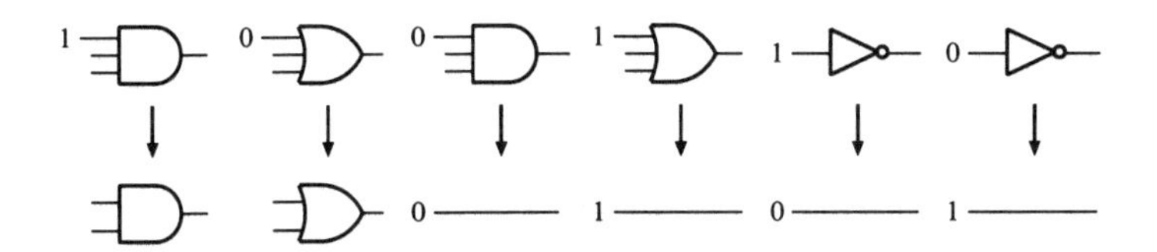

Figure 11.15 Reduction of constants.

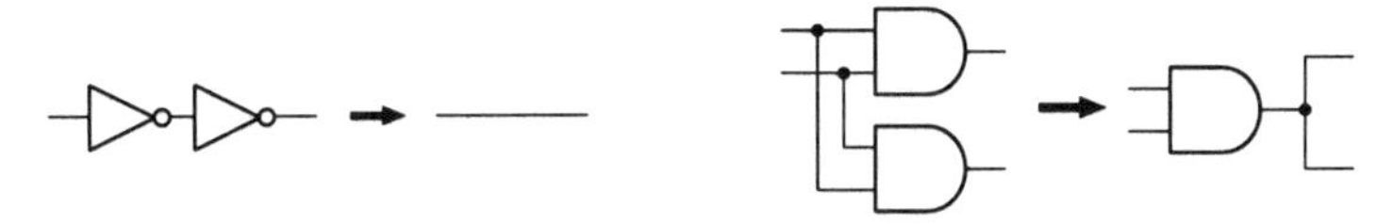

Figure 11.16 Reduction of inverters.

Figure 11.17 Reduction of duplicated gates.

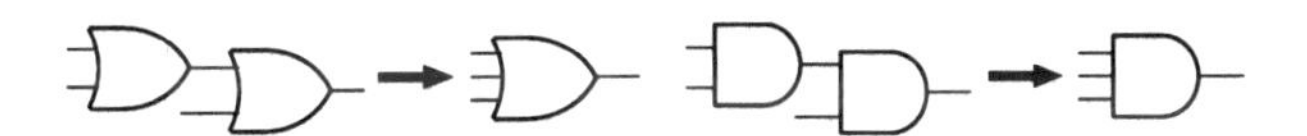

Figure 11.18 Gate merging.

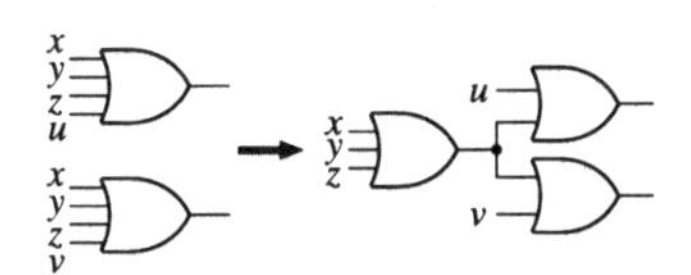

Figure 11.19 Sharing of factors.

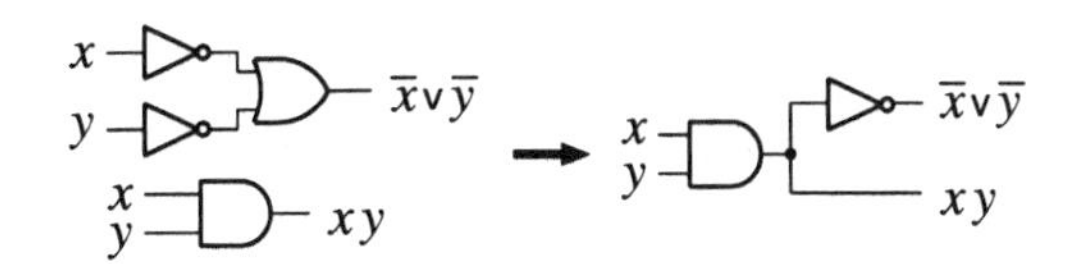

Figure 11.20 Simplification of networks using inverters.

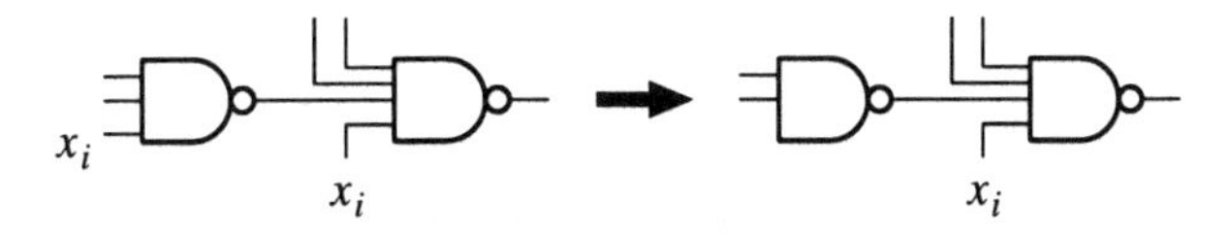

Figure 11.21 Deletion of redundant connections.

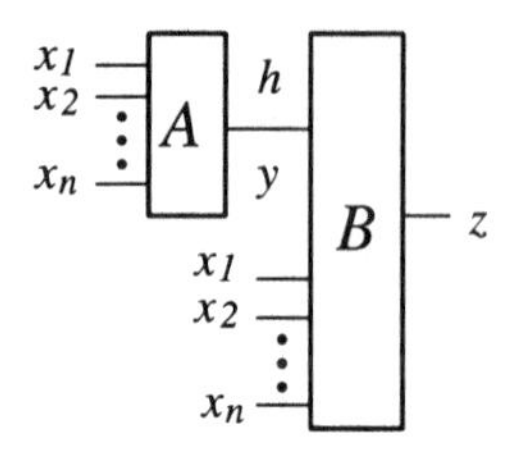

Figure 11.22 Realization of a multi-level logic network.

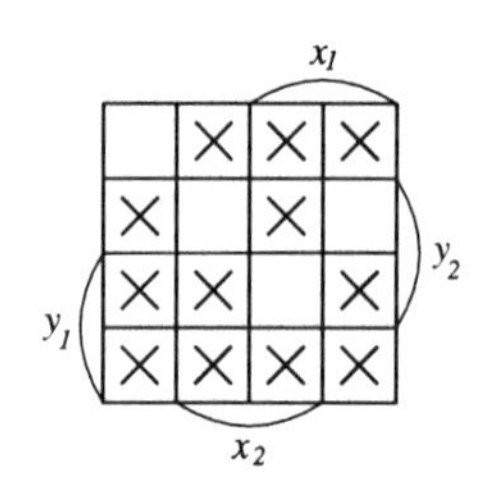

Figure 11.23 Karnaugh map for SDC.

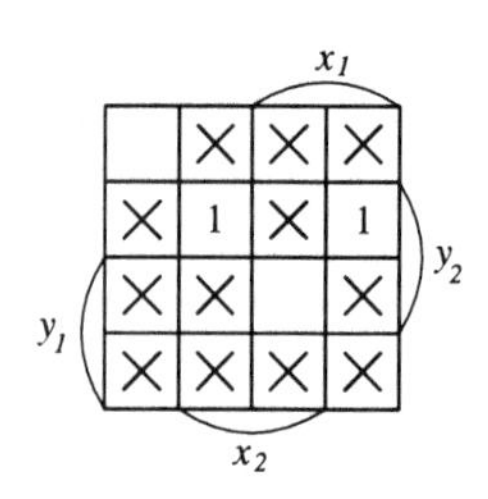

Figure 11.24 Karnaugh map for z.

same inputs are connected to both gates, then remove the connection to the input gate.

Although these methods are simple, they are considerably effective. Some practical logic synthesis systems utilize only local transformations.

11.7 SIMPLIFICATION USING DON'T CARE

When we design multi-level logic networks, we often partition the network into smaller blocks, and design each block independently. In this case, *don't care* conditions happen. By using these *don't cares*, we can simplify the networks.

11.7.1 Satisfiability Don't Care (SDC)

In Fig. 11.22, suppose that the network A is available, and the network B is under design. Let $h = h(x_1, x_2, \ldots, x_n)$ be the output function of the network A. Let $z = z(x_1, x_2, \ldots, x_n, y)$ be the function to be realized in the network B, where $y = h$ is an intermediate variable. Note that y depends on the value of x_1, x_2,..., x_{n-1} and x_n. Thus, if we consider $x_1, x_2, \ldots, x_n$, and y be the input variables for B, then there are combinations that will never occur at the input of B. The set of such combinations is **satisfiability don't care** (SDC). SDC is represented as SDC $= h \oplus y$. This shows the combinations such that h conflicts with y. SDC can be used for the simplification of the network B. When A produces multiple output function $(h_1, h_2, \ldots, h_k)$, the SDC is $\bigvee_{i=1}^{k}(h_i \oplus y_i)$.

When the number of the intermediate variables h_i is large, the SDC becomes very large. In such cases, the use of SDC is not easy.

Example 11.15 (Problem) Let the network A in Fig. 11.22 realize the function $h_1 = x_1 \cdot x_2$ and $h_2 = x_1 \vee x_2$, and the network B realize the function $z = x_1 \oplus x_2$. Obtain the SDC, and simplify the network B.
(Solution) The SDC is obtained as follows:

$$\begin{aligned} \text{SDC} &= (h_1 \oplus y_1) \vee (h_2 \oplus y_2) \\ &= (x_1 x_2)\bar{y}_1 \vee \overline{(x_1 x_2)} y_1 \vee (x_1 \vee x_2)\bar{y}_2 \vee \overline{(x_1 \vee x_2)} y_2. \end{aligned}$$

Fig. 11.23 shows the Karnaugh map of the SDC. Also, Fig. 11.24 shows the Karnaugh map of $z = x_1 \oplus x_2$. From these, we can simplify z as $z = y_2 \cdot \bar{y}_1$. ∎

SDC are useful for designing networks, where some functions are available as inputs in addition to the primary inputs.

11.7.2 Observability Don't Care (ODC)

In Fig. 11.22, suppose that the network B is available and the network A is under design. In this case, the network B often blocks the output of the network A, and the output value z is independent of the outputs of A. The **observability don't care** (ODC) is the set of inputs that are not observable at the output z because of the presence of the network B.

The ODC is represented as

$$\text{ODC} = z(\mid y) \oplus z(\mid \bar{y}) \oplus 1.$$

The ODC can be used to simplify the network A. It is also related to the **permissible function**, which will be explained in the next section.

Example 11.16 (Problem) In Fig. 11.22, let the network B realize the function $z = x_1 x_2 \vee y \bar{x}_3 \vee \bar{y} \bar{x}_2$, and the network A realize the function $h = \bar{x}_1 x_2 \vee x_1 \bar{x}_3$. Obtain the ODC, and simplify the network A.
(Solution) By the definition of ODC, we have

$$\begin{aligned} \text{ODC} &= z(\mid y) \oplus z(\mid \bar{y}) \oplus 1 = (x_1 x_2 \vee \bar{x}_3) \oplus (x_1 x_2 \vee \bar{x}_2) \oplus 1 \\ &= (x_1 x_2 \oplus \bar{x}_3 \oplus x_1 x_2 \bar{x}_3) \oplus (x_1 x_2 \oplus \bar{x}_2) \oplus 1 \\ &= x_2 \oplus \bar{x}_3 \oplus x_1 x_2 \bar{x}_3. \end{aligned}$$

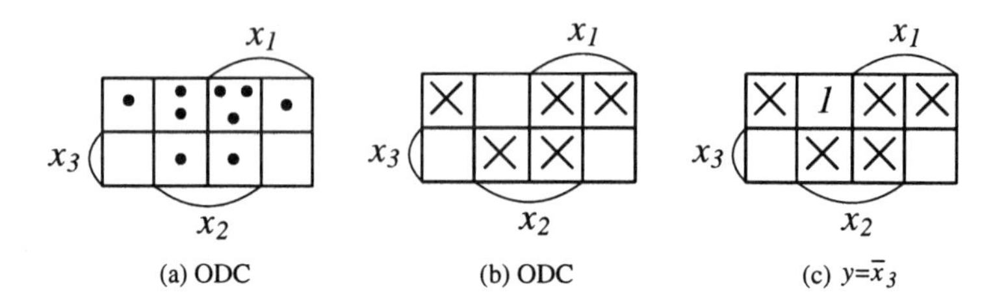

Figure 11.25 Simplification using ODC.

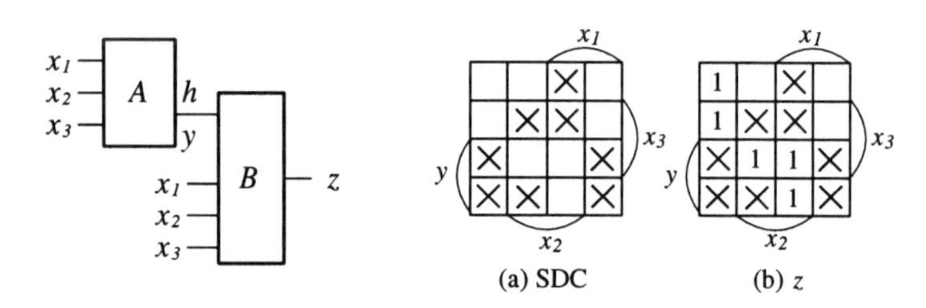

Figure 11.26 Simplification using *don't care*.

Figure 11.27 Simplification using SDC.

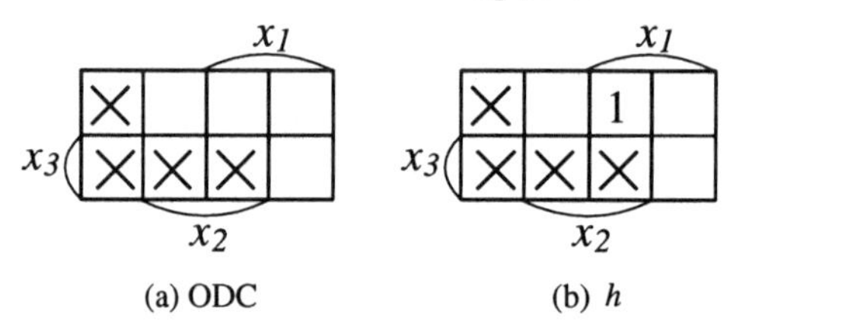

Figure 11.28 Simplification using ODC.

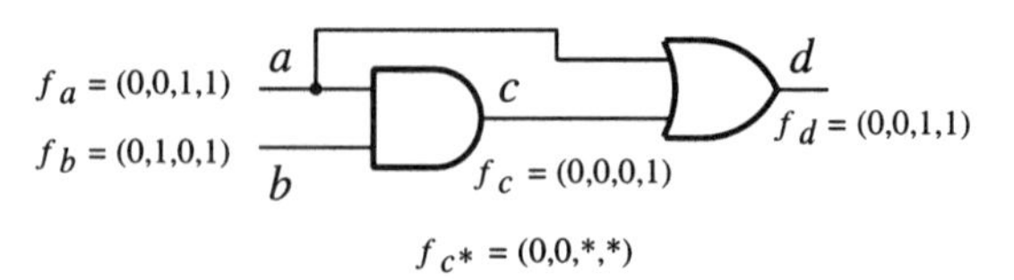

Figure 11.29 Permissible function.

To write the Karnaugh map for the expression with many $\oplus$ operations, we have a convenient way as follows: In the each cell of the Karnaugh map, write one dot each time when a loop covers the cell (Fig. 11.25(a)). The ODC is obtained as shown in Fig. 11.25(b): The cells with the odd numbers of dots constitute the ODC. Thus, in designing the network A, the ODC can be used as *don't cares*. The simplified expression is $h = \bar{x}_3$, as shown in Fig. 11.25(c). ∎

Example 11.17 (Problem) By using *don't care*, simplify the network in Fig. 11.26, where $h = x_1x_2 \vee \bar{x}_1x_2x_3$, $y = h$, and $z = y \vee (x_2x_3 \vee \bar{x}_1\bar{x}_2)$.

(Solution)

1. Let the network A be available, and the network B be under design. Then, we have:
$$\text{SDC} = y \oplus h = y \oplus (x_1x_2 \vee \bar{x}_1x_2x_3).$$
Fig. 11.27(a) shows the SDC, and they can be used as *don't care* in the design of the network B. Fig. 11.27(b) shows the map of z, and the network B is simplified to realize $y \vee \bar{x}_1\bar{x}_2$.
2. Let the network B is available, and the network A is under design. The ODC is obtained as follows:

$$\text{ODC} = f(\mid y) \oplus f(\mid \bar{y}) \oplus 1 = 1 \oplus (x_2x_3 \vee \bar{x}_1\bar{x}_2) \oplus 1 = x_2x_3 \vee \bar{x}_1\bar{x}_2.$$

Fig. 11.28(a) shows the ODC, and for these inputs, the output is independent of the value of h. Fig. 11.28(b) shows the map of h, and the network A is simplified to realize $h = x_1x_2$. ∎

11.7.3 Transduction Method

Permissible function is a similar concept to the ODC. For example, consider the two-variable logic network in Fig. 11.29. Let the logic functions in the points a, b, c, and d be f_a, f_b, f_c, and f_d, respectively. Let the truth table representation of the logic functions in the inputs a and b be $f_a = (0,0,1,1)$ and $f_b = (0,1,0,1)$, respectively. In this case, f_c and f_d are represented as $f_c = (0,0,0,1)$ and $f_d = (0,0,1,1)$, respectively. In the output OR gate, the input from a is $f_a = (0,0,1,1)$. Thus, in order to make the output $f_d = (0,0,1,1)$, the function in c need not be $f_c = (0,0,0,1)$. It can be $f_c* = (0,0,*,*)$, where $*$ denotes *don't care*. In this case, f_c* is the permissible function on c. Let the both values of $*$ be 0. Then, we have $f_c = (0,0,0,0)$, i.e., the constant 0. Thus, in Fig. 11.29, the output function does not change if the AND gate is replaced by the constant 0. As shown in this example, the **transduction method**

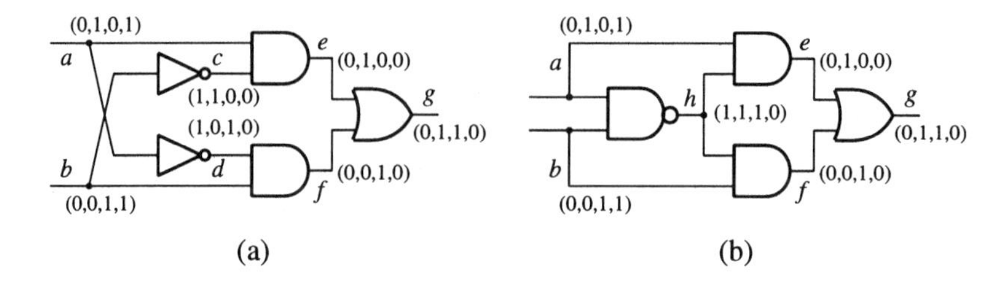

Figure 11.30 Transduction method.

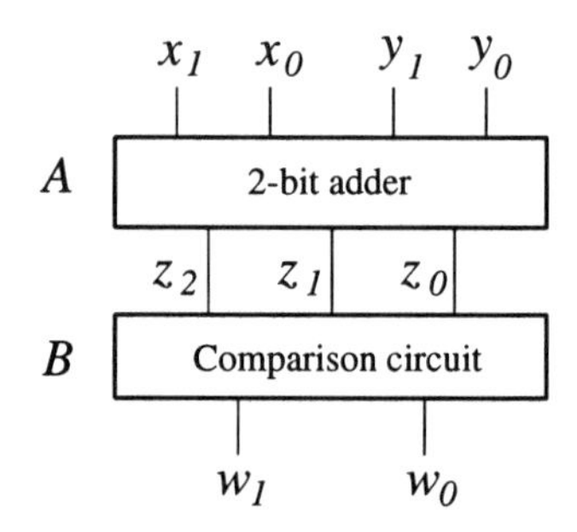

Figure 11.31 Example showing Boolean relation.

simplifies the multi-level logic networks by using the concept of permissible functions.

Example 11.18 In Fig. 11.30(a), consider the output functions of two inverters. The permissible function on c is $f_c = (*, 1, *, 0)$, and the permissible function on d is $f_d = (*, *, 1, 0)$. $f_h = (1, 1, 1, 0)$ is the common function to the both permissible functions, and it can be realized by a single NAND gate. Thus, as shown in Fig. 11.30(b), replacing two inverters with the NAND gate will not change the output function. ∎

11.8 BOOLEAN RELATION

For example, consider the network that adds two 2-bit numbers, and produces the outputs as follows: If the result of the addition is less than 3, then (w_1,w_0)=(0,1). If the result of the addition is equal to 3, then (w_1,w_0)=(0,0).

If the result of the addition is greater than 3, then (w_1,w_0)=(1,0). As shown in Fig. 11.31, the 2-bit adder followed by the comparison circuit (**comparator**) realizes (w_1,w_0). In this case, the comparator realizes the function in Table 11.4, and the adder realizes the function in Table 11.5. The comparator works as follows: If the value of the input is 000, 001, or 010, then the comparator produces the output (0,1). Thus, the set of inputs {000,001,010} forms an equivalence class. Similarly, the sets {011} and {100,101,110,111} form other equivalence classes. When we design the 2-bit adder A, even if the output 000 is replaced by 001 or 010, the output function of B does not change. Similarly, even if the output 100 is replaced by 101, 110, or 111, the output function of B does not change. Table 11.6 shows this relation. For example, the first row shows that when the input is 0000, the output can be either 000, 001, or 010. This condition cannot be described by an ordinary *don't care*. Since, the output is not unique for the input, this is called as a **Boolean relation**. A heuristic algorithm to obtain the near minimum representation that satisfies the given Boolean relation has been developed. Table 11.7 is obtained by that method. If we design A by using ordinary *don't cares*, the resulting network will be more complicated.

11.9 TIMING OPTIMIZATION

In logic design, the delay time is important as well as the cost of hardware (i.e., number of gates and the number of connections). In practical applications, we often need to reduce the delay time, even if we must increase the amount of hardware. The delay time of the logic circuit depends on the number of logic levels, the properties of gates, fan-outs, and the wiring length. If we assume that "the delay time of the gates are the same, and the delay time of the interconnections are zero," then the delay time of the network is the maximum number of gates in all the paths from the inputs to the output. Among the paths from the inputs to the outputs, the one that has the maximum number of gates is the **critical path**. And, the number of gates in the critical path is the **level of the network**. However, the level of the network is not always equal to the delay time of the network. For example, in the network in Fig. 11.32(a), the path consisting of gates 1, 2, 3, 5, and 6 is the maximum, and the number of levels is 5. In order to propagate the signal x through this path to the output, we have to set $y = w = 1$. In a NAND gate, constant 1 can be deleted without changing the output function, and we can modify the network as shown in Fig. 11.32(b). However, in this network, the output value of the gate 3 is always 1, independently of the value of x. Thus, the path of the gates 1, 2, 3, 5,

Table 11.5 Specification of 2-bit adder by logic function.

x_1	x_0	y_1	y_0	z_2	z_1	z_0
0	0	0	0	0	0	0
0	0	0	1	0	0	1
0	0	1	0	0	1	0
0	0	1	1	0	1	1
0	1	0	0	0	0	1
0	1	0	1	0	1	0
0	1	1	0	0	1	1
0	1	1	1	1	0	0
1	0	0	0	0	1	0
1	0	0	1	0	1	1
1	0	1	0	1	0	0
1	0	1	1	1	0	1
1	1	0	0	0	1	1
1	1	0	1	1	0	0
1	1	1	0	1	0	1
1	1	1	1	1	1	0

Table 11.4 Specification for comparison circuit.

z_2	z_1	z_0	w_1	w_0
0	0	0	0	1
0	0	1	0	1
0	1	0	0	1
0	1	1	0	0
1	0	0	1	0
1	0	1	1	0
1	1	0	1	0
1	1	1	1	0

Table 11.6 Specification of 2-bit adder by Boolean relation.

x_1	x_0	y_1	y_0	$z_2\ z_1\ z_0$
0	0	0	0	{000,001,010}
0	0	0	1	{000,001,010}
0	0	1	0	{000,001,010}
0	0	1	1	{011}
0	1	0	0	{000,001,010}
0	1	0	1	{000,001,010}
0	1	1	0	{011}
0	1	1	1	{100,101,110,111}
1	0	0	0	{000,001,010}
1	0	0	1	{011}
1	0	1	0	{100,101,110,111}
1	0	1	1	{100,101,110,111}
1	1	0	0	{011}
1	1	0	1	{100,101,110,111}
1	1	1	0	{100,101,110,111}
1	1	1	1	{100,101,110,111}

Table 11.7 Simplified representation that satisfies the Boolean relation.

x_1	x_0	y_1	y_0	z_2	z_1	z_0
-	1	1	1	1	0	0
1	1	-	1	1	0	0
1	-	1	-	1	0	0
-	-	-	1	0	1	0
-	1	-	-	0	1	0
-	-	1	-	0	0	1
1	-	-	-	0	0	1

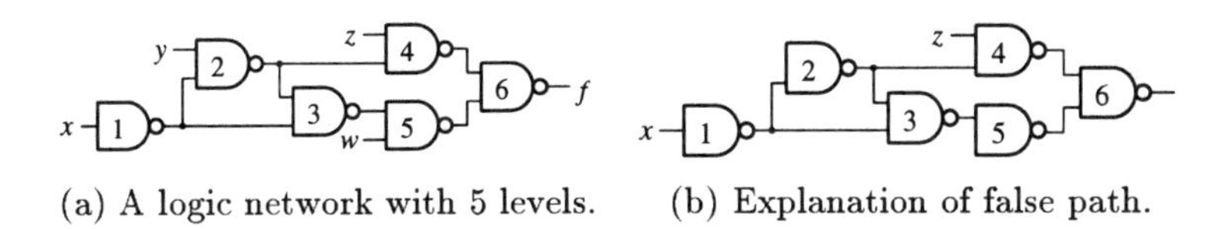

(a) A logic network with 5 levels. (b) Explanation of false path.

Figure 11.32 Network with false path.

and 6, will never be activated. In other words, the delay time of this network is not 5, although the level of the network is 5. In general, we have the following relation:

delay time of the network $\leq$ level of the network $\times$ delay time for each gate.

As shown in the above example, the path that will never be activated is a **false path.** If we assume that the delay time of all the gates are equal, then the (AND-OR) two-level logic networks are the fastest network. However, in two-level networks, fan-in of the gates will be very large, and the interconnections will be longer. Also, the elements with large fan-in are very difficult to implement. We have to synthesize the gates with large fan-in by using several gates. Thus, in practice, well-designed multi-level logic networks are often faster than two-level ones. In the VLSI, due to the wiring resistance and capacitance, the interconnections delay dominates the total delay time.

Bibliographical Notes

Recommended textbooks on logic synthesis are [42, 97, 101, 153], and survey papers are [41, 43]. Examples of logic synthesis systems are: ALERT [124]; LSS [85, 86]; ASYL [374]; SOCRATES [94, 148] (rule based local transformation); CONES [393]; MIS [39]; and CATHEDRAL [299]. As for design language, DDL [111] and VHDL [227] are famous. Detailed discussion on other topics are available as follows: Technology mapping [22, 97, 204]; logic simulation in the presence of unknown inputs [184, 191, 208]; multi-level logic synthesis using integer programming [79, 181, 230, 274, 276]; multi-level logic synthesis using branch and bound methods [88]; factoring [60, 102]; TVFG [348]; weak devision [37]; Roth decomposition [330]; functional decomposition theory [9, 69, 80, 99]; functional decomposition using BDDs [216, 356]; fast methods for functional decomposition [384, 385]; bi-decomposition [370]; local transformation [85, 86, 94, 117, 289]; global flow method [23]; multi-level logic optimization using don't cares [33, 15, 83]; transduction method [280, 281];

encoding [238]; communication complexity [179, 180]; Boolean relation [40, 420]; timing optimization [14, 205, 223, 224, 237]; extension of the QM method to multi-level logic [219]; NAND three-level networks (TANT) [139]; OR-AND-OR three-level networks [342, 367]; AND-AND-OR-OR four-level networks [200]; AND-OR-AND three-level networks [237]; multi-level networks from BDD [187]; cascade [115]; other multi-level networks [147, 214]; decomposition of a function into negative functions [181, 240, 287]; verification of sequential networks [50, 81]; and high-level logic synthesis [74, 134].

Exercises

11.1 Reduce the number of literals in the following expression by factoring with product terms:

$$F = ac \vee ad \vee ae \vee ag \vee bc \vee bd \vee be \vee bf \vee ce \vee cf \vee df \vee dg.$$

11.2 Obtain all the kernels for the each of the following expressions:

$$\begin{aligned} F_1 &= ab \vee ae \vee bd \vee be \vee cd \vee ce \vee g, \text{ and} \\ F_2 &= abcd \vee abce \vee adfg \vee aefg \vee abde \vee beg. \end{aligned}$$

11.3 Obtain all the kernels for the following expression. Suppose that the a and $\bar{a}$, etc., are different literals.

$$F = ab\bar{e}\bar{f} \vee cd\bar{e}\bar{f} \vee ab\bar{c}f \vee \bar{c}\bar{d}ef \vee \bar{a}\bar{b}cd \vee \bar{a}\bar{b}ef.$$

11.4 Reduce the number of literals of the following expression by using parenthesis:

$$F = adf \vee aef \vee bdf \vee bef \vee cdf \vee cef \vee g.$$

11.5 Reduce the number of literals of the following expression by using parenthesis:

$$F = ace \vee ade \vee bce \vee bde \vee acf \vee adf \vee bcf \vee bdf \vee gh.$$

11.6 Reduce the total number of literals in the following expressions by using the common divisor:

$$\begin{aligned} F &= ade \vee bde \vee cde \vee f, \\ G &= bg \vee cg \vee dg \vee aef. \end{aligned}$$

11.7 Prove the following:

1. There exists a function h such that $f \subseteq g \Leftrightarrow f = gh$.
2. There exist functions h and r, such that $fg \neq 0 \Rightarrow f = gh \vee r$.

11.8 Can the network structure shown in Fig. 11.11 realize the following function?

$$f = x_3x_4 \vee x_1x_3 \vee x_2\bar{x}_3 \vee \bar{x}_1\bar{x}_2x_3\bar{x}_4.$$

11.9 Realize the function in Fig. 11.33 by using functional decomposition.

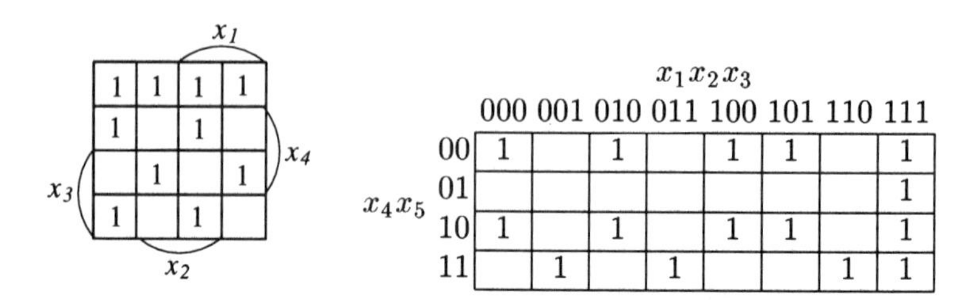

Figure 11.33 Functional decomposition.

Figure 11.34 Functional decomposition.

	0	0	0	0	1	1	1	1
	0	0	1	1	0	0	1	1
	0	1	0	1	0	1	0	1
00	1	0	*	1	*	0	1	*
01	1	0	1	0	*	*	*	0
10	1	*	*	*	1	*	1	0
11	*	*	1	*	1	1	1	0

Figure 11.35 Decomposition chart with *don't cares.*

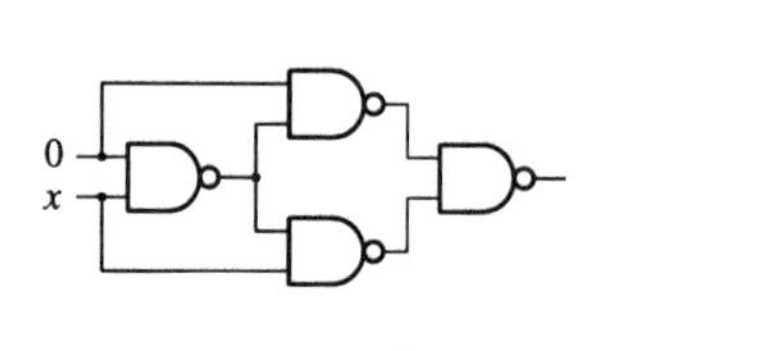

Figure 11.36 Application of local transformations.

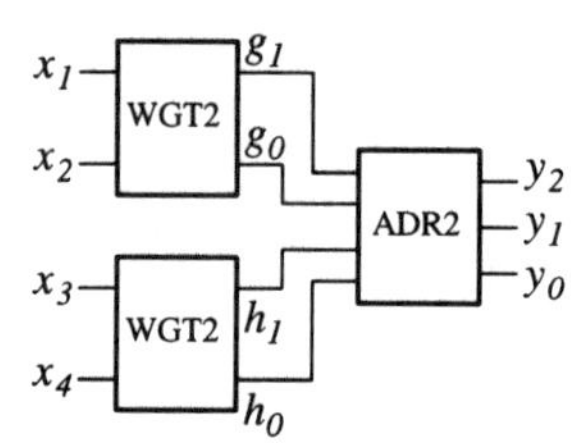

Figure 11.37 Realization of WGT4.

11.10 Realize the function in Fig. 11.34 by using functional decomposition.

11.11 Fig. 11.35 shows the decomposition chart with *don't care* for a 5-variable function. Discuss the column multiplicity of this function.

11.12 Simplify the network in Fig. 11.36 by using local transformation.

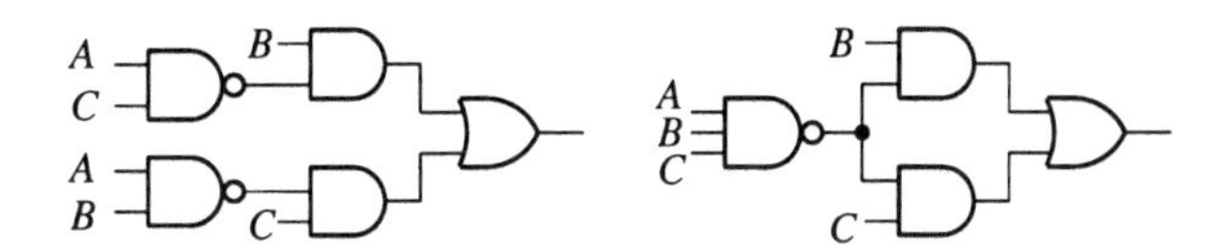

Figure 11.38 Transduction method.

11.13 Consider the 4-input bit count circuit (WGT4), where the input is (x_1, x_2, x_3, x_4), and the output is (y_2, y_1, y_0), the binary representation of $\sum_{i=1}^{4} x_i$. Let Fig. 11.37 be the network structure. Design WGT4 as follows:

1. First, count the numbers of 1's in (x_1, x_2) and (x_3, x_4), by 2-bit counting circuits (WGT2). Let the outputs be (g_1, g_0) and (h_1, h_0), respectively.
2. Next, obtain the sum of two binary numbers (g_1, g_0) and (h_1, h_0) by using a 2-bit adder (ADR2). In this case, (g_1, g_0) and (h_1, h_0) have the combinations that will never appear (SDC). By using this SDC as *don't care*, design the ADR2 by a two-level logic network.

11.14 Prove Lemma 11.1.

11.15 Let the network A realize two functions $h_1 = x_1 \oplus x_2$ and $h_2 = x_1 x_2$, and let the network B realize two functions $z_1 = x_1 \oplus x_2 \oplus x_3$ and $z_2 = x_1 x_2 \vee x_2 x_3 \vee x_3 x_1$. Simplify the network B by using the outputs of A (i.e., by using SDC).

11.16 Let the network A realize the function $h = x_1 x_2 \vee x_2 x_3 \vee x_3 x_1$, and let the network B realize the function $z = x_1 \oplus x_2 \oplus x_3$. Simplify the network B by using the output of A (i.e., by using SDC).

11.17 Let the network A realize the function $y = \bar{x}_1 x_2 \vee x_1 \bar{x}_3$, and let the network B realize the function $z = x_1 x_2 \vee y \bar{x}_3 \vee \bar{y} x_2$. Simplify the network A by using ODC.

11.18 By using transduction method, prove that two networks in Fig. 11.38 produce the same functions.

11.19 By using transduction method, explain that the network in Fig. 11.39(a) can be modified into the networks in Figs. 11.39(b), 11.39(c), and 11.39(d).

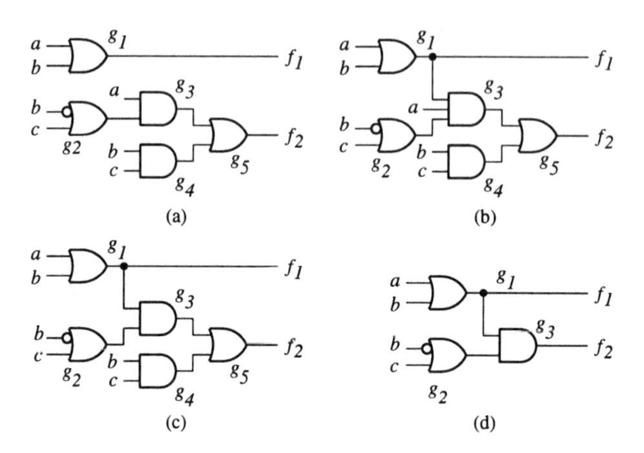

Figure 11.39 Simplification of logic network by transduction method.

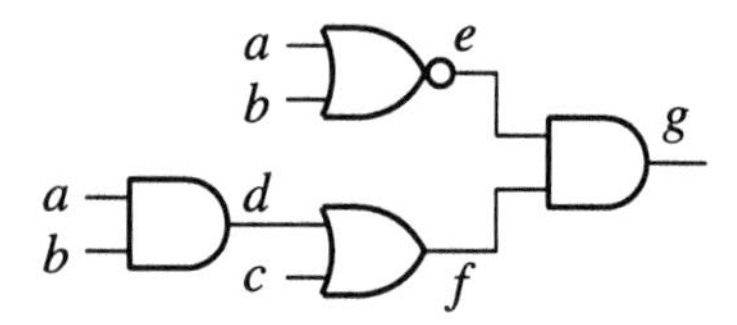

Figure 11.40 A network with false path.

11.20 In Fig. 11.40, show that $adfg$ and $bdfg$ are false paths.

11.21 Design a multi-level logic network for the following function by using AND and OR gates. Describe the design step by step from the original specification to the final realization. Explain the methods used to transform the logical expressions or logic networks. Also, consider the case where EXOR gates are available in addition to AND and OR gates. Suppose that variables and their complements are available as inputs.
ADR3 : 3-bit adder ($Z = X + Y$).

11.22 Do the same thing as Exercise 11.21 for the following function.
INC4 : The circuit that adds 1 to the 4-bit number ($Y = X + 1$).

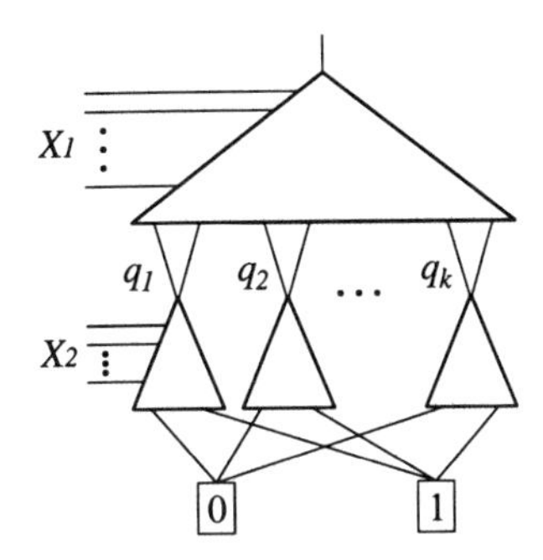

Figure 11.41 Computation of column multiplicity for decomposition $f(X_1, X_2) = g(h(X_1), X_2)$.

11.23 Do the same thing as Exercise 11.21 for the following function.
RDM4 : 4-bit random number generator ($Y = (5X + 1)(\text{mod}16)$).

11.24 Do the same thing as Exercise 11.21 for the following function.
SQR4 : 4-bit square circuit ($Y = X \times X$).

11.25 Do the same thing as Exercise 11.21 for the following function.
WGT4 : 4-input bit-counting circuit (Y = the binary number showing the number of 1's in the inputs).

11.26 Do the same thing as Exercise 11.21 for the following function.
SYM6 : 6-variable symmetric function, where $f = 1$ iff the number of 1's in the inputs is 2, 3, or 4.

11.27 Let $X = (X_1, X_2)$ be a partition of X, and consider the decomposition table for $f(X)$. Let the number of different column patterns in the decomposition table be μ. Let the number of different row patterns in the decomposition table be ν. Then, show that $\mu \leq 2^{\nu}$ and $\nu \leq 2^{\mu}$.

11.28 (M) Let $X = (X_1, X_2)$ be a partition of X. Suppose that the ROBDD for $f(X)$ is partitioned into two blocks as shown in Fig. 11.41. Let k be the number of nodes in the lower block that are adjacent to the upper block, and μ be the column multiplicity of the decomposition table for $f = g(h(X_1), X_2)$. Then, show that $k = \mu$.

12

LOGIC DESIGN USING MODULES

Logic networks are often designed by using logic elements (gates) such as ANDs, ORs, NOTs, etc. However, in some cases, more complex logic elements called **modules** are used. In this chapter, we will consider the logic design using **PLAs** (programmable logic arrays), **MUXs** (multiplexes), and **ROMs** (read only memories).

12.1 LOGIC DESIGN USING PLAS

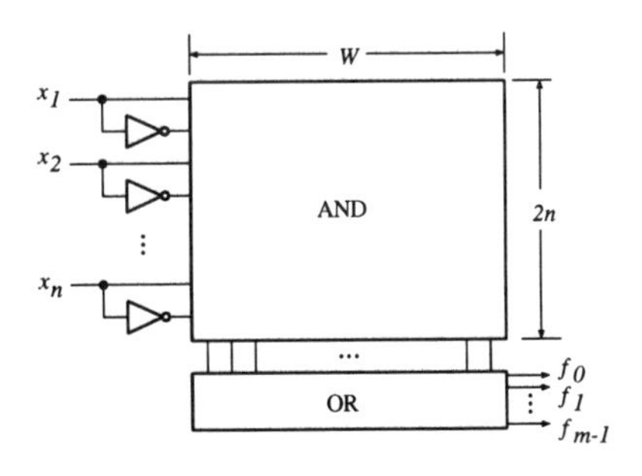

Figure 12.1 n-input m-output PLA.

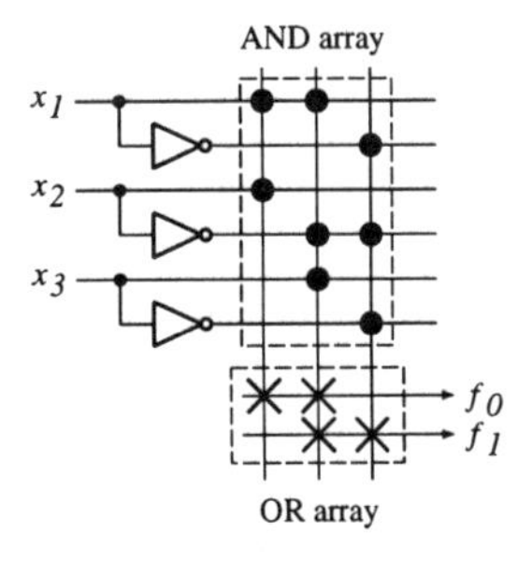

Figure 12.2 Example of 3-input 2-output PLA.

12.1.1 Programmable Logic Array

A logic function can be directly realized by a ROM. However, a ROM requires 2^n words to represent an n-variable function. Thus, when n is large, a realization by ROM is impractical. For control networks for computers, sum-of-products expressions (SOPs) are usually much more compact than truth tables. Fig. 12.1 shows an n-input m-output PLA, which realizes an AND-OR two-level network. In the PLA, the AND array realizes product terms, and the OR array realizes the logical sum of the products. For example, the PLA in Fig. 12.2 realizes SOPs $f_0 = x_1x_2 \vee x_1\bar{x}_2x_3$, $f_1 = \bar{x}_1\bar{x}_2\bar{x}_3 \vee x_1\bar{x}_2x_3$. Vertical lines in the AND array are **product lines**, horizontal lines in the AND array are **literal lines**, and horizontal lines in the OR array are **output lines**. In this PLA, the product lines in the AND array realize, x_1x_2, $x_1\bar{x}_2x_3$, and $\bar{x}_1\bar{x}_2\bar{x}_3$. The OR array realizes the logical sum of product terms. Note that the second product line is connected to the output lines for f_0 and f_1. The **size of the PLA** is defined as $(2n+m) \cdot W$, where W denotes the number of product lines. If we minimize f_0 and f_1 independently, we need four products. To minimize the PLA, we need multi-output simplification (Section 10.5) that considers all the outputs at the same time.

The regular structure of the PLAs has the following features:

1) Logic designs are easy.
2) Modifications are easy. They do not influence the peripheral networks.
3) Layouts are easy.

In the following, we will show optimization methods for PLAs by output phase optimization, and by input decoders.

Two-valued variables that represent the inputs for PLA are denoted by the lower case letters $x_1, x_2, \ldots, x_n$. The output variables are denoted by $f_0, f_1, \ldots, f_{m-1}$. Multi-valued variables that are composed of several binary variables are denoted by the upper case letters $X_1, X_2, \ldots, X_r$.

12.1.2 Output Phase Optimization

Networks can often be simplified by realizing $\bar{f}$, the complement of f, instead of the original function f.

Definition 12.1 *$f = x_1x_2x_3 \vee x_4x_5x_6 \vee x_7x_8x_9 \vee \cdots \vee x_{n-2}x_{n-1}x_n$ ($n = 3r$) is an n-variable* **Achilles' heel function.**

Lemma 12.1 *Let f be an n-variable Achilles' heel function ($n = 3r$). Let $t(f)$ be the number of products in a minimum SOP for f. Then, $t(f) = r$, and $t(\bar{f}) = 3^r$.*

The above lemma shows the function whose complexity is quite different from that of the complement of the function. In designing a PLA, we often can realize $\bar{f}$, the complement of the output function, instead of the original function f. For example, in most PLAs, the outputs are drived by buffers. In that case, we have an option to use a non-inverting buffer or an inverting buffer for each output. When we realize a multi-output function, the combination of the output phases (output polarities) that minimize the total number of products in the PLA is the **optimum output phase**, or the **optimum output polarity**. This option also happens when PLAs do not use inverting buffers. For example, when the outputs of a PLA are connected to the inputs of other PLAs, or the outputs of a PLA are fed back to the PLA, or the outputs of the PLA are connected to flip-flops. Then, also in these cases, the output phase (i.e., f or $\bar{f}$) are chosen freely. To obtain the optimum output phase for m-output function, we can minimize all the 2^m different combinations, and then select the combination that requires the fewest products. However, this method require large amount of computation time when m is large. Fig. 12.3 shows average numbers of product terms for an 8-variable two-valued input single output function. By using pseudo-random numbers, we generated many logic functions whose **density** is r, where the density denotes the number of input a $\boldsymbol{a}$ such that $f(\boldsymbol{a}) = 1$. Then, we minimized the expressions for these functions. In Fig. 12.3, the horizontal axis denotes the density, and the vertical axis denotes the average number of products. This curve has a bell shape, and the number of products takes its maximum when the density is 128. For the cases of different number of input variables n, the graphs have similar shape. Statistically, we have the following:

Property 12.1 *For n-variable functions whose densities are r, if $r \leq 2^{n-1}$, then the average number of products in the SOPs increase as r increases.*

Property 12.2 *For functions whose densities are greater than 2^{n-1}, the number of prime implicants is usually very large, and minimization is difficult. So, usually, the SOPs for the complements of the functions are simpler than SOPs for the original functions.*

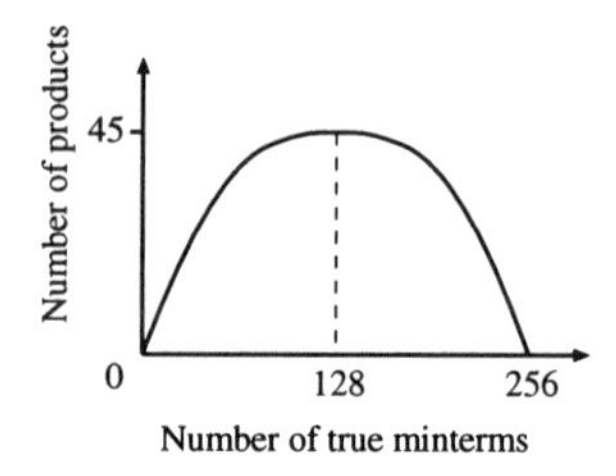

Figure 12.3 Average number of products in minimum SOPs.

In short, statistically speaking, " the more the 0's in the truth table, the simpler the PLAs." Of course, this does not hold for some functions, but is a rough measure. In the truth tables for multi-output functions, we can delete the terms whose outputs are all 0's. So, we have the following:

Theorem 12.1 *In a truth table for an n-input multi-output function, if there are t input combinations that make all the outputs 0, then the functions can be realized by a PLA with at most* $(2^n - t)$ *products.*

From Properties 12.1 and 12.2, and Theorem 12.1, we have the following: A good output phase has "large number of input combinations that make all the outputs 0's," and "large number of 0's in the truth table."

Example 12.1 Design a PLA for 4-input **bit-counting circuit** (the network representing the number of 1's in the inputs by a binary number: WGT4) by considering the output phase.
Table 12.1 shows the truth table for WGT4. A straightforward PLA realization requires 15 products. We can delete only one term whose outputs are all 0's. Next, let us simplify this PLA by considering the output phase. In Table 12.1, the patterns (0,1,0) appear the most frequently in the output (f_2, f_1, f_0). Thus, instead of realizing (f_2, f_1, f_0), if we realize $(f_2, \bar{f}_1, f_0)$, then the WGT4 becomes as shown in Table 12.1. Note that there are 6 combinations that make all the outputs 0's. By deleting these terms, we have the PLA with 10 products. We cannot minimize the PLA any more, and 10 is the minimum solution. ∎

By generalizing Example 12.1, we have the following:

Table 12.1 WGT4.

Input				Output								
				Original			Optimized			Complemented		
x_4	x_3	x_2	x_1	f_2	f_1	f_0	f_2	$\bar{f}_1$	f_0	$\bar{f}_2$	$\bar{f}_1$	$\bar{f}_0$
0	0	0	0	0	0	0	0	1	0	1	1	1
0	0	0	1	0	0	1	0	1	1	1	1	0
0	0	1	0	0	0	1	0	1	1	1	1	0
0	0	1	1	0	1	0	0	0	0	1	0	1
0	1	0	0	0	0	1	0	1	1	1	1	0
0	1	0	1	0	1	0	0	0	0	1	0	1
0	1	1	0	0	1	0	0	0	0	1	0	1
0	1	1	1	0	1	1	0	0	1	1	0	0
1	0	0	0	0	0	1	0	1	1	1	1	0
1	0	0	1	0	1	0	0	0	0	1	0	1
1	0	1	0	0	1	0	0	0	0	1	0	1
1	0	1	1	0	1	1	0	0	1	1	0	0
1	1	0	0	0	1	0	0	0	0	1	0	1
1	1	0	1	0	1	1	0	0	1	1	0	0
1	1	1	0	0	1	1	0	0	1	1	0	0
1	1	1	1	1	0	0	1	1	0	0	1	1

Theorem 12.2 *The minimum PLA for the n-input bit-counting circuit ($WGTn$, $n = 2r$) requires $(2^n - 1)$ products when the output phase is original, and $(2^n - {}_nC_{n/2})$ products when the output phase is optimum.*

Example 12.1 shows the case where the optimum output phase was found. However, for other functions, we cannot find the optimum output phase by using this method. A heuristic algorithm that produces good solutions in a short time has been developed.

Algorithm 12.1 *(Near optimum output phase assignment)*

1. *For a given m-output function $(f_0, f_1, \ldots, f_{m-1})$, let PLA1 be obtained by simplifying the PLA that realizes 2m-output function $(f_0, f_1, \ldots, f_{m-1}, \bar{f}_0, \bar{f}_1, \ldots, \bar{f}_{m-1})$. For the simplification of PLA1, we need a high performance program such as MINI2 or ESPRESSO. PRESTO is unsuitable for this simplification.*
2. *Let the output part of the PLA1 be G. Attach the labels $P_1, P_2, \ldots, P_t$ to the rows of G, where t is the number of products in PLA1. For each output $f_i (i = 0, \ldots, m-1)$, make an SOP: $L_i = P_{a_1} P_{a_2} \cdots \cdot P_{a_r} \vee P_{b_1} P_{b_2} \cdots \cdot P_{b_s}$,*

where P_{a_1}, P_{a_2},..., and P_{a_r} denote the rows whose $(i+1)$th column of G (i.e., the column for the output f_i) are 1's. P_{b_1}, P_{b_2},..., and P_{b_s} denote the rows whose $(i+m+1)$th column of G (i.e., the column for the output $\bar{f}_i$) are 1's.

3. *Expand the logical expression $Q(P_1, P_2, \ldots, P_t) = L_0 \cdot L_1 \cdot \cdots \cdot L_{m-1}$ into an SOP, and obtain the product with the fewest literals.*
4. *Obtain the output phase corresponding to the product obtained in 3.*

Example 12.2 Let us obtain the output phase for a 4-input bit-counting circuit (WGT4), using Algorithm 12.1.

1. Table 12.1 is the truth table for $(f_2, f_1, f_0, \bar{f}_2, \bar{f}_1, \bar{f}_0)$. By simplifying this PLA by MINI2, we have a PLA shown in Table 12.2.
2. Let G be the output PLA1. Then,

f_2	f_1	f_0	$\bar{f}_2$	$\bar{f}_1$	$\bar{f}_0$	
1	0	0	0	1	1	P_1
0	0	1	0	1	0	P_2
0	0	1	0	1	0	P_3
0	0	1	0	1	0	P_4
0	0	0	0	1	1	P_5
0	0	1	0	1	0	P_6
0	1	1	0	0	0	P_7
0	1	0	0	0	1	P_8
0	1	0	0	0	1	P_9
0	1	0	0	0	1	P_{10}
0	1	0	0	0	1	P_{11}
0	1	1	0	0	0	P_{12}
0	1	0	0	0	1	P_{13}
0	1	0	1	0	1	P_{14}
0	1	1	1	0	0	P_{15}
0	1	1	1	0	0	P_{16}
0	0	0	1	0	0	P_{17}
0	0	0	1	0	0	P_{18}

Let L_2, L_1, and L_0 be the expressions for outputs f_2, f_1, and f_0, respectively. Then, we have
$L_2 = P_1 \vee P_{14}P_{15}P_{16}P_{17}P_{18}$,
$L_1 = P_7P_8P_9P_{10}P_{11}P_{12}P_{13}P_{14}P_{15}P_{16} \vee P_1P_2P_3P_4P_5P_6$, and
$L_0 = P_2P_3P_4P_6P_7P_{12}P_{15}P_{16} \vee P_1P_5P_8P_9P_{10}P_{11}P_{13}P_{14}$.

3. By expanding the logical expression $Q(P_1, P_2, \ldots, P_{18}) = L_0 \cdot L_1 \cdot L_2$, we have the following expression:
$P_1P_2P_3P_4P_5P_6P_8P_9P_{10}P_{11}P_{13}P_{14}P_{15}P_{16}P_{17}P_{18}$
$\vee P_1P_2P_3P_4P_5P_6P_7P_{12}P_{14}P_{15}P_{16}P_{17}P_{18}$
$\vee P_1P_5P_7P_8P_9P_{10}P_{11}P_{12}P_{13}P_{14}P_{15}P_{16}P_{17}P_{18}$
$\vee P_2P_3P_4P_6P_7P_8P_9P_{10}P_{11}P_{12}P_{13}P_{14}P_{15}P_{16}P_{17}P_{18}$

Table 12.2 WGT4 simplified by MINI2.

Input part				Output part
x_4	x_3	x_2	x_1	
01	01	01	01	100011
01	10	10	10	001010
10	10	10	01	001010
10	10	01	10	001010
10	10	10	10	000011
10	01	10	10	001010
01	10	01	01	011000
01	10	10	01	010001
01	10	01	10	010001
10	10	01	01	010001
10	01	01	10	010001
10	01	01	01	011000
10	01	10	01	010001
01	01	10	10	010101
01	01	01	10	011100
01	01	10	01	011100
10	11	11	11	000100
11	10	11	11	000100

$\vee\, P_1P_2P_3P_4P_5P_6P_8P_9P_{10}P_{11}P_{13}P_{14}$
$\vee\, P_1P_2P_3P_4P_5P_6P_7P_{12}P_{15}P_{16}$
$\vee\, P_1P_5P_7P_8P_9P_{10}P_{11}P_{12}P_{13}P_{14}P_{15}P_{16}$
$\vee\, P_1P_2P_3P_4P_6P_7P_8P_9P_{10}P_{11}P_{12}P_{13}P_{14}P_{15}P_{16}$.
Note that the first product corresponds to $(\bar{f}_2, \bar{f}_1, \bar{f}_0)$, the second one corresponds to $(\bar{f}_2, \bar{f}_1, f_0)$,..., and the last term corresponds to (f_2, f_1, f_0). The product with the fewest literals is the 6-th, and it corresponds to the output $(f_2, \bar{f}_1, f_0)$. ∎

As shown in the above example, the logical expression L_i shows the product terms necessary to realize either f_i or $\bar{f}_i$. And the logical expression $Q(P_1, \ldots, P_t)$ denotes product terms to realize either f_i or $\bar{f}_i$ for all i ($i = 0, 1, \ldots, m-1$). For example, the 6-th term shows that products P_1, P_2, P_3, P_4, P_5, P_6, P_7, P_{12}, P_{15}, and P_{16} are sufficient to realize the output $(f_2, \bar{f}_1, f_0)$.

Table 12.3 shows the number of products for the PLAs to realize various arithmetic functions. These PLAs are simplified by using Algorithm 12.1. Details of PLAs are shown below:
ADR n has $2n$ inputs and $n+1$ outputs, and computes $f = x + y$.

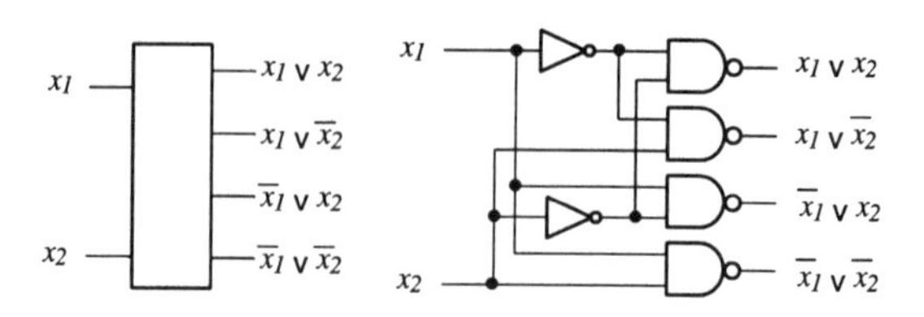

Figure 12.4 2-bit input decoders.

INC n has n inputs and $n+1$ outputs, and computes $f = x + 1$.
LOG n has n inputs and n outputs, and computes $f = \lfloor \frac{(2^n-1)\log(x+1)}{n \log 2} \rfloor$.
MLP n has $2n$ inputs and $2n$ outputs, and computes $f = x \times y$.
NRM n has $2n$ inputs and $n+1$ outputs, and computes $f = \lfloor \sqrt{x^2 + y^2} + 0.5 \rfloor$.
RDM n has n inputs and n outputs, and computes $f = 5x + 1(\mathrm{mod} 2^n)$.
ROT n has n inputs and $\lfloor n/2 \rfloor + 1$ outputs, and computes $f = \lfloor \sqrt{x} + 0.5 \rfloor$.
SQR n has n inputs and $2n$ outputs, and computes $f = x^2$.
WGT n is a bit counting circuit with n inputs.

12.1.3 PLAs with Input Decoders

In the standard PLA, if we delete the input buffers and inverters, and add the input decoders shown in Fig. 12.4, we have a **PLA with 2-bit input decoders** shown in Fig. 12.5. On the other hand, a standard PLA is a "**PLA with 1-bit input decoders.**" Table 12.4 compares the number of products to realize various functions in standard PLAs and PLAs with 2-bit input decoders. This table shows that PLAs with 2-bit input decoders require many fewer products than standard PLAs for symmetric functions. For randomly generated functions of 10 variables, PLAs with 2-bit input decoders require 20 to 30 percent fewer products.

Principle of PLAs with input decoders

A 2-bit input decoder realizes all the maxterms of two variables: $x_1 \vee x_2$, $x_1 \vee \bar{x}_2$, $\bar{x}_1 \vee x_2$, and $\bar{x}_1 \vee \bar{x}_2$. An arbitrary two-variable function is represented by the logical product of some of these maxterms. This is the canonical product-of-sums expression of two variables:

$$f(x_1, x_2) = (c_0 \vee x_1 \vee x_2)(c_1 \vee x_1 \vee \bar{x}_2)(c_2 \vee \bar{x}_1 \vee x_2)(c_3 \vee \bar{x}_1 \vee \bar{x}_2),$$

Table 12.3 Numbers of products to realize arithmetic functions by PLAs.

Input data					Standard PLA		PLAs with input decoders	
Function name		In	Out	Product term	Output phase		Output phase	
					Original	Optimized	Original	Optimized
Adder	ADR2	4	3	15	11	9	5	4
	ADR3	6	4	63	31	25	10	8
	ADR4	8	5	255	75	61	17	14
Add one	INC4	4	5	16	11	10	7	6
	INC6	6	7	64	22	21	13	12
	INC8	8	9	256	37	36	21	20
Logarithm	LOG4	4	4	15	11	10	9	8
	LOG6	6	6	63	40	33	30	27
	LOG8	8	8	255	123	111	93	89
Multiplier	MLP2	4	4	9	7	7	6	5
	MLP3	6	6	49	30	28	21	19
	MLP4	8	8	225	119	108	85	74
Distance	NRM2	4	3	15	7	6	5	5
	NRM3	6	4	63	36	29	19	17
	NRM4	8	5	255	120	101	70	64
Random number generator	RDM4	4	4	15	11	11	7	7
	RDM6	6	6	63	35	35	20	20
	RDM8	8	8	255	76	76	47	47
Square root	ROT4	4	3	15	9	8	6	4
	ROT6	6	4	63	23	20	17	14
	ROT8	8	5	255	57	48	37	32
Square	SQR4	4	8	15	13	10	11	9
	SQR6	6	12	63	47	41	39	36
	SQR8	8	16	255	180	172	146	142
Bit-count circuit	WGT4	4	3	15	15	10	6	5
	WGT6	6	3	63	63	44	17	12
	WGT8	8	4	255	255	186	54	38

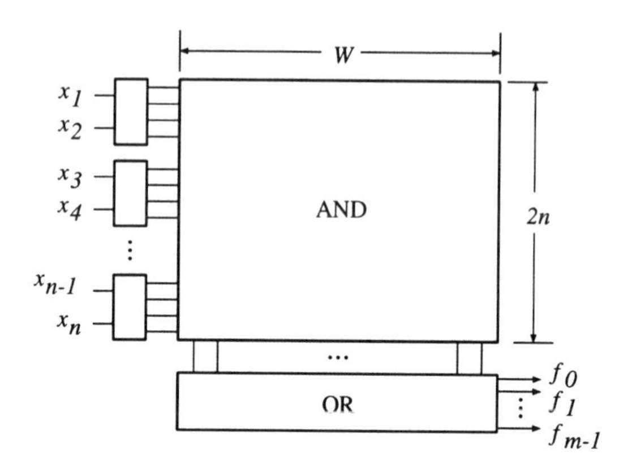

Figure 12.5 PLA with 2-bit input decoders.

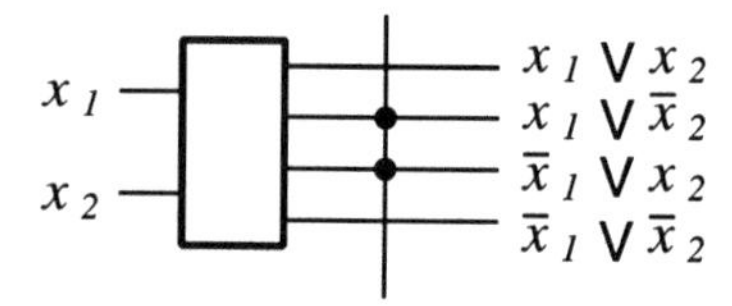

Figure 12.6 Realization of $f = x_1x_2 \vee \bar{x}_1\bar{x}_2$.

where $c_i(i = 0, 1, 2, 3)$ are constants 0 or 1. For example, when $(c_0, c_1, c_2, c_3) = (1, 0, 0, 1)$, the maxterms with $c_i = 0$ remain and we have

$$f(x_1, x_2) = (x_1 \vee \bar{x}_2)(\bar{x}_1 \vee x_2) = x_1x_2 \vee \bar{x}_1\bar{x}_2.$$

Since c_i is either 0 or 1, there are $2^4 = 16$ different combinations to specify the values of (c_0, c_1, c_2, c_3). Table 12.5 shows the combinations of c_i and the functions realized by decoders. By specifying the values of c_i appropriately, we can represent an arbitrary two-variable function.

Example 12.3 Consider Fig. 12.6. In the figure, ● denotes the AND, and the product lines realizes the function $f(x_1, x_2) = (x_1 \vee \bar{x}_2)(\bar{x}_1 \vee x_2) = x_1x_2 \vee \bar{x}_1\bar{x}_2$. Next, consider the PLA shown in Fig. 12.7. This PLA realizes the function $f = f_1(x_1, x_2)f_2(x_3, x_4)f_3(x_5, x_6)$, where

$$\begin{aligned} f_1(x_1, x_2) &= x_1x_2 \vee \bar{x}_1\bar{x}_2, \\ f_2(x_3, x_4) &= x_3x_4 \vee \bar{x}_3\bar{x}_4, \text{ and} \\ f_3(x_5, x_6) &= x_5x_6 \vee \bar{x}_5\bar{x}_6. \end{aligned}$$

Table 12.4 Number of products to realize n-variable functions.

	PLAs with 1-bit decoders	PLAs with 2-bit decoders
Arbitrary function (worst case)	2^{n-1}	2^{n-2}
Symmetric function (worst case)	2^{n-1}	$(\sqrt{3})^{n-2}$
Parity function	2^{n-1}	$(\sqrt{2})^{n-2}$
Adder	$6 \cdot 2^n - 4n - 5$	$n^2 + 1$
10-variable random function (average)	163	120

Table 12.5 Function realized by a 2-bit input decoder.

c_0 c_1 c_2 c_3	Logical expression	c_0 c_1 c_2 c_3	Logical expression
0 0 0 0	0	1 0 0 0	$\bar{x}_1\bar{x}_2$
0 0 0 1	x_1x_2	1 0 0 1	$x_1 \oplus \bar{x}_2$
0 0 1 0	$x_1\bar{x}_2$	1 0 1 0	$\bar{x}_2$
0 0 1 1	x_1	1 0 1 1	$x_1 \vee \bar{x}_2$
0 1 0 0	$\bar{x}_1x_2$	1 1 0 0	$\bar{x}_1$
0 1 0 1	x_2	1 1 0 1	$\bar{x}_1 \vee x_2$
0 1 1 0	$x_1 \oplus x_2$	1 1 1 0	$\bar{x}_1 \vee \bar{x}_2$
0 1 1 1	$x_1 \vee x_2$	1 1 1 1	1

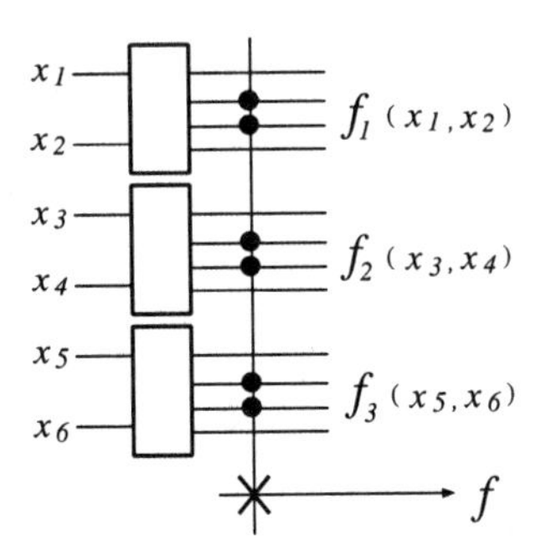

Figure 12.7 Coincidence function for three bits.

In other words,

$$f = (x_1x_2 \vee \bar{x}_1\bar{x}_2)(x_3x_4 \vee \bar{x}_3\bar{x}_4)(x_5x_6 \vee \bar{x}_5\bar{x}_6).$$

This is the 3-bit **coincidence function**. A standard PLA require 8 products. On the other hand, a PLA with 2-bit input decoders require only one product. ∎

In the above example, the PLA with 2-bit input decoders requires one eighth products of the standard PLA (the PLA with 1-bit input decoders). For any function, a PLA with 2-bit input decoders never require more products than the standard PLA. On the average, PLAs with 2-bit decoders require 20-30 percent fewer products. However, we need additional circuit for the input decoders.

Representation for PLAs with Input Decoders

PLAs with input decoders can be specified by bit representation, which is a generalization of the representation for standard PLAs (PLAs with 1-bit input decoders). For example, consider the function f in Example 12.3. Let $X_1 = (x_1, x_2)$, $X_2 = (x_3, x_4)$, and $X_3 = (x_5, x_6)$, where X_i takes one of the values from (0,0), (0,1), (1,0), and (1,1). For simplicity, we denote them by 00, 01, 10, and 11, respectively. In this case, f is represented as

X_1					X_2					X_3			
00,	01,	10,	11		00,	01,	10,	11		00,	01,	10,	11
1	0	0	1	-	1	0	0	1	-	1	0	0	1

This expression shows that the value of f becomes 1, if (X_1=00 or 11) and (X_2=00 or 11) and (X_3=00 or 11).

Theorem 12.3

1) Each product line in a PLA with 2-bit input decoders realizes a product

$$f_1(x_1, x_2) \cdot f_2(x_3, x_4) \cdots f_r(x_{n-1}, x_n).$$

2) The function realized by the product lines is represented by

$$c_0^1c_1^1c_2^1c_3^1 \text{ - } c_0^2c_1^2c_2^2c_3^2 \text{ - } \cdots \text{ - } c_0^rc_1^rc_2^rc_3^r \; (n = 2r).$$

Design Method for PLAs with Input Decoders

PLAs with input decoders are designed in the following steps:

Step 1. Transform the logical expression into the bit representation.

1) Partition the input variables into pairs to make 4-valued variables.
2) By using Table 12.5, derive the bit representation of the function with the 4-valued variables.

Example 12.4 Let us obtain the bit representation of the 2-bit adder (ADR2) shown below:

$$\begin{array}{rrr} & x_1 & x_0 \\ +) & y_1 & y_0 \\ \hline z_2 & z_1 & z_0 \end{array}$$

z_0, z_1, and z_2 are represented by logical expressions:

$$\begin{aligned} z_0 &= x_0 \oplus y_0 = x_0\bar{y}_0 \vee \bar{x}_0 y_0, \\ z_1 &= (x_0 y_0) \oplus (x_1 \oplus y_1) \\ &= (\bar{x}_0 \vee \bar{y}_0)(x_1\bar{y}_1 \vee \bar{x}_1 y_1) \vee (x_0 y_0)(x_1 y_1 \vee \bar{x}_1\bar{y}_1), \text{ and} \\ z_2 &= x_1 y_1 \vee (x_0 y_0)(x_1 \vee y_1). \end{aligned}$$

Let $X_1 = (x_0, y_0)$, $X_2 = (x_1, y_1)$, and $X_3 = (z_0 z_1 z_2)$. Then, we have the bit representation of the 2-bit adder:

$$\begin{array}{c} \begin{array}{ccccc} X_1 & & X_2 & & X_3 \end{array} \\ \left[\begin{array}{ccccc} 0110 & - & 1111 & - & 100 \\ 1110 & - & 0110 & - & 010 \\ 0001 & - & 1001 & - & 010 \\ 1111 & - & 0001 & - & 001 \\ 0001 & - & 0111 & - & 001 \end{array}\right], \end{array}$$

where, X_1 and X_2 are four-valued variables that take one of the values in $\{00, 01, 10, 11\}$. Also, X_3 is a three-valued variable that represents the output part. If $X_3 = 0$, then $z_0 = 1$; if $X_3 = 1$, then $z_1 = 1$; and if $X_3 = 2$, then $z_2 = 1$. In the above bit representation, the first row shows that the combinations ($X_1 = 01$ or 10) and ($X_2 = 00, 01, 10$ or 11) and ($X_3 = 0$) are permitted. The second row shows that the combinations ($X_1 = 00, 01$ or 10) and ($X_2 = 01$ or 10) and ($X_3 = 1$) are permitted. Similarly, the other three rows show the permitted combinations of inputs and outputs. And, these bit representations denote all the permitted combinations. The above bit representation specifies the 4-input 3-output network by the multi-valued input two-valued output function:

$$\{00, 01, 10, 11\} \times \{00, 01, 10, 11\} \times \{0, 1, 2\} \rightarrow \{0, 1\}.$$

∎

Step 2. Transform the bit representation into a PLA.

1) Input part: For the part 0, make an AND connection.

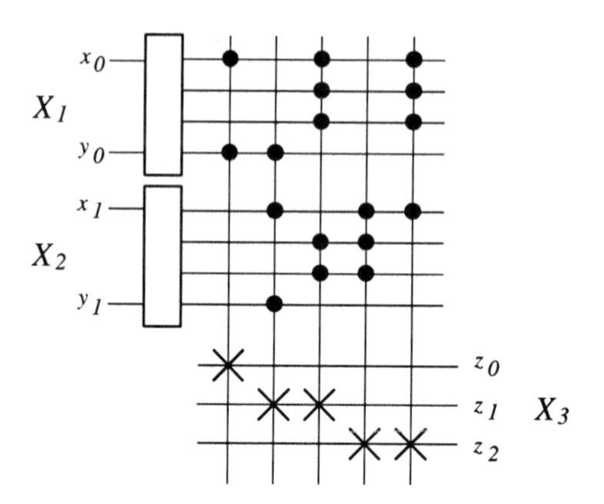

Figure 12.8 Realization of 2-bit adder by a PLA with 2-bit input decoders.

2) Output part: For the part 1, make an OR connection.

Example 12.5 Fig. 12.8 shows the PLA with 2-bit input decoders for a 2-bit adder (ADR2). ∎

Theorem 12.4 *An arbitrary n-variable function ($n = 2r$) is realized by a PLA with 2-bit input decoders using at most 2^{n-2} products.*

(Proof) An arbitrary n-variable function is expanded as

$$f = \bigvee_{\boldsymbol{a}} f(x_1, x_2, \boldsymbol{a}) \cdot (x_3^{a_3} x_4^{a_4}) \cdot (x_5^{a_5} x_6^{a_6}) \cdots (x_{n-1}^{a_{n-1}} x_n^{a_n}),$$

where, $\boldsymbol{a} = (a_3, a_4, \ldots, a_n)$, $a_i \in \{0, 1\}$. The number of products in the above expression is at most 2^{n-2}. On the other hand, by Theorem 12.3, each product line in a PLA with 2-bit input decoders realizes the above product term. Hence, we have the theorem. □

Example 12.6 An arbitrary 4-variable function f is expanded as a sum of four products:

$$\begin{aligned} f &= \bigvee_{(a_3,a_4)} f(x_1, x_2, a_3, a_4)(x_3^{a_3} x_4^{a_4}) \\ &= f(x_1, x_2, 0, 0)x_3^0 x_4^0 \vee f(x_1, x_2, 0, 1)x_3^0 x_4^1 \\ &\quad \vee f(x_1, x_2, 1, 0)x_3^1 x_4^0 \vee f(x_1, x_2, 1, 1)x_3^1 x_4^1. \end{aligned}$$

∎

Theorem 12.5 *An arbitrary n-variable symmetric function ($n = 2r$) is realized by a PLA with 2-bit input decoders using at most 3^{r-1} products.*

(Proof) An arbitrary n-variable symmetric function is expanded as

$$f = \bigvee_{\boldsymbol{b}} S(x_1, x_2, \boldsymbol{b}) \cdot S_{b_2}(x_3, x_4) \cdot S_{b_3}(x_5, x_6) \cdots S_{b_r}(x_{n-1}, x_n),$$

where $\boldsymbol{b} = (b_2, b_3, \ldots, b_r)$, $b_i \in \{0, 1, 2\}$, and

$$S_j(x_i, x_{i+1}) = \begin{cases} 1 & (\text{when } x_i + x_{i+1} = j) \\ 0 & (\text{otherwise}). \end{cases}$$

Note that $S(x_1, x_2, \boldsymbol{b})$ are symmetric with respect to x_1 and x_2. Thus, the number of products in the above expression is 3^{r-1}. On the other hand, by Theorem 12.3, each product line of a PLA with 2-bit input decoders realizes the above products. Thus, we have the theorem. □

Example 12.7 A 4-variable symmetric function f is expanded as a sum of three products:

$$\begin{aligned} f &= \bigvee_{b=0}^{2} S(x_1, x_2, b) S_b(x_3, x_4) \\ &= S(x_1, x_2, 0) S_0(x_3, x_4) \vee S(x_1, x_2, 1) S_1(x_3, x_4) \vee S(x_1, x_2, 2) S_2(x_3, x_4) \\ &= S(x_1, x_2, 0) \bar{x}_3 \bar{x}_4 \vee S(x_1, x_2, 1)(x_3 \oplus x_4) \vee S(x_1, x_2, 2) x_3 x_4. \end{aligned}$$

∎

Theorem 12.6 *An n-variable parity function ($n = 2r$) is realized by a PLA with 2-bit input decoders by using 2^{r-1} products.*

(Proof) An n-variable parity function is represented as

$$f = \bigvee_{\boldsymbol{c}} P(x_1, x_2, \boldsymbol{c}) \cdot P_{c_2}(x_3, x_4) \cdot P_{c_3}(x_5, x_6) \cdots P_{c_r}(x_{n-1}, x_n),$$

where $\boldsymbol{c} = (c_2, c_3, \ldots, c_r)$, $c_i \in \{0, 1\}$, and $P_{c_j}(x_i, x_{i+1}) = x_i \oplus x_{i+1} \oplus c_j$. Note that $P(x_1, x_2, \boldsymbol{c})$ is also a parity function. The number of products in the above expression is 2^{r-1}. On the other hand, by Theorem 12.3, each product line in a PLA with 2-bit input decoders realizes the above product, and we have the theorem. □

Table 12.6 Number of products in PLAs for n-bit adder.

Carry input	PLAs with 1-bit decoders	
	Output phase original	Output phase optimized
No	$6 \cdot 2^n - 4n - 5$	$5 \cdot 2^n - 4n - 3$
Yes	$10 \cdot 2^n - 4n - 9$	$8 \cdot 2^n - 4n - 7$

Carry input	PLAs with 2-bit decoders	
	Output phase original	Output phase optimized
No $(2, 2, \ldots, 2)$	$n^2 + 1$	$n^2 - n + 2$
Yes $(2, 2, \ldots, 1)$	$(n+1)^2$	$n^2 + n + 1$
Yes $(2, 2, \ldots, 3)$	$n^2 + 1$	$n^2 - n + 2$

Example 12.8 A 4-variable parity function f is expanded as a sum of two products:

$$\begin{aligned} f &= \bigvee_{c=0}^{1} P(x_1, x_2, c) P_c(x_3, x_4) \\ &= P(x_1, x_2, 0) P_0(x_3, x_4) \vee P(x_1, x_2, 1) P_1(x_3, x_4) \\ &= P(x_1, x_2, 0)(x_3 \oplus x_4) \vee P(x_1, x_2, 1)(x_3 \oplus \bar{x}_4). \end{aligned}$$

∎

Table 12.6 shows the number of products needed to realize n-bit adders.

$(2, 2, \ldots, 2)$ denotes the case where only 2-bit input decoders are used. $(2, 2, \ldots, 1)$ denotes the case where a 1-bit input decoder is used for the carry input. $(2, 2, \ldots, 3)$ denotes the case where a 3-bit input decoder is used for the carry input, x_0, and y_0.

Example 12.9 In Example 12.4, input variables are assigned as $X_1 = (x_0, y_0)$, and $X_2 = (x_1, y_1)$. On the other hand, if we assign the variables as $X_1 = (x_0, x_1)$ and $X_2 = (y_0, y_1)$, then the PLA requires 9 products. ∎

As shown in Example 12.9, the number of products in the PLA greatly depends on the assignments of the variables to the input decoders. The **optimum assignment of the input variables** is one that minimize the number of products in the PLA. In order to obtain the exact optimum assignment of the input variables, we have to simplify the PLAs for all the combinations. For an $n(=2r)$-variable function, the number of combinations is $n!/(2^r \cdot r!)$. This is 105 when $n = 8$, which is the practical upper bound on the size that we can obtain the optimum solutions by exhaustive method. In order to know how we can reduce the size of the PLA by considering the input assignments of the variables to the decoders, we did the following experiments: For various arithmetic functions, we generated all possible assignments (105 ways), and minimized the expression by using the QM method. Table 12.3 shows the experimental results.

When the number of input variables is large, the exhaustive method requires too much computation time. The heuristic method shown below obtains good solutions in a short time.

Definition 12.2 *Let $I = \{1, 2, \ldots, n\}$ be a set of subscripts for the input variables X. Let Π be a partition of I (corresponding to the partition of X). Let $t(f : \Pi)$ be the number of products in a minimum SOP for f, under the partition Π. Let F be an SOP for the function f. Let $q(i, j)$ be the number of different products in the SOP that are obtained from F by deleting literals for x_i and x_j. Let $t(f : \Pi_{ij})$ be the number of products in a minimum SOP for f, when x_i and x_j are paired.*

Example 12.10 Let F be

$$F = \bar{x}_1\bar{x}_2x_3x_4 \vee \bar{x}_1x_2x_3x_4 \vee x_1\bar{x}_2\bar{x}_3x_4 \vee x_1\bar{x}_2x_3\bar{x}_4 \vee x_1x_2\bar{x}_3\bar{x}_4.$$

The products that are obtained by deleting the literals of x_3 and x_4 from F are: $\bar{x}_1\bar{x}_2, \bar{x}_1x_2, x_1\bar{x}_2, x_1\bar{x}_2$ and x_1x_2. The number of distinct product terms is 4, so we have $q(3, 4) = 4$. Similarly, we have $q(2, 3) = q(2, 4) = 3$, $q(1, 2) = q(1, 4) = q(1, 3) = 4$. ∎

Lemma 12.2 *Let $\Pi_{ij} = \{[1], [2], \ldots, [i, j], \ldots, [n]\}$. Then, $t(f : \Pi_{ij}) \leq q(i, j)$.*

(Proof) Let F be an SOP for f. Without loss of generality, we can assume that $i = 1$ and $j = 2$. F is represented as an SOP:

$$F = \bigvee_{(\boldsymbol{S})} x_1^{S_1} x_2^{S_2} \cdots x_n^{S_n},$$

where $\boldsymbol{S} = (S_1, S_2, \ldots, S_n)$, and $S_i \subseteq \{0,1\}$. Note that

$$x_i^{S_i} = \begin{cases} 1 & (\text{when } S_i = \{0,1\}) \\ x_i & (\text{when } S_i = \{1\}) \\ \bar{x}_i & (\text{when } S_i = \{0\}). \end{cases}$$

Factoring F by $x_3^{S_3} x_4^{S_4} \cdots x_n^{S_n}$, we have

$$F_1 = \bigvee_{(\boldsymbol{S}^*)} G(x_1, x_2, \boldsymbol{S}^*) x_3^{S_3} x_4^{S_4} \cdots x_n^{S_n},$$

where $\boldsymbol{S}^* = (S_3, S_4, \ldots, S_n)$, and $S_i \subseteq \{0,1\}$. Also, $G(x_1, x_2, \boldsymbol{S}^*)$ is an expression that contains only variables x_1 and x_2. Let $t(F_1)$ be the number of products in F_1. Then, $t(F_1)$ is equal to the distinct number of patterns in $x_3^{S_3} x_4^{S_4} \cdots x_n^{S_n}$, which are obtained from F by deleting the literals of x_1 and x_2. From Definition 12.2, we have $t(F_1) = q(1,2)$. Also, by pairing x_1 and x_2 to make a 4-valued variable $X_1 = (x_1, x_2)$, and by replacing $G(x_1, x_2, \boldsymbol{S}^*)$ with literal $X_1^{T_1}, (T_1 \subseteq \{00, 01, 10, 11\})$, we have F_1 which is an SOP under the partition Π_{ij}. Since $t(f : \Pi_{ij})$ denotes the number of products in a minimum SOP, we have $t(f : \Pi_{ij}) \leq t(F_1)$. Thus, we have $t(f : \Pi_{ij}) \leq q(i,j)$. □

The smaller the value of $t(f : \Pi_{ij})$, the simpler the SOP when the variables x_i and x_j are paired. Thus, $t(f : \Pi_{ij})$ is used as a measure of merit when the variables x_i and x_j are paired. However, to compute the value of $t(f : \Pi_{ij})$ is time consuming. In the next algorithm, $q(i,j)$ is used as a measure, which is an upper bound of $t(f : \Pi_{ij})$.

Definition 12.3 *A* **variable assignment graph** *G of an n-variable function $f(x_1, x_2, \ldots, x_n)$ is the complete graph with weights satisfying the following conditions:*

1. *G has n nodes.*
2. *The weight of the edge (i,j) is $q(i,j)$.*

Algorithm 12.2 *(Assignment of the variables for a PLA with 2-bit input decoders)*

1. *Obtain a (near) minimum SOP for f.*
2. *Obtain a variable assignment graph G for f.*
3. *Cover all the nodes of G by a set of edges that have no common elements. In this case, find a set such that the sum of the weights are minimum. This is the* **optimum matching** *of the graph G.*

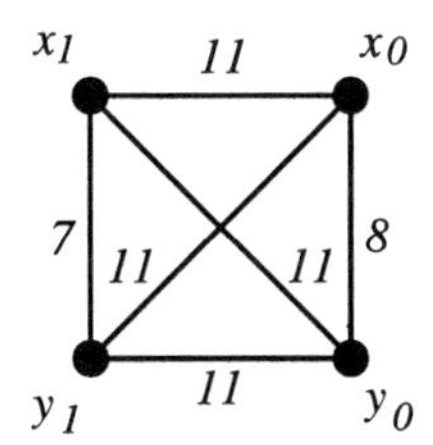

Figure 12.9 Variable assignment graph.

4. Obtain the partition of the variables corresponding to the edges.

Algorithm 12.2 is a heuristic, and it does not always produce the optimal solution. However, experimental results show that it obtains optimal solutions for many functions.

Example 12.11 Let us obtain the input assignment of the variables to the decoder for the 2-bit adder in Example 12.4.

1) The minimum SOP is:

$$
\begin{array}{c}
\begin{matrix} x_1 & x_0 & y_1 & y_0 & z_2 z_1 z_0 \end{matrix} \\
\left[\begin{array}{ccccccccc}
01 & - & 11 & - & 01 & - & 11 & - & 100 \\
11 & - & 01 & - & 01 & - & 01 & - & 100 \\
01 & - & 01 & - & 11 & - & 01 & - & 100 \\
11 & - & 10 & - & 11 & - & 01 & - & 001 \\
11 & - & 01 & - & 11 & - & 10 & - & 001 \\
10 & - & 01 & - & 10 & - & 01 & - & 010 \\
01 & - & 01 & - & 01 & - & 01 & - & 010 \\
10 & - & 11 & - & 01 & - & 10 & - & 010 \\
01 & - & 10 & - & 10 & - & 11 & - & 010 \\
01 & - & 11 & - & 10 & - & 10 & - & 010 \\
10 & - & 10 & - & 01 & - & 11 & - & 010
\end{array}\right].
\end{array}
$$

2) Fig. 12.9 shows the variable assignment graph.
3) Edges (x_1, y_1) and (x_0, y_0) cover all the nodes in G. Also, the sum of weights is $7 + 8 = 15$ and is minimum.
4) The partition of $\{X\}$ is $X = (X_1, X_2)$, where $X_1 = (x_1, y_1)$ and $X_2 = (x_0, y_0)$.
5) Fig. 12.8 is the PLA, where the input assignment of the variables is optimum. ∎

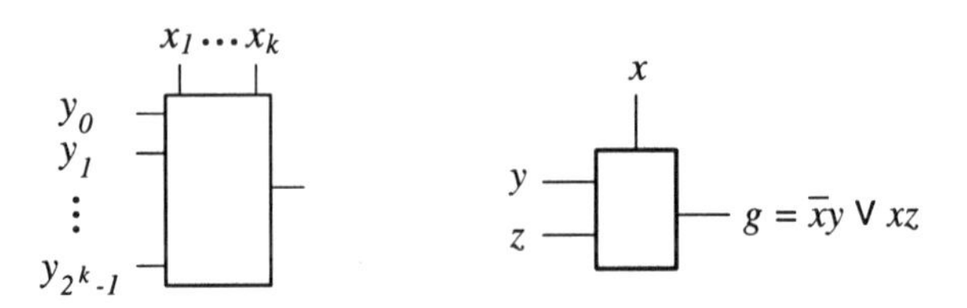

Figure 12.10 k-MUX.

Figure 12.11 1-MUX.

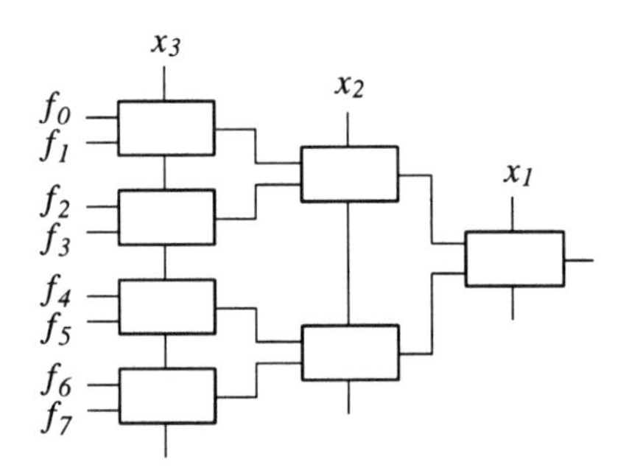

Figure 12.12 Realization of 3-MUX by using 1-MUXs.

12.2 DESIGN USING MULTIPLEXERS

A **multiplexer** (MUX) is the selection circuit shown in Fig. 12.10. It has k control inputs $(x_1, \ldots, x_k)$, and 2^k data inputs $(y_0, y_1, \ldots, y_{2^k-1})$. Let $g(x_1, \ldots, x_k, y_0, y_1, \ldots, y_{2^k-1})$ be the output function. Then, $g = y_a$ when the decimal representation of the control input $(x_1, \ldots, x_k)$ is a. That is, when the control input is $(0, 0, \ldots, 0)$, the top data input y_0 is selected. When the the control input is $(0, 0, \ldots, 1)$, the second data input y_1 is selected. Also, when the control input is $(1, 1, \ldots, 1)$, the last data input y_{2^k-1} is selected. A MUX with k control inputs is a ***k*-MUX**. For an arbitrary n, n-MUX is synthesized by using copies of k-MUXs $(k < n)$. In this section, we assume that constants 0 and 1 are available as inputs.

Example 12.12 A 3-MUX is realized as shown in Fig. 12.12 by using 7 copies of 1-MUXs. ∎

Theorem 12.7 *An arbitrary n-variable function is realized by using at most* $(2^n - 1)$ *copies of 1-MUXs.*

(Proof) By applying the Shannon expansions n times to the n-variable function $f(x_1, x_2, \ldots, x_n)$, we have

$$\bigvee_{(a_1,a_2,\ldots,a_n)} f(a_1, a_2, \ldots, a_n) x_1^{a_1} x_2^{a_2} \cdots x_n^{a_n}, \tag{12.1}$$

where the logical sum is done for 2^n elements. Fig. 12.12 illustrates the case of $n = 3$, and consists of $7 = 2^3 - 1$ copies of 1-MUXs. In this network, when $x_1 = x_2 = x_3 = 0$, the top input f_0 is selected. In (12.1), this corresponds to $f(0,0,0)$, and is a constant 0 or 1. Next, when $x_1 = x_2 = 0$, and $x_3 = 1$, the second input f_1 is selected. Similarly, other inputs are selected. Thus, the network in Fig. 12.12 realizes (12.1), and realizes an arbitrary three-variable function. An arbitrary n-variable function can be realized by connecting constant 0 or 1 to the data inputs of the n-MUX. In order to realize n-MUX, we need $2^n - 1$ copies of 1-MUXs. Thus, we can realize an arbitrary logic function by using at most $2^n - 1$ copies of 1-MUXs. □

The above realization method requires about 2^n copies of 1-MUXs to realize an n-variable function. The next realization method makes this number about a half.

Theorem 12.8 *An arbitrary n-variable function is realized by using at most* 2^{n-1} *copies of 1-MUXs.*

(Proof) Consider the expansion which is obtained by applying the Shannon expansion $(n - 1)$ times:

$$f(x_1, x_2, \ldots, x_n) = \bigvee_{(a_1,a_2,\ldots,a_{n-1})} f(a_1, a_2, \ldots, a_{n-1}, x_n)\, x_1^{a_1} x_2^{a_2} \cdots x_{n-1}^{a_{n-1}},$$

where $f(a_1, a_2, \ldots, a_{n-1}, x_n)$ is 0, 1, x_n, or $\bar{x}_n$. The above expression can be realized by connecting constants 0, 1, variable x_n, or $\bar{x}_n$ to the data inputs of an $(n-1)$-MUX. An $(n-1)$-MUX is synthesized by using $2^{n-1} - 1$ copies of 1-MUXs. If $\bar{x}_n$ is realized by a 1-MUX as shown in Fig. 12.13, then an arbitrary n-variable function is realized with at most 2^{n-1} copies of 1-MUXs. □

Example 12.13 Fig. 12.13 illustrates the realizations of $\bar{x}$, xy, $x \vee y$ and $x \oplus y$ by using 1-MUXs. ∎

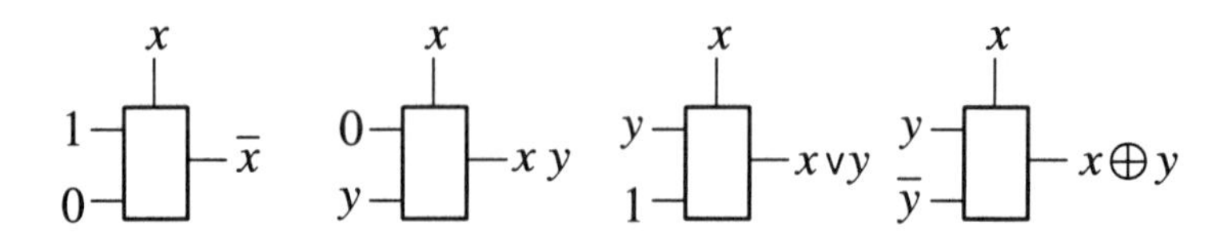

Figure 12.13 Realization of various functions by 1-MUXs.

The next realization method further reduces the number of 1-MUXs.

Theorem 12.9 *An arbitrary n-variable function is realized by using at most $2^{n-2}+11$ copies of 1-MUXs.*

(Proof) Consider the expansion which is obtained by applying the Shannon expansion $(n-2)$ times:

$$f(x_1, x_2, \ldots, x_n) = \bigvee_{(a_1, a_2, \ldots, a_{n-2})} f(a_1, a_2, \ldots, a_{n-2}, x_{n-1}, x_n) x_1^{a_1} x_2^{a_2} \cdots x_{n-2}^{a_{n-2}},$$

where $f(a_1, a_2, \ldots, x_{n-1}, x_n)$ is a two-variable function. The above expression can be realized by connecting two-variable functions to the data inputs of an $(n-2)$-MUX. Also, an $(n-2)$-MUX is synthesized by using $2^{n-2}-1$ copies of 1-MUXs. To realize all the two-variable functions except for constants 0, 1, variable x_n, and x_{n-1}, we need $16-4=12$ copies of 1-MUXs. Thus, this method requires at most $2^{n-2}-1+12 = 2^{n-2}+11$ copies of 1-MUXs to realize an arbitrary n-variable function. □

12.3 LOGIC DESIGN USING ROMS

A **ROM** (read only memory), or an **LUT** (look-up table), is a module that realizes an arbitrary k-variable function. This section shows a method to realize an arbitrary n-variable function by using k-input ROMs $(2 < k < n)$.

Theorem 12.10 *An arbitrary n-variable function is expanded as follows:*

$$f(x_1, x_2, \ldots, x_n) = \bigvee_{(a_{k+1}, a_{k+2}, \ldots, a_n)} g_i(x_1, x_2, \ldots, x_k)\, x_{k+1}^{a_{k+1}} x_{k+2}^{a_{k+2}} \cdots x_n^{a_n}.$$

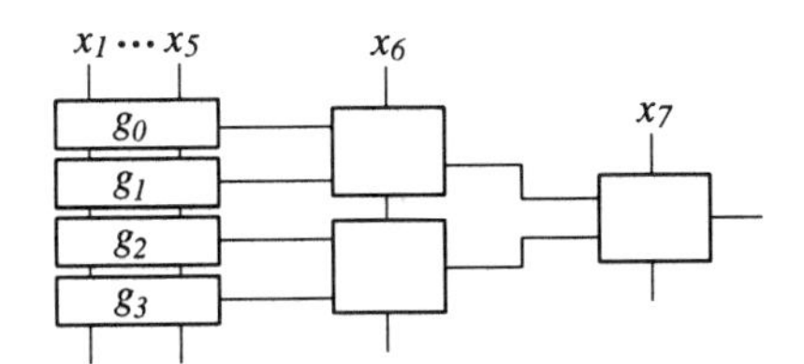

Figure 12.14 Realization of a 7-variable function using ROMs.

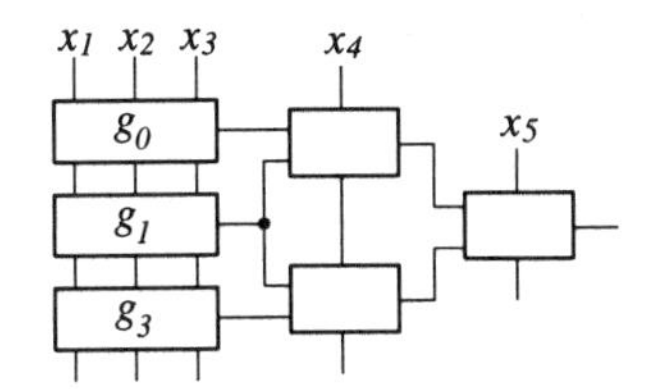

Figure 12.15 Realization of the 5-input majority function.

Example 12.14 In the case of $k = 5$, a 7-variable function f is expanded into a sum of four sub-functions:

$$f(x_1, x_2, \ldots, x_7) = \bigvee_{i=0}^{3} g_i(x_1, x_2, x_3, x_4, x_5)\, x_6^{a_6} x_7^{a_7},$$

where g_0, g_1, g_2, g_3 are 5-variable functions. As Fig. 12.14 shows, one can realize a 2-MUX (the circuit that select one data input out of four data inputs) by using ROMs. The top ROM in the left most column realizes g_0, which is selected when $(x_6, x_7) = (0, 0)$. The second ROM in the left most column realizes g_1, which is selected when $(x_6, x_7) = (1, 0)$. If there are the same functions in g_i, then they can be shared. When the adjacent functions are the same (e.g., $g_0 = g_1$ or $g_2 = g_3$), then the selection circuit can also be simplified. ∎

Theorem 12.11 *An arbitrary n-variable function is realized by using at most $2^{n-k+1} - 1$ copies of k-input ROMs.*

(Proof) First, realize 1-MUXs by ROMs. Next, by combining 1-MUXs, synthesize an $(n - k)$-MUX. By connecting the necessary k-variable function to the data inputs of the $(n-k)$-MUX, we can realize an arbitrary n-variable function. This realization requires $2^{n-k} + (2^{n-k} - 1) = 2^{n-k+1} - 1$ copies of ROMs. In the MUX circuit, all the ROMs are treated as 3-input ROMs. □

When $k = 6$ and $n = 2r$, a ROM can realize a 2-MUX. Thus, the number of necessary ROMs is $(4^{r-2} - 1)/3$. In the above realization, the ROMs in the selection circuit are used as MUXs. Since, a ROM can realize an arbitrary function, number of ROMs can be further reduced.

Example 12.15 Let us realize a 5-variable majority function f by using 3-input ROMs. The function is expanded as follows:

$$\begin{aligned} f(x_1,x_2,x_3,x_4,x_5) &= g_0(x_1,x_2,x_3)\bar{x}_4\bar{x}_5 \vee g_1(x_1,x_2,x_3)\bar{x}_4x_5 \\ &\quad \vee g_2(x_1,x_2,x_3)x_4\bar{x}_5 \vee g_3(x_1,x_2,x_3)x_4x_5, \end{aligned}$$

where

$$\begin{aligned} g_0(x_1,x_2,x_3) &= x_1x_2x_3, \\ g_1(x_1,x_2,x_3) &= g_2(x_1,x_2,x_3) = x_1x_2 \vee x_2x_3 \vee x_1x_3, \text{ and} \\ g_3(x_1,x_2,x_3) &= x_1 \vee x_2 \vee x_3. \end{aligned}$$

Thus, we can realize the function as shown in Fig. 12.15. ∎

Bibliographical Notes

Detailed treatments on each topic can be found as follows: PLAs [58, 90, 122, 166, 190, 248, 278, 310, 347, 426]; FPLAs [190, 218, 311]; microprocessors using PLAs [64, 136, 220]; optimal output phase assignment [343]; PLAs with input decoders [21, 113, 342, 343, 349, 350]; adders using PLAs [343, 423]; minimization of number of pins for PLAs [154, 194]; fault models of PLAs [389]; easily testable PLAs [129, 130]; multi-valued PLAs [350]; logic synthesis using multiplexers [89, 167, 193, 161, 437]; and FPGAs [47, 76, 190, 271, 440].

Exercises

12.1 Design the PLA that calculates the two's complement of 4-bit numbers.

12.2 WGT4 counts the number of 1's in the inputs, and represents it by a binary number. Design WGT4 by using a PLA with 2-bit input decoders.

12.3 Let $X = (x_1, x_2, x_3, x_4)$ and $Y = (y_1, y_2, y_3, y_4)$. Design a PLA with 2-bit input decoders that realizes the function f, where $f = 1$ iff $X \neq Y$.

12.4 Design two PLAs that realize a 3-input bit-counting circuit (WGT3), where the output phase is optimized and non-optimized.

12.5 Design a PLA with 2-bit input decoders that realizes the 6-input 1-output function f, where $f = 1$ iff the number of 1's in the inputs is exactly 3.

12.6 Design a PLA with 2-bit input decoders that realizes $f = x_1 \oplus x_2 \oplus x_3 \oplus x_4$. The number of the products must be minimum.

12.7 (M) Show that an n-bit adder (ADRn) is realizable by an AND-OR two-level PLA with $6 \cdot 2^n - 4n - 5$ products.

12.8 (M) Show that an n-bit adder (ADRn) is realizable by a PLA with 2-bit input decoders by using $n^2 + 1$ products.

12.9 Prove that the number of products in an optimized PLA with 2-bit input decoders for f is not greater than the number of products in an optimized AND-OR two-level PLA for f.

12.10 Obtain the number of nodes in the ROBDD for the multiplexer function: $f(x_1, x_2, \ldots, x_m, y_0, y_1, \ldots, y_{2^m-1})$.

12.11 By using 1-MUXs, realize the 5-variable majority function $M5(x_1, x_2, \ldots, x_5)$.

12.12 By using 1-MUXs, realize the 5-variable parity function $f = x_1 \oplus x_2 \oplus x_3 \oplus x_4 \oplus x_5$.

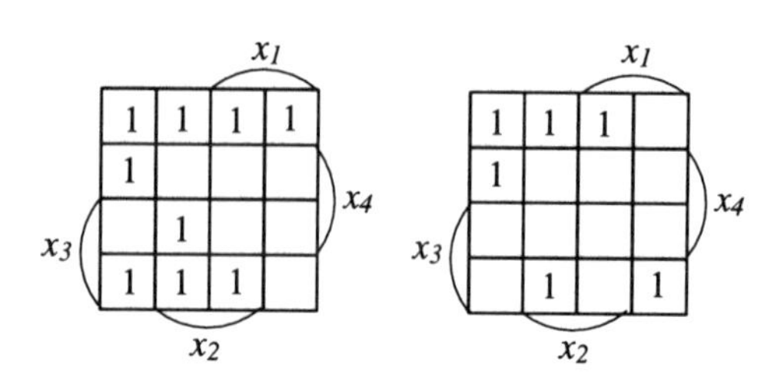

Figure 12.16 Representation using 1-MUX.

12.13 Design a 1-MUX network that realizes two functions shown in Fig. 12.16.

12.14 Prove Lemma 12.1.

12.15 By using 5-input 3-output ROMs, design a 4-bit adder with a carry input (a network with 9-input and 5-output).

12.16 By using 8-input 8-output ROMs, and 4-bit adders with a carry input (networks with 9-input and 5-output), design an 8-bit multiplier (MLP8).

12.17 Let $X = (x_7, x_6, \ldots, x_1, x_0)$ and $Y = (y_7, y_6, \ldots, y_1, y_0)$. Then, an 8-bit comparator works as follows:

$$\begin{aligned} z_1 &= 1 \text{ iff } X > Y, \\ z_2 &= 1 \text{ iff } X = Y, \text{ and} \\ z_3 &= 1 \text{ iff } X < Y. \end{aligned}$$

By using 5-input 3-output ROMs, design the 8-bit comparator.

12.18 Show that an n-bit comparator is realized by a PLA with 2-bit decoders using $2n + 1$ products.

13

LOGIC DESIGN USING EXORS

The logic design methods described in the preceding chapters are based on ANDs, ORs, and NOTs. However, for arithmetic circuits, error correcting circuits, and telecommunication circuits, effective use of EXOR gates can reduce the complexity of networks. In this chapter, we will consider AND-EXOR two-level expressions and their simplifications. These methods will be a base for logic synthesis using EXOR gates. In addition, this chapter introduces Boolean difference and fault detection, as applications of EXOR operations.

13.1 CLASSIFICATION OF AND-EXOR EXPRESSIONS

In this section, we will define various classes of **AND-EXOR expressions**, and show relations among them.

Theorem 13.1 *(Expansion theorem) An arbitrary logic function* $f(x_1, x_2, \ldots, x_n)$ *can be expanded as*

$$f(x_1, x_2, \ldots, x_n) = f_0 \oplus x_1 f_2 \quad (13.1)$$

$$f(x_1, x_2, \ldots, x_n) = \bar{x}_1 f_2 \oplus f_1 \quad (13.2)$$

$$f(x_1, x_2, \ldots, x_n) = \bar{x}_1 f_0 \oplus x_1 f_1, \quad (13.3)$$

where $f_0 = f(0, x_2, \ldots, x_n)$*,* $f_1 = f(1, x_2, \ldots, x_n)$*, and* $f_2 = f_0 \oplus f_1$*.*

Equations (13.1)–(13.3) are the **positive Davio expansion**, the **negative Davio expansion**, and the **Shannon expansion**, respectively. The names of the expansions (13.1) and (13.2) come from Prof. M. Davio who did pioneering work on AND-EXOR logical expressions.

Definition 13.1 *By expanding the function f using (13.1) recursively, we have a logical expression with only un-complemented literals:*

$$a_0 \oplus a_1x_1 \oplus \cdots \oplus a_nx_n \oplus a_{12}x_1x_2 \oplus a_{13}x_1x_3 \oplus \cdots$$
$$\oplus a_{n-1,n}x_{n-1}x_n \oplus \cdots\cdots \oplus a_{12\cdots n}x_1x_2\cdots x_n. \qquad (13.4)$$

This is a **positive polarity Reed-Muller expression** *(***PPRM***).*

In Section 3.5, we called it a Reed-Muller canonical expression. For a logic function, the PPRM is unique and is a canonical expression. Thus, no minimization problem exists. The average number of products in PPRMs for n-variable functions is 2^{n-1}.

Example 13.1 Let us represent the function $f = \bar{x}_1\bar{x}_2\bar{x}_3$ by a PPRM.
By substituting $\bar{x}_1 = x_1 \oplus 1$, $\bar{x}_2 = x_2 \oplus 1$, $\bar{x}_3 = x_3 \oplus 1$, we have

$$\begin{aligned} f &= (x_1 \oplus 1)(x_2 \oplus 1)(x_3 \oplus 1) \\ &= 1 \oplus x_1 \oplus x_2 \oplus x_3 \oplus x_1x_2 \oplus x_2x_3 \oplus x_1x_3 \oplus x_1x_2x_3. \end{aligned}$$

Note that this expression uses un-complemented literals only. ▮

In general, $\bar{x}_1\bar{x}_2\cdots\bar{x}_n$ requires 2^n products in a PPRM.

Definition 13.2 *By applying the positive Davio expansion or the negative Davio expansion to the given function f, we have a logical expression which has similar form as a PPRM. In this case, assume that we can use either un-complemented literals or complemented literals but not both for each variable. Such logical expression is a* **fixed polarity Reed-Muller expression** *(***FPRM***).*

For an n-variable function, there are at most 2^n different FPRMs. The minimization problem is to find one with the minimum products among 2^n possible FPRMs. Among the minimization algorithms of FPRMs, there is one that requires $O(3^n)$ memory storage and computation time.

Example 13.2 Let us represent the function $f = x_1x_2x_3x_4 \vee \bar{x}_1\bar{x}_2\bar{x}_3\bar{x}_4$ by an FPRM.
From the Lemma 3.2, f can be represented as $f = x_1x_2x_3x_4 \oplus \bar{x}_1\bar{x}_2\bar{x}_3\bar{x}_4$. By applying the positive Davio expansion to x_1 and x_2, and the negative Davio

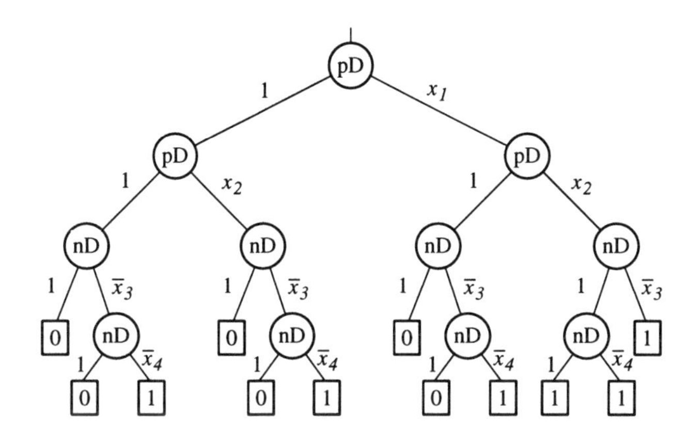

Figure 13.1 Expansion tree for FPRM of $x_1x_2x_3x_4 \vee \bar{x}_1\bar{x}_2\bar{x}_3\bar{x}_4$.

expansion to x_3 and x_4, we have the expression where x_1 and x_2 appear as uncomplemented literals, and x_3 and x_4 appear as complemented literals. Thus, by substituting $\bar{x}_1 = x_1 \oplus 1$, $\bar{x}_2 = x_2 \oplus 1$, $x_3 = \bar{x}_3 \oplus 1$, and $x_4 = \bar{x}_4 \oplus 1$ to f, we have

$$\begin{aligned} f &= x_1x_2(\bar{x}_3 \oplus 1)(\bar{x}_4 \oplus 1) \oplus (x_1 \oplus 1)(x_2 \oplus 1)\bar{x}_3\bar{x}_4 \\ &= x_1x_2(1 \oplus \bar{x}_3 \oplus \bar{x}_4 \oplus \bar{x}_3\bar{x}_4) \oplus (1 \oplus x_1 \oplus x_2 \oplus x_1x_2)\bar{x}_3\bar{x}_4 \\ &= x_1x_2 \oplus x_1x_2\bar{x}_3 \oplus x_1x_2\bar{x}_4 \oplus \bar{x}_3\bar{x}_4 \oplus x_1\bar{x}_3\bar{x}_4 \oplus x_2\bar{x}_3\bar{x}_4. \end{aligned}$$

Note that the last expression is an FPRM. Fig. 13.1 shows the expansion tree. In the tree, the nodes with pD denote the positive Davio expansions, and the nodes with nD denote the negative Davio expansions. In each path from the root node to the constant 1, the logical products of the labels in the path corresponds to a product term in an FPRM. Note that each variable uses the same type of expansion in an FPRM. ∎

In general, $x_1x_2 \cdots x_n \vee \bar{x}_1\bar{x}_2 \cdots \bar{x}_n$ $(n = 2r)$ requires $2^{r+1} - 2$ products in an FPRM.

Definition 13.3 *When the given function f is expanded by either the positive Davio expansion, the negative Davio expansion, or the Shannon expansion, we have a logical expression which is a generalization of FPRMs. Such an expression is a* **Kronecker expression** *(***KRO***).*

For an n-variable function, there are at most 3^n different KROs.

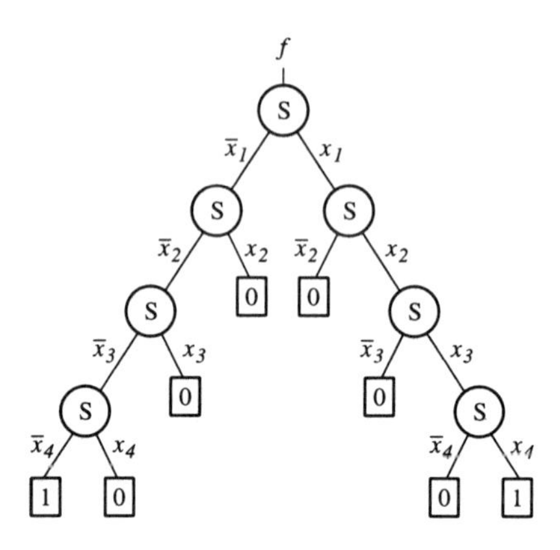

Figure 13.2 Expansion tree of KRO for $x_1x_2x_3x_4 \vee \bar{x}_1\bar{x}_2\bar{x}_3\bar{x}_4$.

Example 13.3 Let us represent the function $f = x_1x_2x_3x_4 \vee \bar{x}_1\bar{x}_2\bar{x}_3\bar{x}_4$ by a KRO. By applying the Shannon expansions for all the variables, f is represented as $f = x_1x_2x_3x_4 \oplus \bar{x}_1\bar{x}_2\bar{x}_3\bar{x}_4$. Fig. 13.2 is the expansion tree for this function. In this tree, the nodes with S denote the Shannon expansion. There are two paths from the root node to the constant 1 nodes. They correspond to $\bar{x}_1\bar{x}_2\bar{x}_3\bar{x}_4$ and $x_1x_2x_3x_4$. Note that all the variables use the same type of expansions. ∎

Definition 13.4 *For a given function f, if f is expanded by the positive Davio expansion or the negative Davio expansion, then we have two sub-functions. For each of these sub-functions, if we expand it by the positive Davio expansion or the negative Davio expansion, where we may use different expansion method for each sub-function. Then, we have a logical expression which is a generalization of an FPRM. Such an expression is a* **pseudo Reed-Muller expression (PSDRM)**.

In a PSDRM, both the complemented and un-complemented literals may appear at the same time. For a given order of variables for expansion, an n-variable function have at most 2^{2^n-1} different PSDRMs.

Example 13.4 Let us represent the function $f = x_1x_2x_3x_4 \vee \bar{x}_1\bar{x}_2\bar{x}_3\bar{x}_4$ by a PSDRM. First, if we use the positive Davio expansion with respect to x_1, we have

$$f = 1 \cdot \bar{x}_2\bar{x}_3\bar{x}_4 \oplus x_1(x_2x_3x_4 \oplus \bar{x}_2\bar{x}_3\bar{x}_4).$$

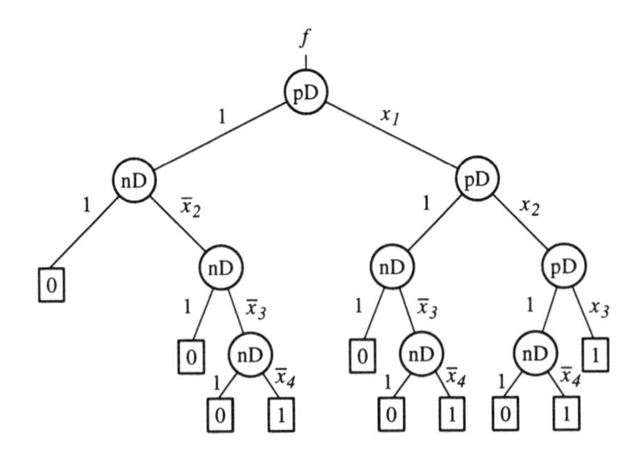

Figure 13.3 Expansion tree of PSDKRO for $x_1x_2x_3x_4 \vee \bar{x}_1\bar{x}_2\bar{x}_3\bar{x}_4$.

$f_0 = \bar{x}_2\bar{x}_3\bar{x}_4$ is a PSDRM, where all the variables are expanded with the negative Davio expansion. Next, consider the expansion of $f_2 = x_2x_3x_4 \oplus \bar{x}_2\bar{x}_3\bar{x}_4$. Since this expression has a similar form to the original one, we can expand it as

$$f_2 = 1 \cdot \bar{x}_3\bar{x}_4 \oplus x_2(x_3x_4 \oplus \bar{x}_3\bar{x}_4) = 1 \cdot f_{20} \oplus x_2 f_{21}.$$

Clearly, $f_{20} = \bar{x}_3\bar{x}_4$ is a PSDRM. Then, we can expand f_{21} as $f_{21} = x_3x_4 \oplus \bar{x}_3\bar{x}_4 = 1 \cdot \bar{x}_4 \oplus x_3 \cdot 1$. Therefore, the expansion is

$$\begin{aligned} f &= 1 \cdot \bar{x}_2\bar{x}_3\bar{x}_4 \oplus x_1(1 \cdot \bar{x}_3\bar{x}_4 \oplus x_2(1 \cdot \bar{x}_4 \oplus x_3 \cdot 1)) \\ &= \bar{x}_2\bar{x}_3\bar{x}_4 \oplus x_1\bar{x}_3\bar{x}_4 \oplus x_1x_2\bar{x}_4 \oplus x_1x_2x_3. \end{aligned}$$

Fig. 13.3 is the expansion tree for this function. Note that x_2 and x_3 use both pD and nD expansions. ∎

In general, $x_1x_2 \cdots x_n \vee \bar{x}_1\bar{x}_2 \cdots \bar{x}_n$ requires n products in a PSDRM.

Definition 13.5 *For a given function f, if f is expanded by the positive Davio expansion, the negative Davio expansion or the Shannon expansion, we have two sub-functions. For each of these sub-functions, suppose that we can use any one of the three expansions, then we have a KRO. In this case, if we can choose the expansion method independently for each sub-function, we have an expression which is a generalization of KRO. Such an expression is a* **pseudo Kronecker expression** *(***PSDKRO***).*

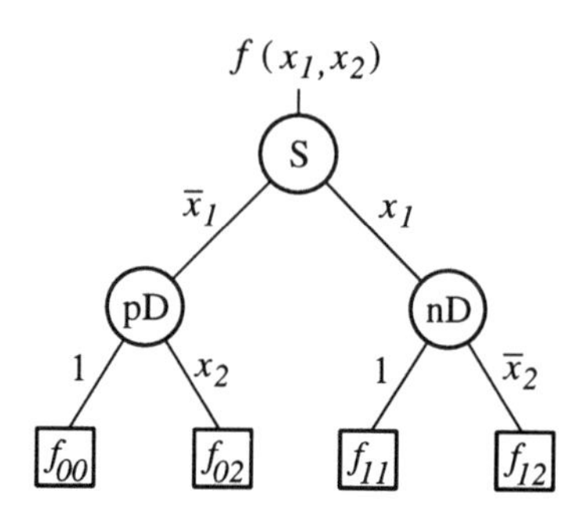

Figure 13.4 Expansion tree of a two-variable function by a PSDKRO.

In a PSDKRO, both complemented and un-complemented literals of the same variable may appear. If we fix the order of the expansion of the variables, there are at most 3^{2^n-1} different PSDKROs for an n-variable function. In general, the number of the products in the minimum PSDKRO depends on the order of the expansion.

Example 13.5 Let us represent a two-variable function $f(x_1, x_2)$ by a PSD-KRO.
For example, if we use the Shannon expansion with respect to x_1, we have

$$f(x_1, x_2) = \bar{x}_1 f_0(x_2) \oplus x_1 f_1(x_2).$$

For f_0, if we use the positive Davio expansion, we have

$$f_0(x_2) = 1 \cdot f_{00} \oplus x_2 f_{02}.$$

For f_1, if we use the negative Davio expansion, we have

$$f_1(x_2) = 1 \cdot f_{11} \oplus \bar{x}_2 f_{12}.$$

From these, we have the expansion for f as follows:

$$\begin{aligned} f(x_1, x_2) &= \bar{x}_1(1 \cdot f_{00} \oplus x_2 f_{02}) \oplus x_1(1 \cdot f_{11} \oplus \bar{x}_2 f_{12}) \\ &= f_{00}\bar{x}_1 \oplus f_{02}\bar{x}_1 x_2 \oplus f_{11} x_1 \oplus f_{12} x_1 \bar{x}_2, \end{aligned}$$

where f_{ij} is a constant 0 or 1. Fig. 13.4 is the expansion tree for the function. In a PSDKRO, three different types of expansions S, pD, and nD may appear at the same time for the same variables. ∎

Definition 13.6 *In an expansion (13.4) for a PPRM, suppose that the polarity of each literal at each product may be chosen arbitrary. In this case, we have the logical expression which is generalization of a PPRM. Such an expression is a* **generalized Reed-Muller expression** *(***GRM***).*

There are at most $2^{n2^{n-1}}$ different GRMs for an n-variable function. We do not know an efficient minimization algorithm for GRMs. Note that some researchers use GRMs to mean different expressions, so we should be careful not to confuse them.

Example 13.6 Let us represent the two-variable function $f(x_1, x_2)$ by a GRM. The PPRM for f is as follows:

$$f = a_0 \oplus a_1 x_1 \oplus a_2 x_2 \oplus a_{12} x_1 x_2.$$

For example, by complementing the polarities of the variables in $x_1 x_2$, we have

$$f = b_0 \oplus b_1 x_1 \oplus b_2 x_2 \oplus b_{12} \bar{x}_1 \bar{x}_2.$$

This is a GRM. However, it is not a PSDKRO. Because, if we set $b_0 = 0$ and $b_1 = b_2 = b_{12} = 1$, we have

$$f = x_1 \oplus x_2 \oplus \bar{x}_1 \bar{x}_2.$$

It is not a PSDKRO. Since, we cannot represent it by combining the positive Davio expansion, the negative Davio expansion, and the Shannon expansion. ∎

Definition 13.7 *A logical expression that combines arbitrary product terms by EXORs is an* **exclusive-OR sum-of-products expression** *(***ESOP***).*

An ESOP is the most general AND-EXOR logical expression. For an n-variable function, there are at most 3^{tn} different ESOPs with t products. No efficient minimization algorithm is known. So, heuristic simplification algorithms are developed.

Example 13.7 There are 9 ESOPs for two-variable functions with one product: $\bar{x}\bar{y}$, $\bar{x}y$, $\bar{x} \cdot 1$, $x\bar{y}$, xy, $x \cdot 1$, $1 \cdot \bar{y}$, $1 \cdot y$, 1. ∎

Theorem 13.2 *An arbitrary n-variable symmetric function ($n = 2r$) is represented by an ESOP with at most $2 \cdot 3^{r-1}$ products.*

(Proof) We use the mathematical induction.
When $r = 1$. There are 8 different two-variable symmetric function: 0, 1, x_1x_2, $\bar{x}_1\bar{x}_2$, $x_1 \oplus x_2$, $x_1 \oplus \bar{x}_2$, $x_1 \vee x_2 = 1 \oplus \bar{x}_1\bar{x}_2$, $\bar{x}_1 \vee \bar{x}_2 = 1 \oplus x_1x_2$. Each of them can be represented by an ESOP with at most two products.
When $r \geq 2$. Suppose that an arbitrary $(n-2)$-variable symmetric function $(n = 2r)$ is represented by an ESOP with at most $2 \cdot 3^{r-2}$ products. An arbitrary n-variable symmetric function can be represented by

$$f(x_1, x_2, \ldots, x_n) = \bar{x}_1\bar{x}_2 f_{00} \oplus \bar{x}_1 x_2 f_{01} \oplus x_1\bar{x}_2 f_{10} \oplus x_1x_2 f_{11},$$

where

$$\begin{aligned} f_{00} &= f(0,0,x_3,\ldots,x_n), \\ f_{01} &= f(0,1,x_3,\ldots,x_n), \\ f_{10} &= f(1,0,x_3,\ldots,x_n), \text{ and} \\ f_{11} &= f(1,1,x_3,\ldots,x_n) \end{aligned}$$

are $(n-2)$-variable symmetric functions. Since f is symmetric, we have $f_{01} = f_{10}$. Thus, we have

$$f(x_1, x_2, \ldots, x_n) = \bar{x}_1\bar{x}_2 f_{00} \oplus (x_1 \oplus x_2) f_{01} \oplus x_1x_2 f_{11}.$$

By assigning $x_1 \oplus x_2 = 1 \oplus x_1x_2 \oplus \bar{x}_1\bar{x}_2$ to the above expression, we have

$$f(x_1, x_2, \ldots, x_n) = \bar{x}_1\bar{x}_2(f_{00} \oplus f_{01}) \oplus f_{01} \oplus x_1x_2(f_{01} \oplus f_{11}).$$

f_{00}, f_{01}, f_{11} are $(n-2)$-variable symmetric functions, and $f_{00} \oplus f_{01}$ and $f_{01} \oplus f_{11}$ are also symmetric functions (Theorem 5.11). Note that from the hypothesis of the mathematical induction, each sub-function can be represented by an ESOP with at most $2 \cdot 3^{r-2}$ products. Thus, the number of products in the left-hand side of the above equation is at most

$$3 \cdot (2 \cdot 3^{r-2}) = 2 \cdot 3^{r-1}.$$

Form the above discussions, we have the theorem. □

Relations among Various Classes

Theorem 13.3 *Suppose that $\mathcal{PPRM}$, $\mathcal{FPRM}$, $\mathcal{PSDRM}$, $\mathcal{PSDKRO}$, $\mathcal{GRM}$, and $\mathcal{ESOP}$ represent the set of all PPRMs, FPRMs, PSDRMs, PSDKROs,*

GRMs, and ESOPs, respectively. Then, we have the following relations:
(1) $\mathcal{PPRM} \subset \mathcal{FPRM}$
(2) $\mathcal{FPRM} \subset \mathcal{PSDRM}$
(3) $\mathcal{FPRM} \subset \mathcal{KRO}$
(4) $\mathcal{KRO} \subset \mathcal{PSDKRO}$
(5) $\mathcal{PSDRM} \subset \mathcal{PSDKRO}$
(6) $\mathcal{PSDRM} \subset \mathcal{GRM}$

(Proof) By definitions, (1)–(5) clearly hold. If we consider an expansion of a PSDRM, then it is also a GRM. Thus, (6) holds. □

Example 13.8

$xy \oplus yz \oplus zx$	PPRM. (All the literals are positive.)
$x\bar{y} \oplus \bar{y}z \oplus zx$	FPRM. It is not a PPRM. (x and z have positive literals, y has negative literals.)
$xy \oplus \bar{y}z \oplus \bar{z}x$	PSDRM. It is not an FPRM. (y and z have positive and negative literals.)
$\bar{x} \oplus xy \oplus x\bar{y}$	PSDKRO. It is not a KRO. (It cannot be represented by a KRO.)
$\bar{x} \oplus xy \oplus x\bar{y}$	PSDKRO. It is not a PSDRM. (y has three different expansions.)
$x \oplus y \oplus \bar{x}\bar{y}$	GRM. It is not a PSDKRO. (It cannot be represented by a PSDKRO.)
$xyz \oplus \bar{x}\bar{y}\bar{z}$	KRO. It is not a GRM. (x, y, and z have the positive and the negative literals, It contains two product terms consisting of the same set of variables.)
$x \oplus y \oplus xy \oplus \bar{x}\bar{y}$	ESOP. It is not a GRM nor a PSDKRO.

∎

Fig. 13.5 shows the relation of Theorem 13.3 and Example 13.8.

Complexity for Various Logic Functions

Arbitrary product terms combined with ORs is a sum-of-products expression (SOP). To represent $x_1 \oplus x_2 \oplus \cdots \oplus x_n$, a PPRM requires n products, and an SOP requires 2^{n-1} products. To represent $\bar{x}_1\bar{x}_2\cdots\bar{x}_n$, a PPRM requires 2^n products, but other expressions require only one product. To represent $x_1x_2\cdots x_n \vee \bar{x}_1\bar{x}_2\cdots\bar{x}_n$ $(n = 2r)$, a PPRM requires $2^n - 1$ products, an FPRM requires $2^{r+1} - 2$ products, a PSDRM requires n products. Other expressions require only two products. To represent $x_1x_2 \vee x_3x_4 \vee \cdots \vee x_{n-1}x_n$ $(n = 2r)$, an SOP

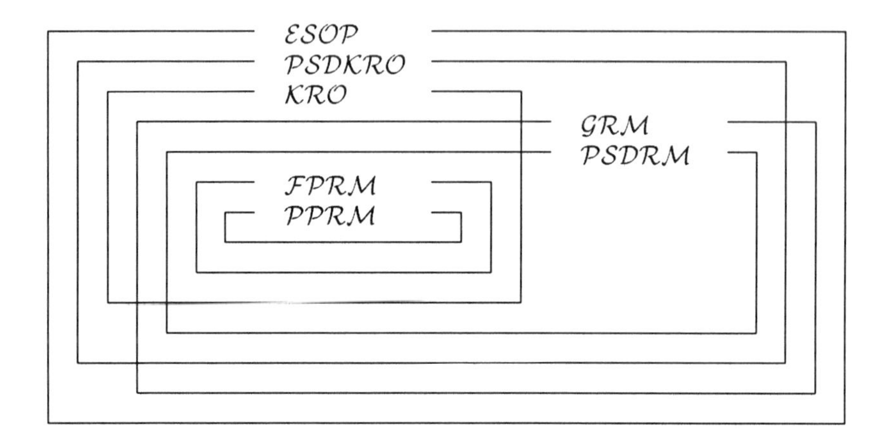

Figure 13.5 Relations among various classes of AND-EXOR expressions.

Table 13.1 Number of 4-variable functions that require t products.

t	PPRM	FPRM	KRO	PSD RM	PSD KRO	GRM	ESOP	SOP
0	1	1	1	1	1	1	1	1
1	16	81	81	81	81	81	81	81
2	120	836	2268	2212	2268	2212	2268	1804
3	560	3496	8424	19160	21384	20856	21744	13472
4	1820	8878	15174	35150	36918	37818	37530	28904
5	4368	17884	19260	7404	3564	4512	3888	17032
6	8008	20152	19440	1480	1296	56	24	3704
7	11440	11600	864	24	0			512
8	12870	2336	0	24	24			26
9	11440	240	0					
10	8008	32	24					
11	4368							
12	1820							
13	560							
14	120							
15	16							
16	1							
av	8.00	5.50	4.73	3.80	3.70	3.68	3.66	4.13

av : average

Table 13.2 Number of products for arithmetic functions.

Data Name	$\lvert f\rvert$	PP RM	FP RM	KRO	PSD RM	PSD KRO	GRM	ESOP	SOP
ADR4	255	34	34	34	34	34	34	31	75
LOG8	255	253	193	171	149	119	105	96	123
MLP4	225	97	97	97	82	76	71	61	121
NRM4	255	216	185	157	140	97	96	71	120
RDM8	255	56	56	56	41	38	31	31	76
ROT8	255	225	118	83	74	43	51	35	57
SQR8	255	168	168	168	158	136	121	112	180
WGT8	255	107	107	107	107	107	107	54	255
SYM9	420	210	173	173	127	90	126	52	84

$\lvert f\rvert$ denotes the number of input combinations that produce non-zero outputs.

Table 13.3 Sufficient numbers of products to realize n-variable functions ($n = 2r$).

Function	PP RM	FP RM	KRO	PSD RM	PSD KRO	GRM	ESOP	SOP
$x_1 \oplus x_2 \oplus \cdots \oplus x_n$	n	n	n	n	n	n	n	2^{n-1}
$\bar{x}_1\bar{x}_2\cdots\bar{x}_n$	2^n	1	1	1	1	1	1	1
$x_1x_2\cdots x_n \vee \bar{x}_1\bar{x}_2\cdots\bar{x}_n$	2^n-1	$2^{r+1}-2$	2	n	2	n	2	2
$x_1x_2 \vee \cdots \vee x_{n-1}x_n$	2^r-1	2^r-1	2^r-1	2^r-1	2^r-1	2^r-1	2^r-1	r
n-bit adder	$2^{n+1}+n-2$					$2^{n+1}+n-2$	$2^{n+1}-1$	$6\cdot 2^n-4n-5$

requires r products, but an AND-EXOR expressions requires $2^r - 1$ products. To represent an n-bit adder, $2^{n+1} - 1$ products are sufficient for an ESOP, $2^{n+1} + n - 2$ products are sufficient for other AND-EXOR expressions, and $6 \cdot 2^n - 4n - 5$ products are sufficient for an SOP. Tables 13.1–13.3 show the number of products needed to represent various functions by various expressions.

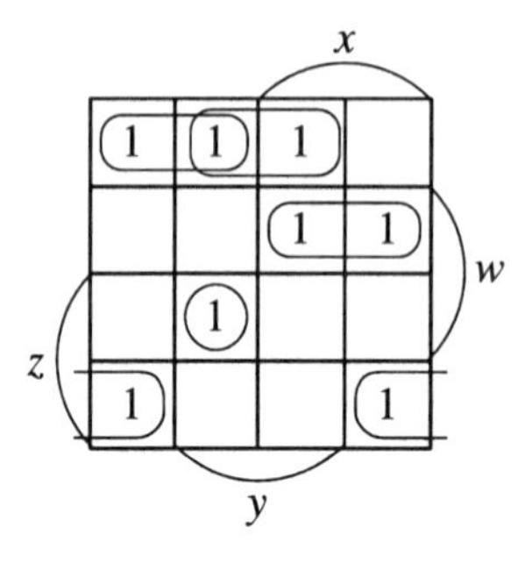

Figure 13.6 Representation using SOP.

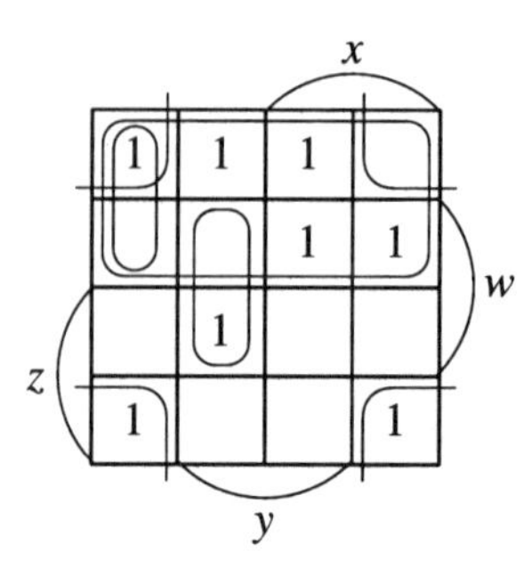

Figure 13.7 Representation using ESOP.

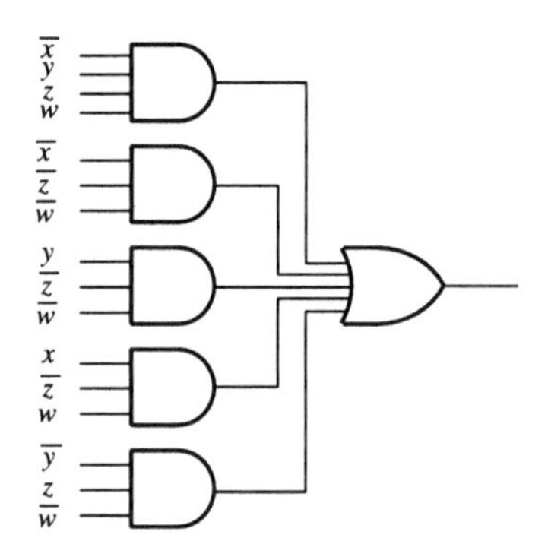

Figure 13.8 AND-OR realization.

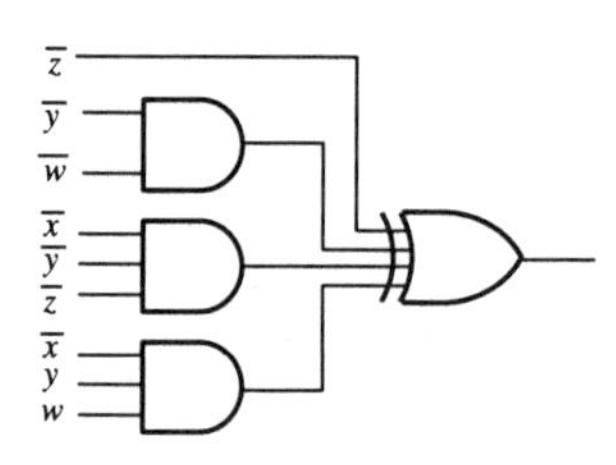

Figure 13.9 AND-EXOR realization.

13.2 SIMPLIFICATION OF ESOPS

The minimization of an SOP is equivalent to the problem of covering all the 1-cells in the Karnaugh map at least once by using a minimum number of loops. On the other hand, the minimization of an ESOP is equivalent to the problem of covering all the 1-cells an odd number of times and all the 0-cells in even number of times by using minimum number of loops. For example, consider the function shown in Fig. 13.6. An SOP requires five products. On the other hand, the ESOP shown in Fig. 13.7 requires only four products. These expressions are realized by two-level logic networks shown in Fig. 13.8 and Fig. 13.9, respectively. Note that the AND-EXOR realization requires fewer gates and fewer connections.

We have no efficient minimization algorithm for ESOPs. In this section, we will show **EXMIN2**, a heuristic algorithm using simplification rules. In EXMIN2, the following rules are iteratively used to simplify ESOPs:

1)	X_MERGE	:	$x \oplus x \rightarrow 0$, $x \oplus \bar{x} \rightarrow 1$, $x \oplus 1 \rightarrow \bar{x}$, $\bar{x} \oplus 1 \rightarrow x$
2)	RESHAPE	:	$xy \oplus \bar{y} \rightarrow \bar{x}\bar{y} \oplus x$
3)	DUAL_COMPLEMENT	:	$x \oplus y \rightarrow \bar{x} \oplus \bar{y}$
4)	X_EXPAND_1	:	$x\bar{y} \oplus \bar{x}y \rightarrow x \oplus y$
5)	X_EXPAND_2	:	$xy \oplus \bar{y} \rightarrow 1 \oplus \bar{x}y$
6)	X_REDUCE_1	:	$x \oplus y \rightarrow x\bar{y} \oplus \bar{x}y$
7)	X_REDUCE_2	:	$1 \oplus \bar{x}y \rightarrow xy \oplus \bar{y}$
8)	SPLIT	:	$1 \rightarrow x \oplus \bar{x}$

Example 13.9 Let us simplify the ESOP for the 4-variable function shown in Fig. 13.6. First, as shown in Fig. 13.10(a), transform the expression into a **disjoint sum-of-products expression** (DSOP), so that the product terms have no common minterms. In the DSOP, the operators $\vee$ can be replaced with $\oplus$ without changing the function represented by the expression.

1) Since, X_MERGE cannot be applied in Fig. (a), we apply the RESHAPE operation to the pair of product terms (①, ②) to obtain Fig. (b).
2) For the pair of products (③, ④) in Fig. (b), we apply the X_EXPAND_2 operation to obtain Fig. (c).
3) For the pair of products (⑤, ⑥) in Fig. (c), we apply the X_EXPAND_2 operation to obtain Fig. (d). Note that 0-cells $x\bar{y}\bar{z}\bar{w}$, $\bar{x}\bar{y}\bar{z}w$ and $\bar{x}y\bar{z}w$ are covered twice by loops.
4) For the pair of products (⑦, ⑧) in Fig. (d), we apply the X_EXPAND_2 operation to obtain Fig. (e). Note that the 1-cell $\bar{x}\bar{y}\bar{z}\bar{w}$ is covered by loops three times.
5) For the pair of products (①, ②) in Fig. (e), we apply the X_EXPAND_2 operation to obtain Fig. (f). This is a very complicated map.
6) For the pair of products (③, ④) in Fig. (f), we apply the X_MERGE operation. Since $\bar{x}\bar{z} \oplus \bar{z} = (\bar{x} \oplus 1)\bar{z} = x\bar{z}$, we have Fig. (g). Note that the number of loops is reduced to four.
7) For the pair of products (⑤, ⑥) in Fig. (g), we apply the X_EXPAND_1 operation to obtain Fig. (h).
8) For the pair of products (⑦, ⑧) in Fig. (h), we apply the X_EXPAND_2 operation to obtain the ESOP shown in Fig. 13.7. Note that, we cannot reduce the number of products nor number of literals any more. ∎

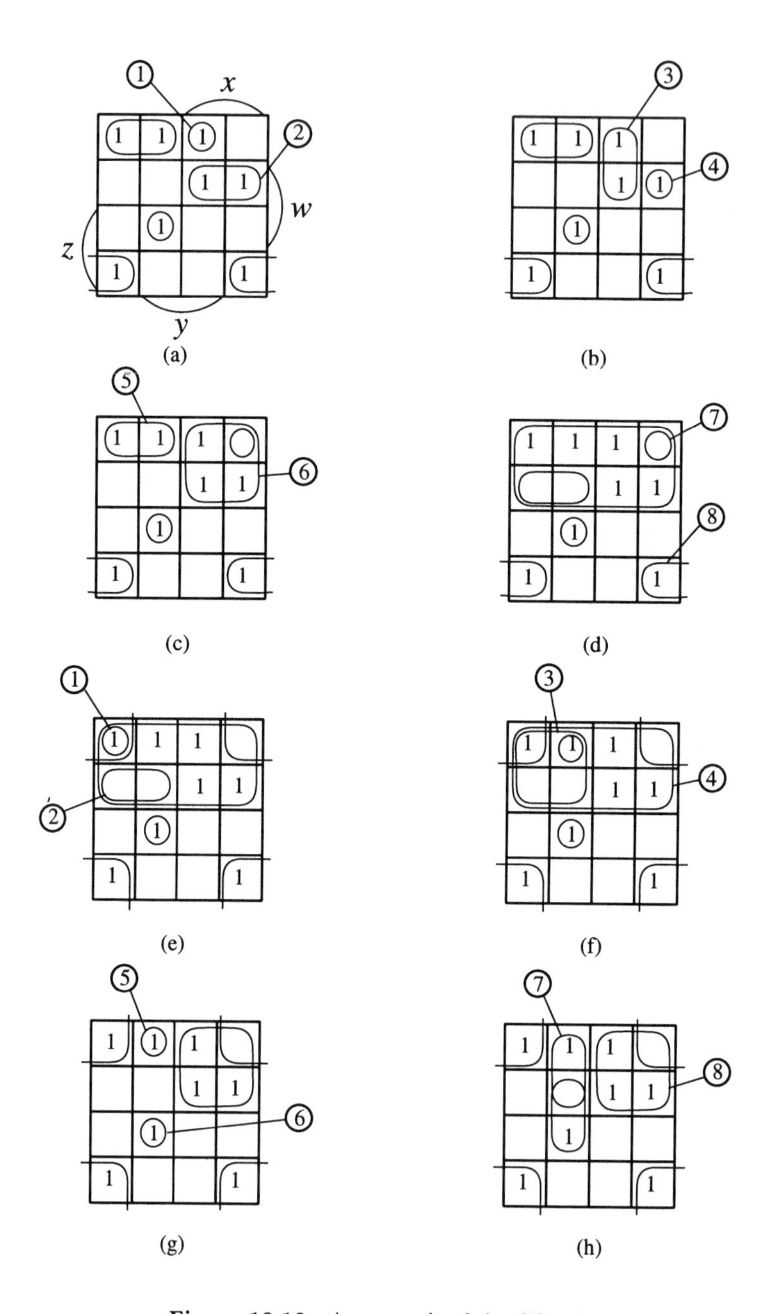

Figure 13.10 An example of simplification.

13.3 FAULT DETECTION AND BOOLEAN DIFFERENCE

Definition 13.8 *Let $f(x_1, x_2, \ldots, x_n)$ be the output function of a fault-free network, and let $f_\alpha(x_1, x_2, \ldots, x_n)$ be the output function of the network with the fault α. Then, the* **fault difference function**, *which distinguishes the two functions, is*

$$h(x_1, x_2, \ldots, x_n) = f(x_1, x_2, \ldots, x_n) \oplus f_\alpha(x_1, x_2, \ldots, x_n).$$

For an input vector $\mathbf{a} = (a_1, a_2, \ldots, a_n)$ that makes $h = 1$, the values of f and f_α are different. Such an input vector $\mathbf{a}$ is a **test input** *for the fault α. A* **stuck-at fault** *is a fault that makes an input or the output terminal to be constant 0 or 1.*

Example 13.10 The network in Fig. 13.11 realizes the function $f = xyz \vee \bar{x}\bar{y}\bar{z}$. If the input terminal of the variable x of the AND gate has a stuck-at-1 fault, then the output function becomes $f_\alpha = 1 \cdot yz \vee \bar{x}\bar{y}\bar{z}$. In this case, the fault difference function is

$$h = (xyz \vee \bar{x}\bar{y}\bar{z}) \oplus (1 \cdot yz \vee \bar{x}\bar{y}\bar{z}).$$

By simplifying this expression, we have

$$h = (xyz \oplus \bar{x}\bar{y}\bar{z}) \oplus (yz \oplus \bar{x}\bar{y}\bar{z}) = xyz \oplus yz = \bar{x}yz.$$

This shows that if $(x, y, z) = (0, 1, 1)$, then the value of f is 0, and the value of f_α is 1. Thus, $(0, 1, 1)$ is a test input for x stuck-at-1. ∎

When the output function of the faulty network is equal to that of the fault-free network, the fault difference function is constantly equal to 0. Such a fault is a **redundant fault** and cannot be detected from the input or output terminals.

Definition 13.9 *Let $f(x_1, x_2, \ldots, x_n)$ be an n-variable logic function. The* **Boolean difference** *of the function f with respect to the variable x_i is*

$$\frac{df}{dx_i} = f(x_1, x_2, \ldots, \bar{x}_i, \ldots, x_n) \oplus f(x_1, x_2, \ldots, x_i, \ldots, x_n). \tag{13.5}$$

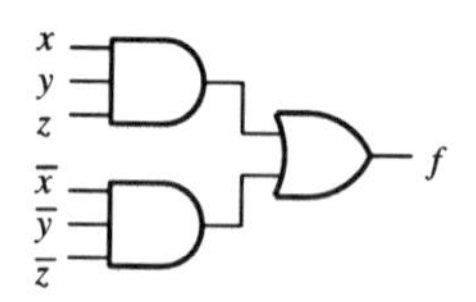

Figure 13.11 Fault model.

From the Shannon's expansion theorem, we have

$$\begin{aligned} f(x_1, x_2, \ldots, \bar{x}_i, \ldots, x_n) &= x_i f_i(0) \oplus \bar{x}_i f_i(1), \text{ and} \\ f(x_1, x_2, \ldots, x_i, \ldots, x_n) &= \bar{x}_i f_i(0) \oplus x_i f_i(1), \end{aligned}$$

where

$$\begin{aligned} f_i(0) &= f_i(x_1, x_2, \ldots, x_{i-1}, 0, x_{i+1}, \ldots, x_n), \text{ and} \\ f_i(1) &= f_i(x_1, x_2, \ldots, x_{i-1}, 1, x_{i+1}, \ldots, x_n). \end{aligned}$$

Therefore, the Boolean difference in (13.5) can be represented as

$$\begin{aligned} \frac{df}{dx_i} &= (x_i f_i(0) \oplus \bar{x}_i f_i(1)) \oplus (\bar{x}_i f_i(0) \oplus x_i f_i(1)) \\ &= f_i(0) \oplus f_i(1). \end{aligned}$$

If $\frac{df}{dx_i} = 0$, f does not depend on x_i. In this case, f is **degenerate.**

In a logic network with the output function $f(x_1, x_2, \ldots, x_n)$, consider a fault α where the input line x_i is stuck-at-0. The fault function f_α produced by the fault α is

$$f_\alpha = f(x_1, x_2, \ldots, x_{i-1}, 0, x_{i+1}, \ldots, x_n) = f_i(0).$$

Also, the fault difference function that detects the fault α is $f \oplus f_i(0)$. By modifying it, we have

$$\begin{aligned} f \oplus f_i(0) &= \bar{x}_i f_i(0) \oplus x_i f_i(1) \oplus f_i(0) \\ &= \bar{x}_i f_i(0) \oplus x_i f_i(1) \oplus (\bar{x}_i \oplus x_i) f_i(0) \\ &= x_i (f_i(0) \oplus f_i(1)) \\ &= x_i \frac{df}{dx_i}. \end{aligned}$$

Therefore, all tests for the stuck-at-0 fault in the input line x_i is represented by

$$\{\boldsymbol{a} \mid x_i \frac{df}{dx_i} = 1\}, \ \boldsymbol{a} \in B^n.$$

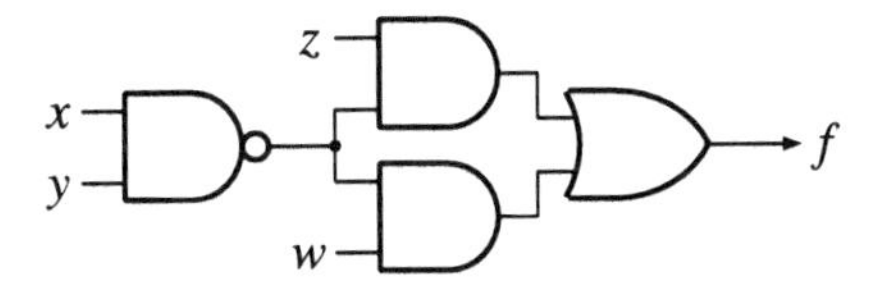

Figure 13.12 Derivation of test inputs.

Similarly, the tests for a stuck-at-1 fault in the input line x_i is represented by

$$\{\boldsymbol{a} \mid \bar{x}_i \frac{df}{dx_i} = 1\}, \ \boldsymbol{a} \in B^n.$$

Example 13.11 Let us obtain the test set for the stuck-at-0 fault in the input line x in Fig. 13.12. First, the output function f is represented as

$$f = \overline{(xy)}(z \vee w).$$

Next, by differentiating f with respect to x, we have

$$\begin{aligned} \frac{df}{dx} &= f(x=0) \oplus f(x=1) = (z \vee w) \oplus \bar{y}(z \vee w) \\ &= y(z \vee w). \end{aligned}$$

The expression showing the test set of a stuck-at-0 fault in x is

$$x\frac{df}{dx} = xy(z \vee w).$$

Thus, $(x, y, z, w) = (1,1,1,0)$, $(1,1,1,1)$, or $(1,1,0,1)$ are the test inputs for x stuck-at-0. ∎

The next example shows that Boolean difference is useful for the simplification of multi-level logic networks.

Example 13.12 In Fig. 13.13, the network A realizes the function $y = h(x_1, x_2, \ldots, x_n)$ and the network B realizes the function $f = z(x_1, x_2, \ldots, x_n, y)$. In this case, the condition that the output of the network B does not depend on the output of the network A is represented by $dz/dy = 0$. In other words, the inputs such that $z(|y) \oplus z(|\bar{y}) \oplus 1 = 1$ will not influence the value of the output of the network B. This is the Observability Don't Care stated in Section 11.7.2. ∎

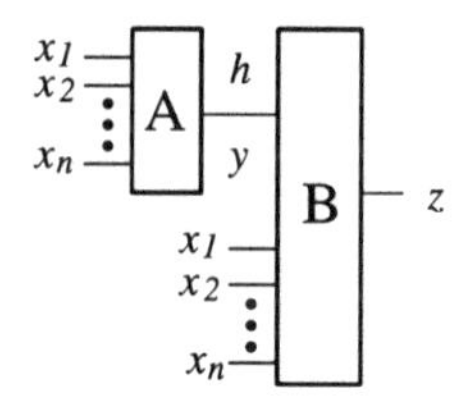

Figure 13.13

Bibliographical Notes

Detailed discussions on each topic can be found as follows: Classification of AND-EXORs [89, 149, 150, 266, 352, 358, 364]; PPRMs [213, 269, 323, 439]; FPRM minimization [25, 27, 107, 108, 232, 365, 415]; PSDKRO minimization [360]; KRO minimization [3, 233, 365]; GRM minimization [73, 78, 91, 368, 369]; ESOP heuristic minimization [35, 119, 123, 165, 163, 329, 357, 359, 336, 391, 403]; ESOP exact minimization [211, 266, 308, 312, 361]; other AND-EXOR expressions [203, 427]; EXOR multi-level logic networks [66, 375, 404]; AND-OR-EXOR networks [92, 93, 109, 362]; AND-EXOR test [317, 322, 369, 335, 378, 409]; textbooks on test [125, 131, 217, 318]; fault detection [6, 133, 242, 331, 332, 399, 432]; Boolean difference [1, 377, 400, 401]; adder test [67, 192]; multi-valued EXORs [110]; decision diagrams using EXORs [106, 203, 364, 371]; and spectral techniques [178, 197, 392, 414, 416].

Exercises

13.1 Prove that the PPRM for $\bar{x}_1\bar{x}_2\cdots\bar{x}_n$ requires 2^n products.

13.2 Consider the n-variable functions that have t products in their PPRMs. Show that there are $\binom{2^n}{t}$ such functions.

13.3 Prove that the average number of products in the PPRMs for n-variable functions is 2^{n-1}.

13.4 Prove that FPRM for $x_1x_2\cdots x_n \vee \bar{x}_1\bar{x}_2\cdots\bar{x}_n$ $(n = 2r)$ requires $2^{r+1} - 2$ products.

13.5 (M) Show that $x_1x_2 \vee x_3x_4 \vee \cdots \vee x_{n-1}x_n$ $(n = 2r)$ can be represented by a PPRM with $2^r - 1$ products.

13.6 Show that there are at most 2^{2^n-1} different PSDRMs for an n-variable function.

13.7 Show that there are at most 3^{2^n-1} different PSDKROs for an n-variable function.

13.8 Show that there are at most $2^{n2^{n-1}}$ different GRMs for an n-variable function.

13.9 Convert the following expressions into ESOPs, and then simplify:

$$\begin{aligned} F &= (1 \oplus xy)(1 \oplus zw) \oplus 1, \\ G &= \bar{x}y\bar{z}\bar{w} \oplus x\bar{z} \oplus \bar{x}w \oplus x\bar{y}zw \oplus z\bar{w}. \end{aligned}$$

13.10 By using the EXMIN2 rules, simplify the following ESOPs:

$$\begin{aligned} F &= \bar{x}\bar{y}\bar{z} \oplus \bar{x}y\bar{z} \oplus x\bar{y}\bar{z} \oplus xyz, \\ G &= \bar{x}\bar{z}\bar{w} \oplus xy\bar{z}\bar{w} \oplus x\bar{y}\bar{z}w \oplus xyzw \oplus x\bar{y}z. \end{aligned}$$

13.11 (M) The **degree** of an AND-EXOR logical expression is the maximum number of literals in the products. Show that the following holds:

(a) An n-variable function $f = x_1x_2\cdots x_n \vee \bar{x}_1\bar{x}_2\cdots\bar{x}_n$ can be represented by a PSDRM with n products and has the degree $n-1$.

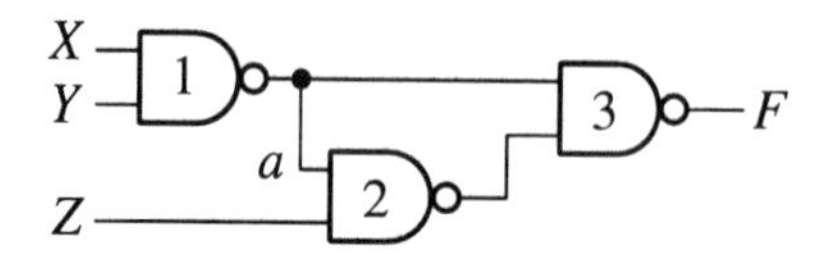

Figure 13.14 Detection of stack-at fault.

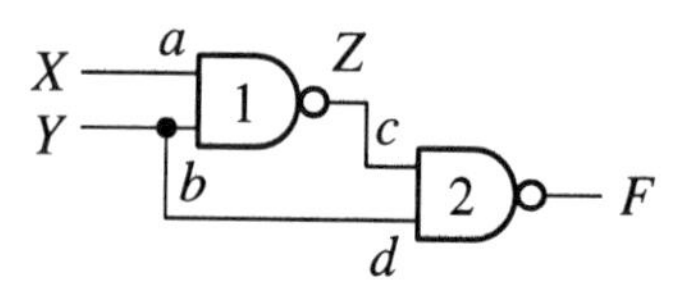

Figure 13.15 Detection of stack-at fault.

(b) Let the degree of an ESOP for a function f be k. Then, the degree of the PPRM for f is at most k.
(c) If the degree of a PPRM for the function f is k, then the degree of any ESOP for f is at least k.
(d) The number of n-variable functions that can be represented by ESOPs whose degree is at most k is $\Phi(n,k) = 2^{\eta(n,k)}$, where $\eta(n,k) = {}_nC_0 + {}_nC_1 + \cdots + {}_nC_k$.

13.12 (M) Consider the logic function f that has an odd number of true minterms. Show that any ESOP for f has at least one minterm. For example, the ESOPs for two-variable function $f = x \vee y$ are $x \oplus \bar{x}y$, $x\bar{y} \oplus y$, $1 \oplus \bar{x} \cdot \bar{y}$, etc. All the ESOPs for f contain minterms of two-variables.

13.13 (M) Let $A, B, C, D, E, F \subseteq P$, and $A \oplus B = (\overline{A} \cap B) \cup (A \cap \overline{B})$. Show the following equations:

$$\begin{aligned} X^A \cdot Y^B \oplus X^C \cdot Y^D &= X^{(A\oplus C)} \cdot Y^B \oplus X^C \cdot Y^{(B\oplus D)} \\ &= X^A \cdot Y^{(B\oplus D)} \oplus X^{(A\oplus C)} \cdot Y^D. \end{aligned}$$

$$\begin{aligned} & X^A \cdot Y^B \cdot Z^C \oplus X^D \cdot Y^E \cdot Z^F \\ = \ & X^{(A\oplus D)} \cdot Y^E \cdot Z^F \oplus X^A \cdot Y^{(B\oplus E)} \cdot Z^F \oplus X^A \cdot Y^B \cdot Z^{(C\oplus F)} \\ = \ & X^{(A\oplus D)} \cdot Y^B \cdot Z^C \oplus X^D \cdot Y^{(B\oplus E)} \cdot Z^C \oplus X^D \cdot Y^E \cdot Z^{(C\oplus F)}. \end{aligned}$$

13.14 Let f and g be functions of $x, y, z, \ldots$. Prove that the following rules for Boolean difference holds:

1. $\dfrac{d\bar{f}}{dx} = \dfrac{df}{dx}$,
2. $\dfrac{df}{dx} = \dfrac{df}{d\bar{x}}$,
3. $\dfrac{d(f\cdot g)}{dx} = f \cdot \dfrac{dg}{dx} \oplus g \cdot \dfrac{df}{dx} \oplus \dfrac{df}{dx} \cdot \dfrac{dg}{dx}$,

4. $\dfrac{d(f \oplus g)}{dx} = \dfrac{dg}{dx} \oplus \dfrac{df}{dx}$,
5. $\dfrac{d(f \vee g)}{dx} = \bar{f} \cdot \dfrac{dg}{dx} \oplus \bar{g} \cdot \dfrac{df}{dx} \oplus \dfrac{df}{dx} \cdot \dfrac{dg}{dx}$,
6. $\dfrac{d}{dx}(\dfrac{df}{dy}) = \dfrac{d}{dy}(\dfrac{df}{dx})$.

13.15 Let $f(x_1, x_2, \ldots, x_n) = x_1 \oplus x_2 \oplus \cdots \oplus x_n$. Show the following:

$$\frac{df}{dx_i} = 1.$$

Let $M(x, y, z) = xy \vee yz \vee zx$. Show the following:

$$\frac{dM}{dx} = y \oplus z.$$

13.16 In Fig. 13.14, are the stack-at faults in the input terminal a of the NAND gate 2 detectable? Discuss the cases for the stack-at-0 fault and the stack-at-1 fault.

13.17 In Fig. 13.15, are the stack-at faults in the terminal b of the NAND gate 1 detectable? Discuss the cases for the stack-at-0 fault and the stack-at-1 fault.

13.18 Let $X = (x_3, x_2, x_1, x_0)$ be a binary representation of an integer, $0 \leq X \leq 15$. Design an AND-EXOR two-level logic network that realizes $Y = X+1$, where $Y = (y_4, y_3, y_2, y_1, y_0)$.

13.19 Let $f = xy \vee \bar{x}\bar{z}$.
1. Represent f by a PPRM.
2. Represent f by a minimum FPRM.

13.20 Show that a function $f(x_1, x_2, \ldots, x_n)$ is linear iff

$$\frac{df}{dx_i} = a_i \quad (a_i \in \{0, 1\})$$

for all i.

13.21 Consider the symmetric function $f = S^5_{\{2,3\}}$.
1. Represent f by a PPRM.

2. Represent f by a minimum ESOP.

13.22 Prove the following:

$$(x_1 \vee \bar{x}_2)(x_2 \vee \bar{x}_3)\cdots(x_n \vee \bar{x}_1) = (x_1 \oplus \bar{x}_2)(x_2 \oplus \bar{x}_3)\cdots(x_n \oplus \bar{x}_1).$$

13.23 (D) Let $SB(n,k)$ be the n-variable function represented by the EXOR sum of all the products consisting of k positive literals:

$$\begin{aligned}
SB(n,0) &= 1, \\
SB(n,1) &= \sum\nolimits^{\oplus} x_i, \\
SB(n,2) &= \sum_{(i<j)}^{\oplus} x_i x_j, \\
SB(n,3) &= \sum_{(i<j<k)}^{\oplus} x_i x_j x_k, \\
\cdots\cdots\cdots & \\
SB(n,n) &= x_1 x_2 \cdots x_n.
\end{aligned}$$

Let $T(n,k)$ be the number of products in the minimum ESOP for $SB(n,k)$. Prove the following [352, 354]:

$$\begin{aligned}
T(n,k) &\leq \binom{n}{k}, \\
T(n,0) &= 1,\ T(n,1) = n,\ T(n,n) = 1, \\
T(n,k) &\leq T(n-1,k) + T(n-1,k-1), \\
T(n,k) &= T(n,n-k).
\end{aligned}$$

14

COMPLEXITY OF LOGIC NETWORKS

This chapter considers the number of gates necessary to realize given logic functions. The necessary number of gates depends on the functions and the realization method. In two-level logic networks, we assume that the gates have no fan-in restrictions. On the other hand, in multi-level logic networks, we assume that gates have fan-in restrictions. The number of gates necessary to realize an arbitrary n-variable function increases exponentially in both cases. The complexity analysis of logic networks is useful for estimating the sizes of networks.

14.1 COMPLEXITY OF TWO-LEVEL LOGIC NETWORKS

This section considers a realization of logic functions using AND-OR two-level logic networks. We assume the following: 1) AND and OR gates have no fan-in restrictions; and 2) the variables, their negation, constants 0's and 1's are available as inputs. First, we will obtain the sufficient number of products in a sum-of-products expression (SOP) to represent an arbitrary logic function.

Definition 14.1 *Let $t(F)$ be the number of products in an SOP F. Let $t(f)$ be the number of products in a minimum sum-of-products expression (MSOP) for the function f.*

Theorem 14.1 *Let f be an arbitrary n-variable logic function. Then, $t(f) \leq 2^{n-1}$.*

(Proof) By applying the Shannon expansions $(n-1)$ times to the n-variable function $f(x_1, x_2, \ldots, x_n)$, we have

$$\bigvee_{(a_2, a_3, \ldots, a_n)} f(x_1, a_2, \ldots, a_n) x_2^{a_2} x_3^{a_3} \cdots x_n^{a_n}, \tag{14.1}$$

where the logical sum is done over 2^{n-1} elements. Thus, the number of products is at most 2^{n-1}. □

Corollary 14.1 *An arbitrary logic function is realized by an AND-OR two-level logic network with at most* $2^{n-1}+1$ *gates.*

Definition 14.2 *Let* $t(n)$ *be the maximum number of products in MSOPs for all the* n*-variable functions. In other words, let* F_n *be the set of the* n*-variable functions, then* $t(n) = \max_{f \in F_n} t(f)$.

The following theorem shows the existence of an n-variable logic function that requires 2^{n-1} products in an MSOP.

Theorem 14.2 $t(n) = 2^{n-1}$.

(Proof) The parity function $f = x_1 \oplus x_2 \oplus \cdots \oplus x_n$ has 2^{n-1} true minterms, and all of them are essential. Thus, $t(f) = 2^{n-1}$. □

The proof of Theorem 14.2 explicitly shows the function that requires the maximum number of products in the SOP (i.e., the worst function). The average numbers of products to represent logic functions by SOPs are plotted in Fig. 12.3. As the value of n increases, the number of products to represent an n-variable function increases exponentially. The following theorem shows a lower bound on the number of products to represent "almost all functions," without showing functions explicitly.

Definition 14.3 *Let* $w(n)$ *be the number of* n*-variable functions that do not satisfy a certain property A. If* $w(n)/2^{2^n} \to 0$ *when* $n \to \infty$*, then "almost all functions have the property A."*

Theorem 14.3 *Almost all* n*-variable functions require at least* $(\log_3 2) \cdot (1-\varepsilon) \cdot 2^n/n$ *products in their MSOPs. Here,* ε *is an arbitrarily small positive number such that* $0 < \varepsilon < 1$.

(Proof) The number of different SOPs of n variables with at most N products is at most $3^{n \cdot N}$. In the following, we will show that if $N = (\log_3 2) \cdot (1-\varepsilon) \cdot 2^n/n$, then the fraction of logic functions that are realized with at most N products approaches zero when $n \to \infty$.

That is, we will show that $\gamma = 3^{n \cdot N}/2^{2^n} \to 0$ when $n \to \infty$.

By taking the logarithm of γ, we have

$$\begin{aligned} \log_2 \gamma &= (n \cdot N) \log_2 3 - 2^n \\ &= (\log_3 2) \cdot (1 - \varepsilon) \cdot 2^n \cdot (\log_2 3) - 2^n = -\varepsilon \cdot 2^n. \end{aligned}$$

Therefore, when $n \to \infty$, we have $\gamma \to 0$. Hence the theorem. □

In other words, almost all functions require $0.63(2^n)/n$ products in their MSOPs.

14.2 COMPLEXITY OF MULTI-LEVEL LOGIC NETWORKS

This section presents the number of gates necessary to realize an arbitrary logic function by using a multi-level logic network. To make the argument simple, we assume the following:

1) Only 1-MUXs (the multiplexer with one control variable in Section 12.2) are used.
2) There is no fan-out restrictions.
3) Constants 0, 1, and variables can be used as inputs.
4) Complemented variables are not available as inputs.

Also in the case of multi-level logic networks, the number of gates necessary to realize an arbitrary n-variable function increases exponentially. Even if we use other logic elements that have fan-in restrictions, the number of gates necessary to realize an arbitrary function increases exponentially. First, we will obtain the number of elements sufficient to realize arbitrary n-variable function. The proof is also an algorithm to realize the given function by using as few elements as possible.

Lemma 14.1 *An n-MUX ($n \geq 1$) is synthesized by using $2^n - 1$ copies of 1-MUXs.*

Lemma 14.2 *All the n-variable logic functions are realized simultaneously by using $2^{2^n} - 2$ copies of 1-MUXs.*

(Proof) The proof is by the mathematical induction on n.

0) For $n = 0$, no MUXs are needed since constant functions do not require any 1-MUX.
1) When $n = 1$, all the 1-variable functions x and $\bar{x}$ are realized by using two 1-MUXs.
2) Suppose that all the functions up to k variables are realized by using $2^{2^k} - 2$ 1-MUXs. Then, we will show that all the functions up to $k+1$ variables are realized by using $2^{2^{k+1}} - 2$ copies of 1-MUXs. An arbitrary $(k+1)$-variable function can be expanded as $f = \bar{x}_{k+1} f_0 \vee x_{k+1} f_1$, where f_0 and f_1 are k-variable functions. There are 2^{2^k} different k-variable functions. Among the $(k+1)$-variable functions, the number of functions that depends on the variable x_{k+1} is $2^{2^{k+1}} - 2^{2^k}$. Each function can be realized by using one 1-MUX as shown in Fig. 14.1. Thus, all the $(k+1)$-variable functions, is realized by using $(2^{2^k} - 2) + (2^{2^{k+1}} - 2^{2^k}) = 2^{2^{k+1}} - 2$ copies of 1-MUXs.

From 0), 1), and 2), we have the lemma. □

Theorem 14.4 *An arbitrary n-variable logic function is realized by using $R(n,k) = 2^{2^k} + 2^{n-k} - 3$ copies of 1-MUXs, where $0 \leq k \leq n-1$.*

(Proof) By applying the Shannon expansion $n - k$ times to the n-variable function, we have the following:

$$\bigvee_{(a_{k+1}, a_{k+2}, \ldots, a_n)} f(x_1, x_2, \ldots, x_k, a_{k+1}, \ldots, a_n) x_{k+1}^{a_{k+1}} x_{k+2}^{a_{k+2}} \cdots x_n^{a_n},$$

where, the logical sum is done over 2^{n-k} elements. Consider the network shown in Fig. 14.2. First, in the network in the left-hand side block, realize all the functions up to k variables. Next, in the network in the right-hand side, realize the $(n-k)$-MUX network. This MUX selects the function $f(x_1, x_2, \ldots, x_k, a_{k+1}, \ldots, a_n)$ when the values of $(x_{k+1}, x_{k+2}, \ldots, x_n)$, are set to $(a_{k+1}, a_{k+2}, \ldots, a_n)$. Therefore, an arbitrary logic function can be realized

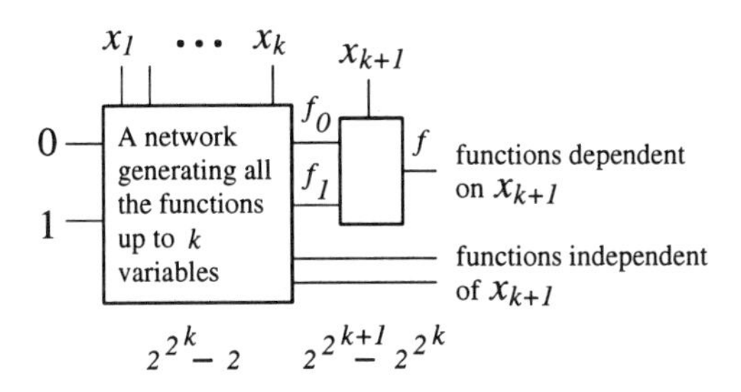

Figure 14.1 Realization of all the $(k+1)$-variable functions.

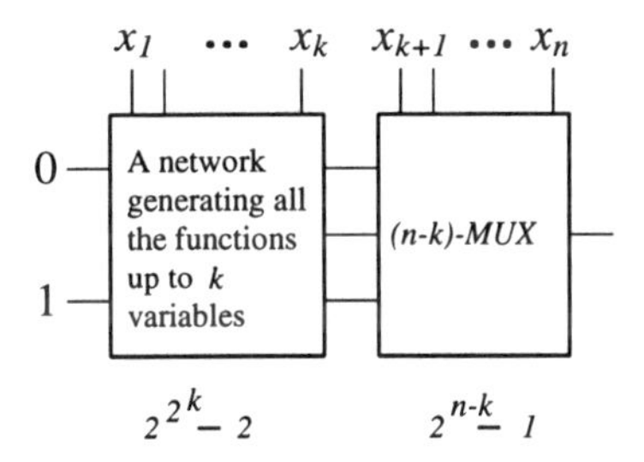

Figure 14.2 Network generating an arbitrary n-variable function.

by the network in Fig. 14.2. The total number of 1-MUXs in this network is

$$2^{2^k} - 2 + 2^{n-k} - 1 = 2^{2^k} + 2^{n-k} - 3.$$

□

Next, consider the number $R(n)$ of 1-MUXs that is sufficient to realize an arbitrary n-variable function.

Theorem 14.5 *When n is sufficiently large, an arbitrary n-variable logic function is realized by using at most $\frac{2^n}{n}(2+\varepsilon)$ copies of 1-MUXs, where, ε is an arbitrarily small positive number such that $0 < \varepsilon < 1$.*

(Proof) Let $k = \lfloor \log_2 n - t \rfloor$ in Theorem 14.4, where t is a small positive number, and $\lfloor x \rfloor$ denotes the largest integer smaller than or equal to x. For this integer, we can make $\dfrac{n}{2^{t+1}} < 2^k \leq \dfrac{n}{2^t}$.

By setting

$$B(n,k) = \frac{1}{2^k} + \frac{2^{2^k}}{2^n},$$

we have the following:

$$B(n,k) \leq \frac{2^{t+1}}{n} + \frac{2^{\frac{n}{2^t}}}{2^n} = \frac{1}{n}(2^{t+1} + \frac{n}{2^{np}}),$$

where $p = 1 - \frac{1}{2^t}$ is a positive number. From this, we have

$$R(n,k) = 2^{2^k} + 2^{n-k} - 3 = 2^n B(n,k) - 3 < 2^n B(n,k).$$

Thus, for sufficiently large integer n, we have

$$R(n) < \frac{2^n}{n}(2^{t+1} + \frac{n}{2^{np}}).$$

Next, when n is sufficiently large, we have $2^{t+1} + \frac{n}{2^{np}} < 2 + \varepsilon$, $0 < \varepsilon$.

Therefore, we have

$$R(n) < \frac{2^n}{n}(2 + \varepsilon).$$

From this, we have the theorem. □

Next, we will obtain a lower bound on the number of elements to realize an arbitrary n-variable function. In the case of two-level logic networks, the parity function is "the most complicated function." However, in the case of multi-level logic networks, it is very difficult to show "the most complicated function" explicitly. Theorem 14.6 shows the existence of the function that requires $2^n/n$ elements. However, it does not show the specific function that require $2^n/n$ elements. The technique of the proof is due to Shannon.

Theorem 14.6 *Let $w(n)$ be the number of n-variable logic functions that are realized by using at most $2^n/2n$ copies of 1-MUXs. Then, we have* $\frac{w(n)}{2^{2^n}} \to 0$ $(n \to \infty)$.

(Proof) Let the number of 1-MUXs in the network be $k = \lfloor 2^n/2n \rfloor$. For each input of a 1-MUX, constants 0, 1, variables, or outputs of other 1-MUXs can be connected. Since a 1-MUX has three inputs, the total number of inputs is $3k$. Also, we can assume that all the MUX cannot be distinguished. Therefore, the number of the connection patterns is at most $\frac{(k+n+1)^{3k}}{k!}$.

Since, $w(n) \leq \frac{(k+n+1)^{3k}}{k!}$, by using Stirling's formula

$$k! \simeq (2\pi)^{\frac{1}{2}} k^{k+\frac{1}{2}} e^{-k}$$

we have

$$\begin{aligned}
\log_2 \frac{w(n)}{2^{2^n}} &\leq 3k \log_2(k+n+1) - (k \log_2 k - k \log_2 e + 2^{-1} \log_2 k) - 2^n \\
&\leq 2k \log_2(k+n+1) + k \log_2 e - 2^{-1} \log_2 k - 2^n \\
&\leq \frac{2^n}{n}(\log_2 \frac{2^n}{2n} + O(1)) - 2^n \\
&\leq \frac{2^n}{n}(n - \log_2 2n + O(1)) - 2^n \\
&\leq \frac{2^n}{n}(-\log_2 2n + O(1)).
\end{aligned}$$

From this, when $n \to \infty$, we have $\log_2 \frac{w(n)}{2^{2^n}} \to -\infty$. Hence, we have the theorem. □

Corollary 14.2 *Almost all functions of n variables require $2^n/2n$ elements in a multi-level network consisting of 1-MUXs.*

Theorem 14.5 shows that an upper bound on the number of 1-MUXs to realize an arbitrary n-variable function is $(1+\varepsilon)2^{n+1}/n$. On the other hand, Corollary 14.2 shows that a lower bound is $2^{n-1}/n$. Note that the upper bound is about four time greater than the lower bound. This is because the realization method in the proof of Theorem 14.5 is simple. If we use more complicated realization method, we can make the upper bound nearly equal to the lower bound.

A n-variable symmetric function can be realized with $O(n^2)$ 1-MUXs as shown in the following:

Theorem 14.7 *An arbitrary n-variable symmetric function can be realized by using at most $n(n+1)/2$ copies of 1-MUXs.*

(Proof) An arbitrary n-variable symmetric function can be represented as $f = a_0 S_0^n \vee a_1 S_1^n \vee \cdots \vee a_n S_n^n$, where, $a_i = 0$ or 1. From here, we will show that the network for this expression can be realized with $n(n+1)/2$ copies of 1-MUXs. We will prove it by mathematical induction on n.

0) When $n = 0$, constants 0 and 1 can be realized without using 1-MUX.

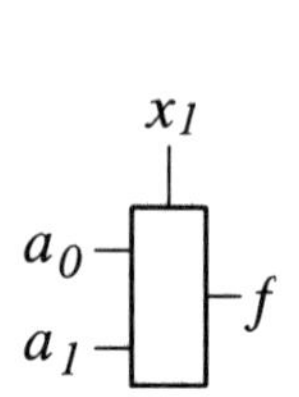

Figure 14.3 Realization of 1-variable symmetric functions.

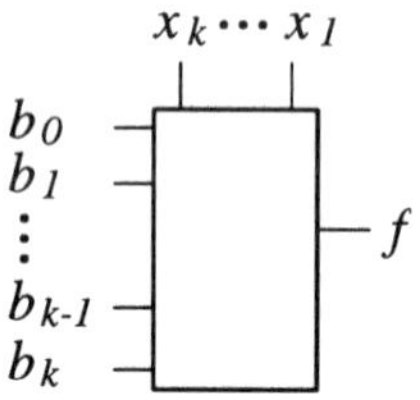

Figure 14.4 Realization of k-variable symmetric functions.

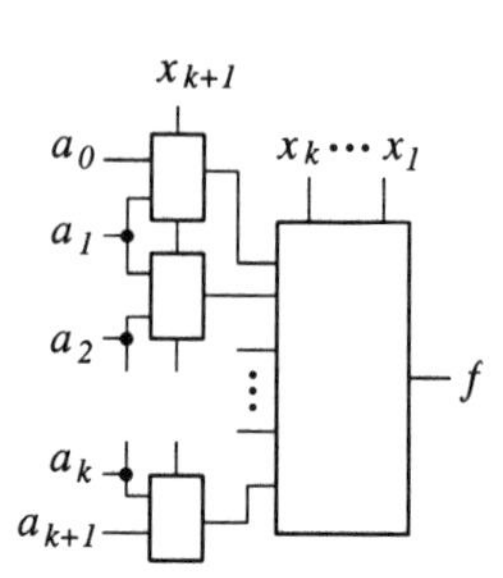

Figure 14.5 Realization of $(k + 1)$-variable symmetric functions.

1) When $n = 1$, all the 1-variable functions are symmetric. Thus, the network in Fig. 14.3 realizes them.
2) Suppose that the theorem holds for k-variable functions. That is, by using $k(k+1)/2$ copies of 1-MUXs, the network in Fig. 14.4 realize arbitrary k-variable symmetric function in the form of $g = b_0 S_0^k \vee b_1 S_1^k \vee \cdots \vee b_k S_k^k$. In this case, by connecting $k+1$ copies of 1-MUXs to the inputs of Fig. 14.4, Fig. 14.5 realizes an arbitrary $(k+1)$-variable symmetric function in the form $f = a_0 S_0^{k+1} \vee a_1 S_1^{k+1} \vee \cdots \vee a_{k+1} S_{k+1}^{k+1}$.
   The total number of 1-MUXs in Fig. 14.5 is

$$\frac{k(k+1)}{2} + (k+1) = \frac{(k+1)(k+2)}{2}.$$

From 0), 1), and 2), the theorem holds. □

By replacing each 1-MUX in the network realized in Theorem 14.5 and Theorem 14.7, with a node of a BDD, we have a BDD representing the logic function f. Therefore, we have the following:

Theorem 14.8 *An arbitrary n-variable function can be represented by a BDD with at most $2^n \cdot (2+\varepsilon)/n$ nodes. An arbitrary n-variable symmetric function can be represented by a BDD with at most $n(n+1)/2$ nodes.*

Bibliographical Notes

Survey and textbooks on complexity of logic networks are [160, 202, 307, 376, 421]. Detailed discussion on each topic can be found as follows: Complexity of AND-OR two-level networks [19, 68, 72, 75, 141, 304, 353, 381]; complexity of OR-AND-OR three-level networks [342, 367]; complexity of AND-AND-OR-OR four-level networks [200]; and complexity of multiple-level networks [270, 381].

Exercises

14.1 Show that an arbitrary logic function of n variables can be realized by using $O(2^n/n)$ 2-input gates.

14.2 Show that an arbitrary symmetric function of n variables can be realized by using $O(n^2)$ 2-input gates.

14.3 Show that a parity function of n variables can be realized by using $O(n)$ 2-input gates.

14.4 (D)

1. Show that the number of different series-parallel contact networks with N contacts is at most 4^N.
2. Show that series-parallel contact networks that realize almost all functions of n-variable require at least $\frac{2^n}{\log_2 n}(1-\varepsilon)$ contacts, where $\varepsilon > 0$.

(Riordan-Shannon 1942 [328]).

A

HISTORY OF SWITCHING THEORY

A.1 OVERVIEW

Switching theory has developed as shown in Table A.1. When C. E. Shannon started his works, the main logic elements were electromechanical, i.e., switches and relays. Afterwards, vacuum tubes, diodes, and transistors were used to make logic elements. In these days, the logic elements were very expensive. Office products, consumer products, toys, etc., began to use a large amount of logic elements, after the appearance of LSIs and microprocessors. These logic elements made influence on switching theory. Besides these elements, the logic elements that made major influence on switching theory include parametrons (realizing majority functions), TTL ICs (realizing NAND gates), PLAs (realizing AND-OR two-level logic networks), and FPGAs (field programmable gate arrays).

When logic elements were expensive, and networks to be realized were relatively small, logic designs were done manually. In this setting, the intuition and the experience of human were more important than mathematical design theory. Many people claimed that 'Switching theory was useless'. However, such criticism did not discourage researchers; instead, it motivated them to do more useful research. Later, VLSIs appeared. VLSIs contain too many gates to design manually. Moreover, efficient and inexpensive computers have become available. Now, logic design using computers (logic synthesis) is indispensable. Thus, the importance of switching theory has increased recently. In logic synthesis, some methods that were developed many years earlier have become practical recently.

Moreover, the objectives of logic design have changed. When the logic elements were expensive, reduction of gate counts was the first objective. However, in the age of VLSIs, higher speed, higher reliability, good testability, quick design, low power, productivity with a small amount with many varieties, modifiability, etc., are important criteria.

A.2 LOGIC ELEMENTS

In 1938, K. Zuse (Germany) [61] completed the Z1, an electromechanical computer. In 1939, J. V. Atanasoff [259] and his student (U.S.A.) completed the prototype of an electronic computer: It was a serial, binary, digital, special-purpose computer with regenerative memory using vacuum tubes. Independently, H. Aiken of Harvard University and G. Stibitz of Bell Telephone Liberators developed an automatic calculator using relays [295]. In 1945, J. Mauchuly and J. P. Eckert Jr. completed ENIAC, which used 18,000 vacuum tubes. In 1947, Bell Telephone Laboratory announced transistors. However, their reliability was quite low, and vacuum tubes were used for a while. In 1958, J. Kilby (Texas Instruments) developed a prototype semiconductor IC. In the same year R. Noyce (Fairchild Semiconductor Corp.) worked separately on ICs. In 1970, the PLA (programmable logic array) was invented by Texas Instruments [114]. PLAs are suitable for integrated circuits and easy to design. In the same year, IBM developed PLAs with decoders [121], however, they called them *writable control storage*. In 1971, INTEL announced 4004, the first microprocessor. In 1984, Motorola introduced 68020 that used many PLAs with built-in self-test (BIST) [84]. In 1985, XILINX announced a LUT (look-up table) type FPGA, where each LUT is rewritable. The logic design of FPGAs is quite different from conventional networks, since each cell realizes an arbitrary function of four or five variables.

A.3 TWO-LEVEL LOGIC NETWORKS

Early developments of switching theory can be found in [46, 202, 293, 327]. George Boole (United Kingdom) introduced Boolean algebra in 1847 [31]. However, it was not practically used until late 1930. I. I. Zhegalkin [439] (Russia) presented a canonical form of logic function, which is called as a positive polarity Reed-Muller expression (PPRM). A. Nakashima and M. Hanzawa (Japan) [286] formulated the design method for switching networks by using an algebra,

Table A.1 History of switching theory.

	Element	Two-level logic network	Multi-level logic network	Sequential network	Language & system
1847		Boole			
1937		Nakashima			
1938		Shannon			
1941			Post		
1947	Transistor				
1949		Shannon	Shannon		
1952		Quine			
1953		Karnaugh			
1954	Parametron	Muller		Huffman	
1956		McCluskey			FORTRAN
1957	IC		Ashenhurst	Mealy	
1958			Lupanov		
			Markov		
1960			Roth	Moore	ALGOL/LISP
1962			Curtis		
1964	SSI		Lawler		
1965		Miller, Harrison			
1967		Tison	TANT		
1968			NAND Synthesis		DDL
1969		Svoboda			ALERT
1970	PLA	Kohavi	Integer Programming	Unger	
1971	4004	Dietmeyer			
1974	8080	MINI	Transduction method		
1978	8086	Davio	LORES		
1979		Muroga			
1980			LSS		
1981				Retiming	
1982		Brayton	Weak division		IDL
1984	68020	ESPRESSO-II			VHDL
		MINI-II			
1985	FPGA	ESPRESSO-MV			SOCRATES
					PARTHENON
1986			Bryant		
			BDD		CONES
1987					MIS
1989		Boolean Relation	Applications of BDDs		
1992		EXMIN2			

which is essentially the same as Boolean algebra. V. I. Shestakov [383] (Russia) and Piesch [314] (Austria) also considered similar problems. C. E. Shannon (U.S.A.) [379] applied Boolean algebra to analyze contact networks, when he was a master's student of MIT. His paper has been the most influential one on switching theory. In addition to this, Shannon developed the theory of communication [380] which is the most important in the area.

A. Blake [30] showed that the set of all the prime implicants for a logic function f is a canonical form for f. M. Karnaugh [195], W. V. Quine [320] and E. J. McCluskey [243] showed simplification methods for SOPs. I. Reed [323] and D. E. Muller [269] showed AND-EXOR canonical forms (PPRMs). R. E. Miller [252] published textbooks on switching theory considering logic synthesis. M. A. Harrison's [157] textbook on switching theory contains enumeration of equivalence classes of logic functions. P. Tison [405] showed a method to generate PIs by using consensus. Later, he presented an algebra of multiple-valued input two-valued output logic functions [406]. A. Mukhopadhyay put together interesting topics on switching theory into a book [267]. Z. Kohavi [212] published a long lasting textbook. S. J. Hong, R. G. Cain, and D. L. Ostapko at IBM developed MINI [174], a multiple-valued logic minimizer to simplify PLA with decoders. A. Svoboda developed the PRESTO logic minimizer, which was published after he passed away [45]. He also developed a special hardware to generate PIs [396]. D. L. Dietmeyer developed logic synthesis system and published a CAD oriented book [103]. M. Davio, J-P Deschamps, and A. Thayse published two books [89, 90]. Although [89] is not so easy to read, it contains good survey of AND-EXOR expressions. S. Muroga contributed to threshold logic, two-level logic minimization, and multi-level logic minimization. Among his three major books [275, 277, 278], [277] describes various two-level logic minimization methods. After 1980, many microprocessors using PLAs were developed [84, 137, 220]. Since they used PLAs as control parts, the minimizations of PLAs or SOPs became very important. R. K. Brayton et al. developed ESPRESSO-II [38], which is an improvement of MINI. MINI-II [343] is a multi-valued logic minimizer, which detects essential prime implicants. R. L. Rudell and A. L. Sangiovanni-Vincentelli developed ESPRESSO-MV [333], a multi-valued logic minimizer, which is written in C and widely used as a standard two-level logic minimizer. R. K. Brayton and F. Somenzi considered applications of Boolean relation [40]. EXMIN2 [359] is the first practical logic minimizer for ESOPs.

Table A.2 Asymptotic complexities of logic networks.

Elements	Network form	Cost	Complexity
AND, OR	k-level AND-OR-AND	one per	
gates	fanout$=1$, $k=2$	input	$n2^{n-1}$
	fanout$=1$, $k \geq 3$		$2^n/\log_2 n$
	fanout≥ 1, $k \geq 3$		$2^n/n$
Contact	series-parallel	one per	$2^n/\log_2 n$
	arbitrary	contact	$2^n/n$
AND, OR	fanout$=1$	one per	$n/2$
NOT	fanout≥ 1	NOT	$\log_2 n$

A.4 MULTI-LEVEL LOGIC NETWORKS

E. L. Post [316] developed a theory on the complete set of logic functions (Theorem 5.17). S. V. Yablonskii [428] showed that an irredundant complete set of logic functions contains at most four primitive functions.

Let $L(n)$ be the least number of contacts to realize any switching function of n-variable by an arbitrary contact network. C. E. Shannon [381] proved that $2^n/n \leq L(n) \leq 4 \cdot 2^n/n$, as $n \to \infty$. O. B. Lupanov [234] improved Shannon's upper bound to $L(n) \sim 2^n/n$, as $n \to \infty$. To prove this, he invented a complete n-variable decoding network with $K(n)$ relay contacts, where $K(n) \sim 2^n$, as $n \to \infty$. This decoding network requires about a half of contacts of the straightforward realization as $n \to \infty$. He also showed that in the case of series-parallel contact networks, $L(n) \sim 2^n/\log_2 n$, as $n \to \infty$. A. A. Markov [240] showed that any logic function of n-variable can be realized by using at most $\lfloor \log_2(n+1) \rfloor$ inverters, where $\lfloor k \rfloor$ denotes the maximum integer not greater than k. (See Problem A.1). Table A.2 summaries asymptotic complexities of various classes of logic networks.

R. L. Ashenhurst [9] and H. A. Curtis [80] developed functional decomposition methods. J. P. Roth [330] showed another method to derive multi-level logic networks. E. L. Lawler [219] formulated an algorithm to derive minimum multi-level network. J. F. Gimpel [139] developed a design method of NAND three-level networks called *TANT*. D. L. Dietmeyer and Y. Su [102] showed a design method for fan-in limited NAND networks: They developed a factoring algorithm for logical expressions. S. Muroga et al. used integer linear programming to find various optimum multi-level networks [16, 274, 276]. His group also proposed the *transduction method* to improve the multi-level networks in the early 70s, but it was practically used after the development of efficient BDD packages [280, 281]. R. K. Brayton and C. McMullen [37] developed an algebraic method to derive multi-level network called *weak division*. R. E. Bryant [48] showed that BDD is a canonical representation of a logic function and it is a very good data structure for computer manipulation. After this, many logic design tools adopted BDDs as data structure.

A.5 SEQUENTIAL NETWORKS

D. A. Huffman [175], G. H. Mealy [247], and E. F. Moore [261] formulated various models of sequential networks. S. H. Unger [413] published a book on asynchronous circuits. Retiming was proposed by Leiserson and Saxe [223, 224]. Implicit finite-state machine traversal was considered by Coudert et al. [81].

A.6 LANGUAGE AND DESIGN SYSTEMS

DDL [111] is a register transfer level design language. IBM developed an experimental logic synthesis system ALERT [124], which converted APL language into logic networks. However, the performance was not acceptable: The major parts of IBM 1800 designed by ALERT required 160 percent more gates than manual design. At that time, there were no good logic minimizers, no good methods to derive multi-level logic networks, nor powerful computers to perform these complicated jobs.

LSS [85, 86] developed by IBM converted TTL logic networks into ECL logic using a local transformation method. LSS was a successful production tool for large-scale design: They designed many LSIs for IBM 3090 computers.

VHDL [227] (VHSIC hardware description language) has become an industry standard. SOCRATES [94, 148] was one of the earliest successful automatic multi-level logic synthesis systems. The keys of the success are; good logic minimizer (ESPRESSO); good method to generate multi-level logic (weak division); and reasonable methods to improve the networks (local transformations and rule-based expert system). It produces networks comparable to manual design. CONES [393] developed by Bell Laboratories converts behavioral models written in C into standard cell, PLAs, or PLDs. MIS [39] developed at U. C. Berkeley was one of the most successful one, and some commercial tools incorporated this technique. SIS developed at U. C. Berkeley is a tool for sequential and combinational logic synthesis. It supports all the features of MIS.

A.7 SWITCHING THEORY IN JAPAN

In 1930, Akira Nakashima, just graduated Tokyo University, entered NEC (Nippon Electric Company). First, he was assigned to the research group for designing special relay network systems. In this group, he designed automatic exchange systems as well as other relay application systems for six years. During that period, he became interested in the transient analysis of relay networks and formulated the design theory of relay networks. Since November 1934, he published his idea in a series of papers "Theory and practice of relay engineering," in the monthly technical journals of NEC. His papers attracted attention of many people, and the Telegraph and Telephone Society of Japan invited him as a guest speaker at its technical meeting in early 1935. His three-hour talk included design theories of relay networks that he formulated through designing various automatic exchange systems. The pre-print of the talk was published as "Synthesis theory of relay networks," [284] in the *Journal of the Telegraph and Telephone Society of Japan* in September 1935. The paper presented basic switching theory in terms of definitions and theorems, including De Morgan's Theorem. However, they were not represented by symbols. The formulation of these theorems using symbols were completed by A. Nakashima and M. Hanzawa in 1936 and 1937. At that time, they were not aware of Boolean algebra. Later, in August 1938, Nakashima found that his algebra was exactly the same as Boolean algebra.

In 1956, Toshio Ikeda at Fujitsu completed FACOM128, the first commercial relay computer in Japan [116].

Eiichi Goto [143] invented the *parametron* in 1954 when he was a graduate student of Tokyo University. A parametron consists of two magnetic cores and one capacitor, and realizes a three-input majority function [144]. In March 1957, NTT (Nippon Telegraph and Telephone Corp.) developed the Musashino-1, the first parametron computer using 519 vacuum tubes and 5,400 parametrons. Since it was an ILLIAC compatible machine, the computer could run the software of the ILLIAC. Parametron computers were manufactured by Hitachi, Fujitsu and NEC. Especially, NEC sold a few hundred sets [116]. Although parametron computers were reliable, they were very slow: They worked only at the speed of 200 kHz. Thus, after the appearance of reliable transistors, parametron computers became obsolete. Although the parametron computers were sold for only for two years, they made a strong influence in the switching theory. The realization of logic functions using majority elements were the first problem [258, 408]. After that the majority elements were generalized to threshold elements [272, 273], and they attracted many Japanese researchers [11, 145, 430].

Mochinori Goto [142] at the Electro-Technical Laboratory (ETL) analyzed operations of contact relay networks by introducing three-valued logic.

In July 1956, ETL developed a transistor computer, the ETL Mark III, that used 130 transistors and 1,700 diodes. This was the first stored program computer using transistors. They used point-contact transistors whose reliability was quite low. In spite of this difficulty, they completed the computer in two years.

Y. Komamiya [213] in ETL derived relation between arithmetic expressions and Boolean functions. He also considered the Reed-Muller transforms of logic functions. (See Problem A.2 and A.3). I. Ninomiya [298] considered invariant properties of Fourier transformations, and applied them to the classification of switching functions. His main results were included in the Harrison's book [157]. K. Kobayashi [209] considered the notion of almost complete set of logic elements, the details are shown in Mukhopadhyay's book [268].

Switching theory was active in the following universities: Tohoku (S. Noguchi, A. Maruoka [241], T. Higuchi [167], M. Kameyama [193]), Meiji (M. Mukaidono [264]), Tokyo Institute of Technology (Y. Thoma [408]), Nagoya (A. Fukumura), Kyoto (S. Yajima [430], H. Mine [257]), Osaka (H. Ozaki [306], T. Kasami [287], K. Kinoshita [206], T. Kitahashi [207], H. Fujiwara [131], T. Sasao), Hiroshima (N. Yoshida [438]), Kyushu (M. Ito [189]). Among them, S. Yajima's group at Kyoto University produced many researchers: Y. Kambayashi [194], T. Ibaraki [181], H. Hiraishi, K. Inagaki [431], H. Yasuura [436], N. Takagi [398], N. Ishiura [186], S. Minato [254, 255], K. Hamaguchi [155], H. Ochi [303], and others.

In addition to the universities, switching theory was also active in Electrical Technical Laboratories (M. Goto [142], Y. Komamiya [213]), NTT (Z. Kiyasu, S. Muroga [272], K. Ibuki [182]), NEC (A. Nakashima [284], T. Watanabe, S. Naito, T. Nanya [290, 291], T. Yamada [432]). LORES [117, 289] developed by MITSUBISHI converted TTL logic networks into ECL logic networks by local transformation method. PARTHENON [74] developed at NTT translated SFL (Structural Functional description Language) into microprocessors.

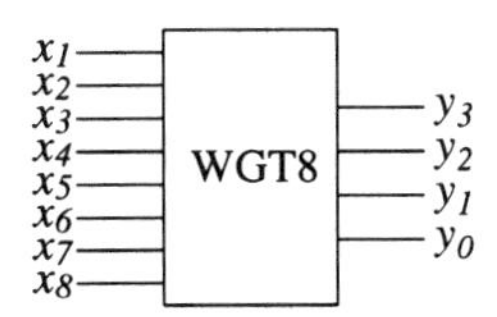

Figure A.1 WGT8.

Problem

A.1 (M) By using only AND gates, OR gates and inverters, realize three functions $f_1 = \bar{x}_1$, $f_2 = \bar{x}_2$, and $f_3 = \bar{x}_3$. You may use any number of AND and OR gates. However, you can use only two inverters.

A.2 (D) Let x_1, x_2,..., x_n be binary variables and r be an integer defined by $r = x_1 + x_2 + \cdots + x_n$, where + is an ordinary integer addition. Let the binary representation of r be

$$(y_k, y_{k-1}, \ldots, y_1, y_0)_2, \; y_j \in \{0, 1\} \; (j = 0, 1, \ldots, k).$$

In other words,

$$x_1 + x_2 + \cdots + x_n = 2^k y_k + 2^{k-1} y_{k-1} + \cdots + 2y_1 + y_0.$$

Show that

$$y_i = SB(n, 2^i).$$

(Komamiya 1959 [213, 301]).

For example, consider the network that represents the sum of 8 inputs as a binary representation. The network has $x_1, x_2, \ldots, x_8$ as inputs and y_3, y_2, y_1, y_0 as outputs (Fig. A.1). Then,

$$\begin{aligned} y_3 &= SB(8,8) = x_1 x_2 \cdots x_8 \\ y_2 &= SB(8,4) = \sum_{i<j<k<l}^{\oplus} x_i x_j x_k x_l \\ y_1 &= SB(8,2) = \sum_{i<j}^{\oplus} x_i x_j \\ y_0 &= SB(8,1) = x_1 \oplus x_2 \oplus \cdots \oplus x_8 \end{aligned}$$

A.3 (D) Let

$$f(x_1, x_2, \ldots, x_n) = \bigvee_{(a_1, a_2, \ldots, a_n)} f(a_1, a_2, \ldots, a_n) x_1^{a_1} x_2^{a_2} \cdots x_n^{a_n}$$

be the minterm expansion of f, where $a_i \in \{0,1\}$. Consider the function

$$f^T(x_1, x_2, \ldots, x_n) = \sum_{(a_1, a_2, \ldots, a_n)}^{\oplus} f(a_1, a_2, \ldots, a_n) x_1^{2-a_1} x_2^{2-a_2} \cdots x_n^{2-a_n},$$

where $x_i^2 = 1$. f^T is called a **Reed-Muller spectrum** of f [392]. Then, prove that $(f^T)^T = f$. (Komamiya 1959 [213, 301]).

REFERENCES

Some publication names are abbreviated as follows: **AIEE** for the American Institute of Electrical Engineers, **IEEE** for the Institute of Electrical and Electronics Engineers, **IRE** for The Institute for Radio Engineers, **IEE** for The Institute of Electrical Engineers (United Kingdom), **IEICE** for The Institute of Electronics, Information and Communication Engineers (Japan), **TX** for Transactions on X, **(E)C** for (Electronic) Computers, **CAD** for Computer-Aided Design of Integrated Circuits and Systems, **IT** for Information Theory, **CE** for Communication and Electronics, **CS** for Circuits and Systems, **DAC** for ACM/IEEE Design Automation Conference, **ISMVL** for IEEE International Symposium on Multiple-Valued Logic, **ICCAD** for IEEE International Conference on Computer Aided Design, **ICCD** for IEEE International Conference on Computer Design, **ISCAS** for IEEE International Symposium on Circuits and Systems, **EDAC** for European Conference on Design Automation.

[1] S. B. Akers, "On a theory of Boolean functions," *J. Soc. Indust. Appl. Math.*, Vol. 7, No. 4, pp. 487-498, Dec. 1959.

[2] S. B. Akers, "Binary decision diagrams," *IEEE TC*, Vol. C-27, No. 6, pp. 509-516, June 1978.

[3] A. E. A. Almaini and L. McKenzie, "Tabular techniques for generating Kronecker expansions," *IEE Proc. Comput. Digit. Tech.*, Vol. 143, No. 4, pp. 205-212, July 1996.

[4] Z. Arevalo and J. G. Bredeson, "A method to simplify a Boolean function into a near minimal sum-of-products for programmable logic arrays," *IEEE TC*, Vol. C-27, pp. 1028-1039, Nov. 1978.

[5] D. B. Armstrong, "A programmed algorithm for assigning internal codes to sequential machines," *IEEE TEC*, Vol. EC-11, pp. 466-472, Aug. 1962.

[6] B. D. Armstrong, "On finding a nearly minimal set of fault detection tests for combinational logic nets," *IEEE TC*, Vol. EC-15, No. 1, pp. 66-73, Feb. 1966.

[7] R. Arrathoon, *Optical Computing: Digital and Symbolic*, Marcel Dekker Inc., 1989.

[8] K. V. Asari and C. Eswaren, "An optimization technique for the design of multiple-valued PLA's," *IEEE TC*, Vol. C-43, No. 1, pp. 118-122, Jan. 1994.

[9] R. L. Ashenhurst, "The decomposition of switching functions," *Proc. of an Int. Symp. on the Theory of Switching*, pp. 74-116, April 1957.

[10] P. Ashar, S. Devadas, and A. Newton, *Sequential Logic Synthesis*, Kluwer Academic Publishers, Boston, MA, 1992.

[11] H. Ataka, "A basic theorem on threshold devices," *IEEE TEC*, Vol. EC-13, No. 5, p. 631, 1964.

[12] M. Auguin, F. Boeri, and C. Andre, "An algorithm for designing multiple Boolean functions: Application to PLA's," *Digital Process*, Vol. 4, No. 3-4, pp. 215-230, 1978.

[13] T. C. Bartee, "Computer designs of multiple-output logical networks," *IRE TEC*, EC-10, No. 1, pp. 21-30, March 1961.

[14] K. Bartlett, W. Cohen, A. De Geus, and G. Hachtel, "Synthesis and optimization of multilevel logic under timing constraints," *IEEE TCAD*, Vol. CAD-5, No. 4, pp. 582-596, Oct. 1986.

[15] K. A. Bartlett, R. K. Brayton, G. D. Hachtel, R. M. Jacoby, C. R. Morrison, R. L. Rudell, A. Sangiovanni-Vincentelli, and A. R. Wang, "Multilevel logic minimization using implicit don't cares," *IEEE TCAD*, Vol. CAD-7, No. 6, pp. 723-740, June 1988.

[16] C. R. Baugh, C. S. Chandersekaran, R. S. Swee, and S. Muroga, "Optimum networks for NOR-OR gates for functions of three variables," *IEEE TC*, Vol. C-21, No. 2, pp. 153-160, Feb. 1972.

[17] B. Becker and R. Drechsler, "Exact minimization of Kronecker expressions for symmetric functions," *IEE Proc. Comput. Digit. Tech.*, Vol. 143, No. 6, pp. 349-354, 1996.

[18] B. Becker and R. Drechsler, "On the expressive power of OKFDDs," *Formal methods in System Design*, Vol. 11, No. 1, pp. 5-21, Kluwer Academic Publishers, July 1997.

[19] E. A. Bender and J. T. Butler, "On the size of PLA's required to realize binary and multiple-valued functions," *IEEE TC*, Vol. 38, No. 1, pp. 82-98, Jan. 1989.

[20] A. Bender and J. T. Butler, "Asymptotic approximations for the number of fanout-free functions," *IEEE TC*, Vol. C-27, pp. 1180-1183, Dec. 1978.

[21] M. S. T. Benten, and S. M. Sait, "GAP: A genetic algorithm approach to optimize two-bit decoder PLAs," *Int. J. Electron.*, Vol. 76, No. 1, pp. 99-106, Jan. 1994.

[22] R. A. Bergamaschi, "SKOL: A System for logic synthesis and technology mapping," *IEEE TCAD*, Vol. CAD-10, No. 11, pp. 1342-1355, Nov. 1991.

[23] L. Berman and L. Trevillyan, "Global flows optimization in automatic logic design," *IEEE TCAD*, Vol. CAD-10, No. 5, pp. 557-564, May 1991.

[24] Ph. W. Besslich and P. Pichlbauer, "Fast transform procedure for the generation of near minimal covers of Boolean functions," *IEE Proc.*, Vol. 128, Part E, No. 6, pp. 250-254, Nov. 1981.

[25] Ph. W. Besslich, "Efficient computer method for ExOR logic design," *IEE Proc.*, Vol. 130, Part E, pp. 203-206, Nov. 1983.

[26] Ph. W. Besslich, "Spectral processing of switching functions using signal-flow transformations," in M. Karpovsky (ed.), *Spectral Techniques and Fault Detection*, Orlando, FL, Academic Press, 1985, pp. 91-141.

[27] G. Bioul, M. Davio, and J. P. Deschamps, "Minimization of ring-sum expansions of Boolean functions," *Philips Res. Rpts.*, Vol. 28, pp. 17-36, 1973.

[28] G. Birkoff, *Lattice Theory*, American Mathematical Society, 1967.

[29] N. N. Biswas, "Computer aided minimization procedure for Boolean functions," *DAC*, pp. 699-702, June 1984.
[30] A. Blake, "Canonical expressions in Boolean algebra," Ph. D. Dissertation, Dept. of Mathematics, Univ. Chicago, 1937. Published by Univ. Chicago, Libraries, 1937.
[31] G. Boole, *An Investigation of the Laws of Thought*, 1854. Reprinted by Dover Publications, Inc., New York, 1954.
[32] K. S. Brace, R. L. Rudell, and R. E. Bryant, "Efficient implementation of a BDD package," *DAC*, pp. 40-45, June 1990.
[33] D. Brand, "Redundancy and don't cares in logic synthesis," *IEEE TC*, Vol. C-32, No. 10, pp. 947-952, Oct. 1983.
[34] D. Brand and V. Iyengar, "Timing analysis using a functional relationship," *ICCAD*, pp. 126-129, 1986.
[35] D. Brand and T. Sasao, "Minimization of AND-EXOR expressions using rewriting rules," *IEEE TC*, Vol. 42, No. 5, pp. 568-576, May 1993.
[36] R. K. Brayton, J. D. Cohen, G. D. Hachtel, B. M. Tragger, and D. Y. Yun, "Fast recursive Boolean function manipulation," *ISCAS*, pp. 58-62, May 1982.
[37] R. K. Brayton and C. McMullen, "The decomposition of factorization of Boolean expressions," *ISCAS*, pp. 49-54, May 1982.
[38] R. K. Brayton, G. D. Hachtel, C. T. McMullen, and A. L. M. Sangiovanni-Vincentelli, *Logic Minimization Algorithms for VLSI Synthesis*, Kluwer Academic Publishers, Boston, 1984.
[39] R. K. Brayton, R. Rudell, A. Sangiovanni-Vincentelli, and A. R. Wang, "MIS: A multi-level logic optimization systems," *IEEE TCAD*, Vol. CAD-6, No. 6, pp. 1062-1081, Nov. 1987.
[40] R. Brayton and F. Somenzi, "An exact minimizer of Boolean relations," *ICCAD*, pp. 316-319, 1989.
[41] R. Brayton, G. Hachtel, and A. Sangiovanni-Vincentelli, "Multilevel logic synthesis," *IEEE Proc.*, Vol. 78, No. 2, pp. 264-300, Feb. 1990.
[42] M. A. Breuer (ed.), *Design Automation of Digital Systems*, Vol. 1: Theory and Techniques, Prentice-Hall, 1972.
[43] M. A. Breuer, "Recent development in the automated design and analysis of digital systems," *Proc. IEEE*, pp. 12-27, Jan. 1972.
[44] F. Brglez, D. Bryan, J. Calhoun, G. Kedem, and R. Lisanke, "Automated synthesis for testability," *IEEE Transactions on Industrial Electronics*, Vol. 36, No. 2, pp. 263-277, May 1989.
[45] D. W. Brown, "A state-machine Synthesizer – SMS," *DAC*, pp. 301-304, June 1981.
[46] F. M. Brown, *Boolean Reasoning: The logic of Boolean equations*, Kluwer Academic Publishers, Boston, 1990.
[47] S. D. Brown, R. J. Francis, J. Rose, and Z. G. Vranesic, *Field Programmable Gate Arrays*, Kluwer Academic Publishers, Boston, 1992.
[48] R. E. Bryant, "Graph-based algorithms for Boolean function manipulation," *IEEE TC*, Vol. C-35, No. 8, pp. 677-691, Aug. 1986.
[49] R. E. Bryant, "On the complexity of VLSI implementations and graph rep-

resentations of Boolean functions with application to integer multiplication," *IEEE TC*, Vol. 40, No. 2, pp. 205-213, Feb. 1991.

[50] J. R. Burch, E. M. Clarke, D. E. Long, K. L. McMillan, and D. L. Dill, "Symbolic model checking for sequential circuit verification," *IEEE TCAD*, Vol. CAD-14, No. 4, pp. 401-424, April 1994.

[51] J. T. Butler and K. J. Breeding, "Some characteristics of universal cell nets," *IEEE TC*, Vol. C-22, pp. 897-903, Oct. 1973.

[52] J. T. Butler, "A note on cellular automata complexity tradeoffs," *Information and Control*, 26, pp. 286-295, Nov. 1974.

[53] J. T. Butler, "On the number of functions realized by cascades and disjunctive networks," *IEEE TC*, Vol. C-24, pp. 681-690, July 1975.

[54] J. T. Butler, "Analysis and design of fanout-free networks of positive symmetric gates," *Journal of the Association for Computing Machinery*, Vol. 24, pp. 481-498, July 1978.

[55] J. T. Butler, "Tandem networks of universal cells," *IEEE TC*, Vol. C-27, pp. 785-800, Sept. 1978.

[56] J. T. Butler, D. S. Herscovici, T. Sasao, and R. J. Barton, "Average and worst case number of nodes in decision diagrams of symmetric multiple-valued functions," *IEEE TC*, Vol. 46, No. 4, pp. 491-494, April 1997.

[57] K. M. Butler, D. E. Ross, R. Kapur, and M. R. Mercer, "Heuristics to compute variable ordering for efficient manipulation of ordered binary decision diagrams," *ICCAD*, pp. 417-420, June 1991.

[58] W. N. Carr and J. P. Mize, *MOS/LSI Design and Application*, McGraw-Hill, New York, 1972.

[59] G. Caruso, "A selection algorithm for switching function minimization," *IEEE TC*, Vol. C-33, No. 1, pp. 91-97, Jan. 1984.

[60] G. Caruso, "Near optimal factorization of Boolean functions," *IEEE TCAD*, Vol. 10, No. 8, pp. 1072-1078, Aug. 1991.

[61] B. Carlson, A. Burgees, and C. Miller, "Timeline of computing history," *IEEE Computer*, Oct. 1996.

[62] E. Cerny and M. A. Martin, "An approach to unified methodology of combinational switching circuits," *IEEE TC*, Vol. C-26, No. 8, pp. 745-756, Aug. 1977.

[63] A. H. Chan, "Using decision trees to derive the complement of a binary function with multiple-valued inputs," *IEEE TC*, Vol. C-36, No. 2, Feb. 1987.

[64] H. H. Chao et. al., "Designing the Micro/370," *IEEE Design & Test*, pp. 32-40, June 1987.

[65] M. Chatterjee, D. K. Pradhan, and W. Kunz, "LOT: Logic optimization with testability: New transformations using recursive learning," *ICCAD*, pp. 318-325, 1995.

[66] S. Chattopadhyay, S. Royt, and P. P. Chaudhuri, "KGPMIN: An efficient multilevel multi-output AND-OR-XOR minimizer," *IEEE TCAD*, Vol. 16, No. 3, pp. 257-265, March 1997.

[67] W.-T. Cheng and J. H. Patel, "A minimum test set for multiple fault detection in ripple carry adders," *IEEE TC*, Vol. C-36, No. 7, 891-895, July 1987.

[68] K. Cheng and V. D. Agrawal, "An entropy measure for the complexity of multi-output Boolean functions," *DAC*, pp. 302-305, June 1990.
[69] M. J. Ciesielski and S. Yang, "PLADE: A two-stage PLA decomposition," *IEEE TCAD*, Vol. 11, No. 8, pp. 943-954, Aug. 1992.
[70] C. H. Clare, *Designing Logic Systems Using State Machines*, McGraw-Hill, New York, 1973.
[71] E. M. Clarke, M. Fujita, and X. Zhao, "Multi-terminal binary decision diagrams and hybrid decision diagrams," Chapter 4 in [363].
[72] A. Cobham, R. Fridshal, and J. H. North, "A statistical study of the minimization of Boolean functions using integer linear programming," *IBM Research Report RC-756*, June 1962.
[73] M. Cohn, "Inconsistent canonical forms of switching functions," *IRE Trans.*, EC-11, p.284, 1962.
[74] R. Composano and W. Wolf (ed.), *High-Level VLSI Synthesis*, Kluwer Academic Publishers, Boston, MA, 1991.
[75] R. W. Cook and M. J. Flynn, "Logical network cost and entropy," *IEEE TC*, Vol. C-22, Vol. 9, pp. 823-826, Sept. 1973.
[76] J. Cong and Y. Ding, "FlowMap: An optimal technology mapping algorithm for delay optimization in lookup-table based FPGA design," *IEEE TCAD*, Vol. 13, No. 1, pp. 1-12, Jan. 1994.
[77] O. Coudert and J. C. Madre, "Implicit and incremental computation of primes and essential primes of Boolean functions," *DAC*, June 1992. Also, Chapter 2 in [355].
[78] L. Csanky, M. A. Perkowski, and I. Schäefer, "Canonical restricted mixed polarity exclusive-or sums of products and the efficient algorithm for their minimization," *Proc. IEE*, Part E, Vol. 140, No. 1, pp. 69-77, Oct. 1993.
[79] J. N. Culliney, M. H. Young, T. Nakagawa, and S. Muroga, "Results of the synthesis of optimal networks of AND and OR gates for four-variable switching functions," *IEEE TC*, Vol. C-27, No. 1, pp. 76-85, Jan. 1979.
[80] H. A. Curtis, *A New Approach to The Design of Switching Circuits*, D. Van Nostrand Co., Princeton, NJ, 1962.
[81] O. Coudert, C. Berthet, and J. C. Madre, "Verification of sequential machines based on symbolic execution," in J. Sifakis (ed.), *Lecture Notes in Computer Science*, Springer-Verlag, Berlin, Germany, 1990.
[82] M. R. Dagenais, V. K. Agarwal, and N. C. Rumin, "McBoole: A new procedure for exact logic minimization," *IEEE TCAD*, Vol. CAD-5, No. 1, pp. 229-233, Jan. 1986.
[83] M. Damiani and G. De Micheli, "Don't care specifications in combinational and synchronous logic circuits," *IEEE TCAD*, Vol. CAD-12, No. 3, pp. 365-388, March 1993.
[84] R. G. Daniels and W. C. Bruce, "Built-in self-test trends in Motorola microprocessors," *IEEE Design & Test*, pp. 64-71, April 1985.
[85] J. Darringer, W. Joyner, L. Berman, and L. Trevillyan, "LSS: Logic synthesis through local transformations," *IBM J. Res. and Develop.*, Vol. 25, No. 4, pp. 272-280, July 1981.

[86] J. Darringer, D. Brand, J. Gerbi, W. Joyner Jr., and L. Trevillyan, "LSS: A system for production logic synthesis," *IBM J. Res. and Develop.*, Vol. 28, No. 5, pp. 537-545, Sept. 1984.

[87] S. R. Das, "A new algorithm for generating prime implicants," *IEEE TC*, Vol. C-20, No. 12, pp. 1614-1615, 1971.

[88] E. Davidson, "An algorithm for NAND decomposition under network constraints," *IEEE TC*, Vol. C-18, No. 12, pp. 1098-1109, Dec. 1969.

[89] M. Davio, J-P Deschamps, and A. Thayse, *Discrete and Switching Functions*, McGraw-Hill International, 1978.

[90] M. Davio, J-P Deschamps, and A. Thayse, *Digital Systems with Algorithm Implementation*, John Wiley and Sons, New York, 1983.

[91] D. Debnath and T. Sasao, "GRMIN2: A heuristic simplification algorithm for generalized Reed-Muller expressions," *IEE Proceedings-Computers and Digital Techniques*, Vol. 143, No. 6, pp. 376-384, Nov. 1996.

[92] D. Debnath and T. Sasao, "Minimization of AND-OR-EXOR three-level networks with AND gate sharing," *IEICE Trans. Information and Systems*, Vol. E80-D, No. 10, pp. 1001-1008, Oct. 1997.

[93] D. Debnath and T. Sasao, "A heuristic algorithm to design AND-OR-EXOR three-level networks," *Proc. Asia and South Pacific Design Automation Conference (ASP-DAC'98)*, pp. 69-74, Yokohama, Japan, Feb. 1998.

[94] A. J. De Geus and W. Cohen, "A rule-based system for optimizing combinational logic," *IEEE Design and Test*, pp. 22-32, Aug. 1985.

[95] G. De Micheli, R. K. Brayton, and A. L. Sangiovanni-Vincentelli, "Optimal state assignment of finite state machine," *IEEE TCAD*, Vol. CAD-4, No. 3, pp. 262-285, July 1985.

[96] G. De Micheli, "Symbolic design of combinational and sequential circuit implemented by two-level logic macros," *IEEE TCAD*, Vol. CAD-5, No. 4, pp. 597-616, 1986.

[97] G. De Micheli, *Synthesis and Optimization of Digital Circuits*, McGraw-Hill, 1994.

[98] S. Devadas, T. Ma, R. Newton, and A. Sangiovanni-Vincentelli, "MUSTANG: State assignment of finite-state machines targeting multi-level logic implementations," *IEEE TCAD*, Vol. CAD-7, No. 12, pp. 1290-1300, Dec. 1988.

[99] S. Devadas, A. Wang, A. R. Newton, and A. Sangiovanni-Vincentelli, "Boolean decomposition in multilevel logic optimization," *IEEE Journal of Solid-State Circuits*, Vol. 24, pp. 399-408, April 1989.

[100] S. Devadas and A. R. Newton, "Exact algorithms for output encoding, state assignment, and four-level Boolean minimization," *IEEE TCAD*, Vol. 10, No. 1, pp. 13-27, Jan. 1991.

[101] S. Devadas, K. Keutzer, and A. Ghosh, *Logic Synthesis*, McGraw Hill, 1994.

[102] D. Dietmeyer and Y. Su, "Logic design automation of fan-in limited NAND networks," *IEEE TC*, Vol. C-18, No. 1, pp. 11-22, Jan. 1969.

[103] D. L. Dietmeyer, *Logic Design of Digital Systems (Second Edition)*, Allyn and Bacon Inc., Boston, 1978.

[104] W. E. Donath, "Equivalence of memory to 'random logic'," *IBM J. Res. and*

Develop., pp. 401-407, Sept. 1974.

[105] T. Downs and M. F. Schulz, *Logic Design with Pascal: Computer-Aided Design Techniques*, Van Nostrand Reinhold, New York, 1988.

[106] R. Drechsler, A. Sarabi, M. Theobald, B. Becker, and M. A. Perkowski, "Efficient representation and manipulation of switching functions based on ordered Kronecker functional decision diagrams," *DAC*, pp. 415-419, June 1994.

[107] R. Drechsler, M. Theobald, and B. Becker, "Fast OFDD-based minimization of fixed polarity Reed-Muller expressions," *IEEE TC*, Vol. 45, No. 11, pp. 1294-1299, Nov. 1996.

[108] R. Drechsler and B. Becker, "Sympahy: Fast exact minimization of fixed polarity Reed-Muller expressions for symmetric functions," *IEEE TCAD*, Vol. 16, No. 1, pp. 1-5, Jan. 1997.

[109] E. V. Dubrova, D. M. Miller, and J. C. Muzio, "Upper bounds on the number of products in AN-OR-EXOR expansion of logic functions," *Electronics Letters*, Vol. 31, No. 7, pp. 541-542, March 1995.

[110] G. Dueck and D. M. Miller, "A 4-valued PLA using the MOD SUM," *ISMVL*, pp. 232-240, May 1986.

[111] J. R. Duley and D. L. Dietmeyer, "A digital system design language (DDL)," *IEEE TC*, Vol. C-17, pp. 850-861, Sept. 1968.

[112] B. Dunham and R. Fridshal, "The problem of simplifying logical expressions," *Journal of Symbolic Logic*, Vol. 24, pp. 17-19, 1959.

[113] A. B. Ektare and M. H. Al-Sheakhly, "Function symmetries and decoded-PLA realization," *Int. J. Electronics*, Vol. 60, No. 6, pp. 691-707, 1986.

[114] Electronics, "MOS array is custom-programmable," *Electronics*, pp. 133-134, March 30, 1970.

[115] B. Elspas, "The theory of multirail cascades," Chapter VIII of [267].

[116] S. Endo, *Early Computer Engineers in Japan*, (in Japanese) ASCII, (ISBN4-7561-0607-2), 1996.

[117] K. Enomoto, S. Nakamura, T. Ogihara, and S. Murai, "LORES-2: A logic reorganization system," *IEEE Design and Test*, pp. 35-42, Oct. 1985.

[118] G. Epstein, *Multiple-Valued Logic Design: An introduction*, Institute of Physics Publishing, Bristol and Philadelphia, 1993.

[119] S. Even, I. Kohavi, and A. Paz, "On minimal modulo-2 sum of products for switching functions," *IEEE TEC*, Vol. EC-16, pp. 671-674, Oct. 1967.

[120] E. D. Fabricius, *Modern Digital Design and Switching Theory*, CRC Press, 1992.

[121] H. Fleisher, A. Weinberger, and V. Winkler, "The writable personalized chip," *Computer Design*, Vol. 9, No. 6, pp. 59-66, June 1970.

[122] H. Fleisher and L. I. Maissel, "An introduction to array logic," *IBM J. Res. and Develop.*, Vol. 19, pp. 98-109, March 1975.

[123] H. Fleisher, M. Tarvel, and J. Yeager, "A computer algorithm for minimizing Reed-Muller canonical forms," *IEEE TC*, Vol. C-36, No. 2, Feb. 1987.

[124] T. D. Friedman and S. C. Yang, "Methods used in an automated logic design generator (ALERT)," *IEEE TC*, Vol. C-18, No. 7, pp. 593-614, July 1969.

[125] A. D. Friedman and P. R. Menon, *Fault Detection in Digital Circuits*, Prentice

Hall, Englewood Cliffs, NJ, 1971.
[126] A. D. Friedman, *Logical Design of Digital Systems*, Computer Science Press, Woodland Hills, CA, 1975.
[127] A. D. Friedman, *Fundamentals of Logic Design and Switching Theory*, Computer Science Press, Rockville, MD, 1986.
[128] S. J. Friedman and K. J. Supowit, "Finding the optimal variable ordering for binary decision diagrams," *IEEE TC*, Vol. 39, No. 5, pp. 710-713, May 1990.
[129] H. Fujiwara and K. Kinoshita, "A Design of programmable logic arrays with universal tests," *Joint special issue on Design for Testability IEEE TC*, Vol. C-30, No. 11, pp. 823-828; also *IEEE TCS*, Vol. CAS-28, No. 11, pp. 1027-1032, Nov. 1981.
[130] H. Fujiwara, "A new PLA design for universal testability," *IEEE TC*, Vol. C-33, No. 8, pp. 745-750, Aug. 1984.
[131] H. Fujiwara, *Logic Testing and Design for Testability*, The MIT Press, Cambridge, MA, 1985.
[132] M. Fujita, H. Fujisawa, and Y. Matsunaga, "Variable ordering algorithms for ordered binary decision diagrams and their evaluation," *IEEE TCAD*, Vol. 12, No. 1, pp. 6-12, Jan. 1993.
[133] K. Furuya, "A probabilistic approach to locally exhaustive testing," *FTCS 17: Digest of Papers. The Seventeenth Int. Symp. on Fault-Tolerant Computing*, pp. 62-65, 1987.
[134] D. Gajski, N. Dutt, and S. Lin, *High-Level Synthesis: Introduction to Chip and System Design*, Kluwer Academic Publishers, Boston, 1992.
[135] M. R. Garey and D. S. Johnson, *Computers and Intractability*, W. H. Freeman and Company, San Francisco, l979.
[136] P. P. Gelsinger, "Built in self test of the 80386," *ICCD*, pp. 169-173, 1986.
[137] P. P. Gelsinger, "Design and test of the 80386," *IEEE Design & Test*, pp. 42-50, June 1987.
[138] J. F. Gimpel, "A reduction technique for prime implicant tables," *IEEE TEC*, Vol. EC-14, pp. 535-541, Aug. 1965.
[139] J. F. Gimpel, "The minimization of TANT networks," *IEEE TEC*, Vol. EC-16, No. 2, pp. 12-38, Feb. 1967.
[140] S. Ginsburg, "A synthesis technique for minimal state sequential machines," *IRE TEC*, Vol. EC-8, pp. 13-24, March 1959.
[141] V. V. Glagolev, "Some bounds for disjunctive normal forms of the algebra of logic," *Problemi Kibernetiki 19*, pp. 74-93, 1970. (English translation, System Theory Research, Consultants Bureau, New York).
[142] M. Goto, "Applications of logical equations to the theory of relay contact networks," *Electric Soc. of Japan* (in Japanese), Vol. 69, p. 125, April, 1949.
[143] E. Goto, Patent.
[144] E. Goto, "The parametron: A digital computer element which utilizes parametric oscillation," *Proc. IRE*, Vol. 47, No. 8, pp. 1304-1316, 1959.
[145] E. Goto and H. Takahashi, "Some theorems in threshold logic for enumerating Boolean functions," *Proc. IFIP Congress*, pp. 747-752, Noth-Holland, 1962.
[146] A. Grasselli and F. Luccio, "A method for minimizing the number of internal

states in incompletely specified sequential networks," *IEEE TEC*, Vol. EC-14, pp. 350-359, June 1965.

[147] D. L. Greer, "An associative logic matrix," *IEEE Journal of Solid-State Circuits*, Vol. SC-11, No. 5, pp. 679-691, Oct. 1976.

[148] D. Gregory, K. Bartlett, A. de Geus, and G. Hachtel, "Socrates: A system for automatically synthesizing and optimizing combinational logic," *DAC*, pp. 79-85, June 1986.

[149] D. Green, *Modern Logic Design*, Addison-Wesley Publishing Company, 1986.

[150] D. H. Green, "Families of Reed-Muller canonical forms," *Int. Journal of Electronics*, Vol. 63, No. 2, pp. 259-280, Jan. 1991.

[151] D. H. Green and G. A. Khuwja, "Tabular simplification method for switching functions expressed in Reed-Muller algebraic form," *Int. Journal of Electronics*, Vol. 75, No. 2, pp. 297-314, 1993.

[152] G. D. Hachtel and R. M. Jacoby, "Verification algorithms for VLSI synthesis," *IEEE TCAD*, Vol. CAD-7, No. 5, pp. 616-640, May 1988.

[153] G. D. Hachtel and F. Somenzi, *Logic Synthesis and Verification Algorithms*, Kluwer Academic Publishers, Boston, 1996.

[154] C. Halatsis and N. Gaitanis, "Irredundant normal forms and minimal dependence set of a Boolean function," *IEEE TC*, Vol. C-27, No. 11, pp. 1064-1068, Nov. 1978.

[155] K. Hamaguchi, A. Morita, and S. Yajima, "Efficient construction of binary moment diagrams for verifying arithmetic circuits," *ICCAD*, pp. 78-82, 1995.

[156] F. Harary, *Graph Theory*, Addison-Wesley, Reading, MA, 1972.

[157] M. A. Harrison, *Introduction to Switching and Automata Theory*, McGraw-Hill, 1965.

[158] M. A. Harrison, "Counting theorems and their application to classifications switching functions," in Chapter 4 of [267].

[159] J. Hartmanis and R. Stearns, *Algebraic Structure Theory of Sequential Machines*, Prentice-Hall, Englewood Cliffs, NJ, 1966.

[160] J. T. Hastad, *Computational Limitations for Small-Depth Circuits*, The MIT Press, Cambridge, 1989.

[161] Hafiz Md. Hasan Babu and T. Sasao, "Design of multiple-output networks using time domain multiplexing and shared multi-terminal multiple-valued decision diagrams," *ISMVL*, pp. 45-51, Fukuoka, Japan, May 1998.

[162] J. P. Hayes, "The fanout structure of switching functions," *J. Ass. Comput. Mach.*, Vol. 22, pp. 551-571, Oct. 1975.

[163] M. A. Heap, W. A. Rogers, and M. R. Mercer, "A synthesis algorithm for two-level XOR based circuits," *ICCD*, pp. 459-462, Oct. 1992.

[164] L. Hellerman, "A catalog of three-variable OR-invert and AND-invert logical circuits," *IEEE TEC*, Vol. EC-12, No. 3, pp. 198-223, June 1963.

[165] M. Helliwell and M. Perkowski, "A fast algorithm to minimize multi-output mixed-polarity generalized Reed-Muller forms," *DAC*, pp. 427-432, 1988.

[166] P. J. Hicks (ed.), *Semi-Custom IC Design and VLSI*, Peter Peregrinus, 1983,

[167] T. Higuchi and M. Kameyama, "Ternary logic system based on T-gate," *ISMVL*, pp. 290-304, May 1975.

[168] T. Higuchi and M. Kameyama, "Static-hazard-free T-gate for ternary memory element and its application to ternary counters," *IEEE TC*, Vol. C-26, No. 12, pp. 1212-1221, Dec. 1977.
[169] F. Hill and G. Peterson, *Introduction to Switching Theory and Logical Design*, Wiley, New York, 1981.
[170] F. J. Hill and G. R. Peterson, *Digital Logic and Microprocessors*, John Wiley & Sons, New York 1984.
[171] F. J. Hill and G. R. Peterson, *Computer Aided Logic Design with Emphases on VLSI*, Wiley, New York, 1993.
[172] B. Holdsworth, *Digital Logic Design*, Butterworths, London, 1987.
[173] S. J. Hong and D. L. Ostapko, "On complementation of Boolean functions," *IEEE TC*, Vol. C-21, p. 1022, Sept. 1972.
[174] S. J. Hong, R. G. Cain, and D. L. Ostapko, "MINI: A heuristic approach for logic minimization," *IBM J. Res. and Develop.*, pp. 443-458, Sept. 1974.
[175] D. A. Huffman, "The synthesis of sequential switching circuits," *J. Franklin Inst.*, Vol. 257, No. 3, pp. 161-190, No. 4, pp. 275-303, March-April, 1954.
[176] S. L. Hurst, *The Logical Processing of Digital Signals*, Crane Russak, and Edward Arnold, 1978.
[177] S. L. Hurst, "Multiple-valued logic—its status and its future," *IEEE TC*, Vol. C-33, No. 12, Dec. 1984.
[178] S. L. Hurst, D. M. Miller, and J. C. Muzio, *Spectral Techniques in Digital Logic*, Academic Press, London, 1985.
[179] T-T. Hwang, R. M. Owens, and M. J. Irwin, "Exploiting communication complexity for multilevel logic synthesis," *IEEE TCAD*, Vol. 9, No. 10, pp. 1017-1027, Oct. 1990.
[180] T-T. Hwang, R. M. Owens, and M. J. Irwin, "Efficiently computing communication complexity for multilevel logic synthesis," *IEEE TCAD*, Vol. 11, No. 5, pp. 545-554, May 1992.
[181] T. Ibaraki and S. Muroga, "Synthesis of networks with a minimum number of negative gates," *IEEE TC*, Vol. C-20, pp. 49-58, Jan. 1971.
[182] K. Ibuki, K. Naemura, and S. Nozaki, "General theory of complete sets of logical functions," *Electron. Commun. Japan (IEEE Translation)*, Vol. 46, No. 7, pp. 55-56, July 1963.
[183] Y. Igarashi, "An improved lower bound on the maximum number of prime implicants," *IEICE Trans.*, Vol. E62, No. 6, pp. 389-394, June 1979.
[184] Y. Iguchi, T. Sasao, and M. Matsuura, "On properties of Kleene TDDs," *Asia and South Pacific Design Automation Conference (ASP-DAC'97)*, pp. 473-476, Jan. 1997.
[185] K. Ishikawa, T. Sasao, and H. Terada, "A minimization algorithm for logical expressions and its bounds of application," (in Japanese), *Trans. IECE Japan*, Vol. J65-D, No. 6, pp. 797-804, June 1982.
[186] N. Ishiura, H. Sawada, and S. Yajima, "Minimization of binary decision diagrams based on exchange of variables," *ICCAD*, pp. 472-475, Nov. 1991.
[187] N. Ishiura, "Synthesis of multilevel logic circuits from binary decision diagrams," *IEICE Trans. Inf. Syst.*, Vol. E76-D, No. 9, 1085-92, Sept. 1993.

[188] O. Ishizuka, "On multivalued multithreshold networks composed of conventional threshold elements," *IEEE TC*, Vol. C-26, No. 12, pp. 1251-1257, Dec. 1977.

[189] M. Ito, "On general solutions of n dimensional Boolean (two-valued) equations," *(in Japanese) Bul. of Fac. of Eng.*, Kyushu University, Vol. 28, No. 4, 1955.

[190] J. H. Jenkins, *Designing with FPGAs and CPLDs*, PTR Prentice Hall, Upper Saddle River, NJ., 1994.

[191] G. Jennings, "Symbolic incompletely specified functions for correct evaluation in the presence of indeterminate input values," *Proc. of the 28th Hawaii Int. Conference on System Sciences*, pp. 23-31, Vol. 1, 1995.

[192] S. Kajihara and T. Sasao, "On adders with minimum test," *IEEE The 6th Asian Test Symp.*, pp. 10-15, Akita, Japan, Nov. 17-19, 1997.

[193] M. Kameyama and T. Higuchi, "Synthesis of multiple-valued logic based on tree-type universal logic module," *IEEE TC*, Vol. C-26, pp. 1297-1302, Dec. 1977.

[194] Y. Kambayashi, "Logic design of programmable logic arrays," *IEEE TC*, Vol. C-28, No. 9, pp. 609-617, Sept. 1979.

[195] M. Karnaugh, "The map method for synthesis of combinational logic circuits," *AIEE Trans. on Comm. and Electronics*, Vol. 9, pp. 593-599, 1953.

[196] K. Karplus, "Amap: A technology mapper for selector-based field-programmable gate arrays," *DAC*, pp. 244-247, June 1991.

[197] M. G. Karpovsky (ed.), *Spectral Techniques and Fault Detection*, Academic Press, 1985.

[198] M. G. Karpovsky, *Finite Orthogonal Series in the Design of Digital Devices*, Jon Wiley & Sons, New York, 1976.

[199] S. Karunanithi and A. D. Friedman, "Some new types of logical completeness," *IEEE TC*, Vol. C-27, No. 11, pp. 998-1005, Nov. 1978.

[200] M. Karpovsky, "Multilevel logical networks," *IEEE TC*, Vol. C-36, No. 2, pp. 215-226, Feb. 1987.

[201] R. H. Katz, *Contemporary Logic Design*, Benjamin/Cummings, 1994.

[202] W. H. Kautz, "A survey and assessment of progress in switching theory and logical design in the Soviet Union," *IEEE TEC*, Vol. EC-15, No. 2, pp. 164-204, April 1966.

[203] U. Kebschull, E. Schubert, and W. Rosenstiel, "Multilevel logic synthesis based on functional decision diagrams," *EDAC 92*, pp. 43-47, 1992.

[204] K. Keutzer, "DAGON: Technology binding and local optimization by DAG matching," *DAC*, pp. 341-347, June 1987.

[205] K. Keutzer, S. Malik, and A. Saldanha, "Is redundancy necessary to reduce delay?" *IEEE TCAD*, Vol. CAD-10, No. 4, pp. 427-435, April 1991.

[206] K. Kinoshita T. Sasao, and J. Matsuda, "On magnetic bubble logic circuits," *IEEE TC*, Vol. C-25, No. 3, pp. 214-221, March 1976.

[207] T. Kitahashi and K. Tanaka, "Orthogonal expansion of many-valued logical functions and its application to their realization with a single threshold element," *IEEE TC*, Vol. C-21, No. 2, pp. 211-218, Feb. 1972.

[208] S. C. Kleene, *Introduction to Metamathematics*, Wolters-Noordhoff, North-

Holland Publishing, 1952.

[209] K. Kobayashi, "Almost complete set of logic primitives," *(in Japanese), Electron Commun. Japan*, Vol. 50, No. 12, 1967.

[210] K. L. Kodandapani, and R. V. Setlur, "A note on minimum Reed-Muller canonical forms of switching functions," *IEEE TC*, Vol. C-26, No. 3, pp. 310-313, March 1977.

[211] N. Koda and T. Sasao, "Four-variable AND-EXOR minimum expressions and their properties" *(in Japanese), IEICE Trans.*, Vol. J74-D-1, No. 11, pp.765-773, Nov., 1991. Also in *System Computer of Japan* (U.S.A.), Vol. 23, No. 10, pp. 27-41, 1992 (English translation).

[212] Z. Kohavi, *Switching and Finite Automata Theory*, McGraw-Hill Book Co., 1970.

[213] Y. Komamiya, *Theory of Computing Networks*, Researches of ETL, Sept. 1959.

[214] W. Kunz, D. Stoffel, and P. R. Menon, "Logic optimization and equivalence checking by implication analysis," *IEEE TCAD*, Vol. 16, No. 3, pp. 266-81, March 1997.

[215] Y. S. Kuo, "Generating essential primes for a Boolean function with multiple-valued inputs," *IEEE TC*, Vol. C-36, No. 3, March 1987.

[216] Y-T. Lai, M. Pedram, S. B. K. Vrudhula, "EVBDD-based algorithm for integer linear programming, spectral transformation, and functional decomposition," *IEEE TCAD*, Vol. 13, No. 8, pp. 959-975, Aug. 1994.

[217] P. K. Lala, *Fault-Tolerant and Fault Testable Hardware Design*, Prentice Hall, 1985.

[218] P. K. Lala, *Practical Digital Logic Design and Testing*, Prentice Hall, Englewood Cliffs, NJ., 1996.

[219] E. L. Lawler, "An approach to multilevel Boolean minimization," *Journal of ACM*, Vol. 11, No. 3, pp. 283-295, July 1964.

[220] H-F. S. Law and M. Shoji, "PLA design for the BELLMAC-32A microprocessor," *ICCC-82*, pp. 161-164, Sept. 1982.

[221] C. Lee, "Representation of switching circuits by binary-decision programs," *Bell System Technical Journal*, Vol. 19, pp. 985-999, July 1959.

[222] S. C. Lee, *Modern Switching Theory and Digital Design*, Prentice Hall, 1978.

[223] C. E. Leiserson and J. B. Saxe, "Optimizing synchronous systems," *Proc. of the Symp. on Foundations of Computer Science*, pp. 23-26, Oct. 1981.

[224] C. Leiserson and J. Saxe, "Retiming synchronous circuitry," *Algorithmica*, Vol. 6, pp. 5-35, 1991.

[225] H-T. Liaw and C-S. Lin, "On the OBDD representation of generalized Boolean functions," *IEEE TC*, Vol. 41, No. 6, pp. 661-664, June 1992.

[226] B. Lin, O. Coudert, and J. C. Madre, "Symbolic prime generation for multiple-valued functions," *DAC*, pp. 40-44, June 1992.

[227] R. Lipsett, C. Schaefer, and C. Ussery, *VHDL: Hardware Description and Design*, Kluwer Academic Publishers, Boston, MA., 1989.

[228] C. L. Liu, *Introduction to Combinatorial Mathematics*, McGraw-Hill, 1968.

[229] C. L. Liu, *Elements of Descreate Mathematics* (second edition), McGraw-Hill, New York, 1985.

[230] T-K. Liu, K. R. Hohlin, L-E. Shiau, and S. Muroga, "Optimal one-bit full adders with different types of gates," *IEEE TC*, Vol. C-23, No. 1, pp. 63-70, Jan. 1974.

[231] F. Luccio, "A method for the selection of prime implicants," *IEEE TEC*, Vol. EC-15, No. 4, pp. 205-212, April 1966.

[232] P. K. Lui and J. Muzio, "Boolean matrix transforms for the parity spectrum and the minimization of modulo-2 canonical expansions," *Proc. IEE*, Vol. 138, Part E, No. 6, pp. 411-417, Nov. 1991.

[233] P. K. Lui and J. C. Muzio, "Boolean matrix transforms for the minimization of modulo-2 canonical expansions," *IEEE TC*, Vol. C-41, No. 3, pp. 342-347, March 1992.

[234] O. B. Lupanov, "Asymptotic bounds on the complexity of formulas which realize the functions of logical algebra," *Doklady A. N.*, Vol. 119, No. 1, pp. 23-26, 1958, *Trans: Automation Express*, International Physical Index Inc., Vol. 2, No. 6, pp. 12-14, 1960.

[235] O. B. Lupanov, "The complexity of realizing functions of logical algebra by means of formula," *Prob. Kiber*, Vol. 3, pp. 61-80, *Trans: Problems of Cybernetics*, Pergamon.

[236] A. A. Malik, D. Harrison, and R. K. Brayton, "Three-level decomposition with application to PLDs," *ICCD*, pp. 628-633, Oct. 1991.

[237] S. Malik, E. Sentovich, R. Brayton, and A. Sangiovanni-Vincentelli, "Retiming and resynthesis; optimizing sequential networks with combinational techniques," *IEEE TCAD*, Vol. CAD-10, No. 1, pp. 74-84, Jan. 1991.

[238] S. Malik, L. Lavagno, R. Brayton, and A. Sangiovanni-Vincentelli, "Symbolic minimization of multiple-level logic and input encoding problem," *IEEE TCAD*, Vol. CAD-11, No. 7, pp. 825-843, July 1992.

[239] D. Mange, *Analysis and Synthesis of Logic Systems*, Artech House, Norwood, MA, 1986.

[240] A. A. Markov, "On the inversion complexity of a system of functions," *J. of ACM.*, Vol. 5, No. 4, pp. 331-334, 1958.

[241] A. Maruoka and N. Honda, "Logical networks of flexible cells," *IEEE TC*, Vol. C-22, No. 4, pp. 347-58, April 1973.

[242] Y. Matsunaga, "MINT: An exact algorithm for finding minimum test set," *IEICE Trans. Fundamentals*, Vol. E76-A, No. 10, Oct. 1993,

[243] E. J. McCluskey Jr., "Minimization of Boolean functions," *Bell System Technical Journal*, Vol. 35, No. 6, pp. 1417-1444, Nov. 1956.

[244] E. J. McCluskey and H. Schorr, "Essential multiple-output prime implicants, mathematical theory of automata", *Proc. Polytechnic Institute of Brooklyn Symp.*, Vol. 12, pp. 437-457, April 1962.

[245] E. J. McCluskey, *Introduction to the Theory of Switching Circuits*, McGraw-Hill, New York, 1965.

[246] R. McNaughton, "Unate truth functions," *IRE TEC*, Vol. EC-10, No. 1, pp. 1-6, March 1961.

[247] G. H. Mealy, "A method for synthesizing sequential circuits," *Bell System Technical Journal*, Vol. 34, No. 5, pp. 1045-1079, Sept. 1955.

[248] C. Mead and L. Conway, *Introduction to VLSI systems*, Addison-Wesley Publishing Co., 1980.

[249] C. Meinel and T. Theobald, *Algorithms and Data Structures in VLSI Design*, Springer, Berlin, 1998.

[250] R. S. Michalski and Z. Kulpa, "A system for programs for the synthesis of switching circuits using the method of disjoint stars," *Proc. of IFIP Congress*, 1971, North-Holland, pp. 61-65, 1972.

[251] F. Mileto and G. Putzolu, "Average values of quantities appearing in Boolean function minimization," *IEEE TEC*, Vol. EC-13, No. 4, pp. 87-92, April 1964.

[252] R. E. Miller, *Switching Theory*, Wiley, New York, 1965.

[253] G. Miles, "Field programmable logic arrays for next generation design," *WESCON 75*.

[254] S. Minato, N. Ishiura, and S. Yajima, "Shared binary decision diagram with attributed edges for efficient Boolean function manipulation," *DAC*, pp. 52-57, June 1990.

[255] S. Minato, "Fast generation of prime-irredundant covers from binary decision diagrams," *IEICE Trans. Fundamentals*, Vol. E76-A, No. 6, pp. 967-973, June 1993.

[256] S. Minato, *Binary Decision Diagrams and Applications for VLSI Synthesis*, Kluwer Academic Publishers, 1996.

[257] H. Mine and Y. Koga, "Basic properties and a construction method for fail-safe logical systems," *IEEE TEC*, Vol. EC-16, No. 3, pp. 282-289, June 1967.

[258] F. Miyata, "Realization of arbitrary logical functions using majority element," *IEEE TEC*, Vol. EC-12, No. 3, pp. 183-191, 1963.

[259] C. R. Mollenhoff, *ATANASOFF: Forgotten Father of the Computer*, Iowa State University Press, 1988.

[260] D. Moller, J. Mohnke, and M. Weber, "Detection of symmetries of Boolean functions represented by ROBDDs," *ICCAD*, pp. 680-684, Nov. 1993.

[261] E. F. Moore, "Gedanken-experiments on sequential machine," *Automata Studies, Annals of Mathematical Studies*, No. 34, pp. 129-153, Princeton University, 1956.

[262] E. Morreale, "Recursive operators for prime implicant and irredundant normal form determination," *IEEE TC*, Vol. C-19, No. 6, pp. 504-509, June 1970.

[263] T. H. Mott Jr., "Determination of irredundant formal forms of a truth function by iterated consensus of the prime implicants," *IRE TEC*, pp. 245-252, June 1960.

[264] M. Mukaidono, "Regular ternary logic functions: Ternary logic functions suitable for treating ambiguity," *IEEE TC*, Vol. C-35, No. 2, pp. 179-183, Feb. 1986.

[265] A. Mukherjee, *Introduction to nMOS and CMOS VLSI Systems Design*, Prentice-Hall, Englewood Cliffs, N.J., 1986.

[266] A. Mukhopadhyay and G. Schmitz, "Minimization of exclusive OR and logical equivalence of switching circuits," *IEEE TC*, Vol. C-19, No. 2, pp. 132-140, Feb. 1970.

[267] A. Mukhopadhyay (ed.), *Recent Developments in Switching Theory*, Academic

Press, New York, 1971.

[268] A. Mukhopadhyay, "Complete sets of logic primitives," in Chapter 1 of [267].

[269] D. E. Muller, "Application of Boolean algebra to switching circuit design and to error detection," *IRE TEC*, Vol. EC-3, No. 3, pp. 6-12, Sept. 1954.

[270] D. E. Muller, "Complexity in electronic switching circuits," *IRE TEC*, Vol. EC-5, No. 1, pp. 15-19, 1956.

[271] R. Murgai, R. K. Brayton, and A. Sangiovanni-Vincentelli, *Logic Synthesis for Field-Programmable Gate Arrays*, Kluwer Academic Publishers, Boston, 1995.

[272] S. Muroga, I. Toda, and S. Takasu, "Theory of majority decision elements," *J. Franklin Inst.*, Vol. 271, pp. 376-418, 1961.

[273] S. Muroga, "Functional forms of majority functions and a necessary and sufficient condition for their realizability," *IEEE TCE*, Vol. 83, No. 74, pp. 474-486, 1964.

[274] S. Muroga, "Logical design of optimal digital networks by integer programming," Chapter 5 in *Advances in Information System Science*, Vol. 3, pp. 283-348, edited by J. T. Tou, Plenum Press, 1970.

[275] S. Muroga, *Threshold Logic and Its Applications*, John Wiley, 1971.

[276] S. Muroga and T. Ibaraki, "Design of optimal switching network by integer programming," *IEEE TC*, Vol. C-21, No. 6, pp. 573-58, June 1972.

[277] S. Muroga, *Logic design and Switching Theory, Wiley-Interscience Publication*, 1979.

[278] S. Muroga, *VLSI System Design, John Wiley & Sons*, New York, 1982.

[279] S. Muroga, "Logic design of VLSI Electronic circuits, Tutorial," *VLSI: Algorithms and Architectures*, P. Bertolazzi and F. Luccio (ed.), Elsevier Science Publishers B.V. 1985.

[280] S. Muroga, Y. Kambayashi, H. C. Lai and J. N. Culliney, "The transduction method — design of logic networks based on permissible functions," *IEEE TC*, Vol. C-38, No. 10, pp. 1404-1424, Oct. 1989.

[281] S. Muroga, "Logic synthesizer, the transduction method and its extension, SYLON," in Chapter 3 of [355], pp. 59-86.

[282] J. C. Muzio and T. C. Wesselkamper, *Multiple-Valued Switching Theory*, Adam Hilger Ltd. Bristol and Boston, 1986.

[283] H. T. Nagle, B. D. Carroll, and J. D. Irwin, *An Introduction to Computer Logic*, Prentice-Hall, New Jersey, 1975.

[284] A. Nakashima, "A realization theory for relay circuits," *(in Japanese), Preprint. J. Inst. Electrical Communication Engineers of Japan*, Sept. 1935.

[285] A. Nakashima and M. Hanzawa, "The theory of equivalent transformation of simple partial paths in the relay circuit," *(in Japanese), J. Inst. Electrical Communication Engineers of Japan*, No. 165, 167, Dec. 1936, Feb. 1937.

[286] A. Nakashima and M. Hanzawa, "Algebraic expressions relative to simple partial paths in the relay circuits," *(in Japanese), J. Inst. Electrical Communication Engineers of Japan*, No. 173, Aug. 1937 (Condensed English translation : Nippon Electrical Comm. Engineering, No. 12, pp. 310-314, Spet. 1938) Section V, "Solutions of acting impedance equations of simple partial paths."

[287] K. Nakamura, N. Tokura, and T. Kasami, "Minimal negative gate networks,"

IEEE TC, Vol. C-21, No. 1, pp. 5-11, Jan. 1972.

[288] K. Nakamura, "Synthesis of gate-minimum multi-output two-level negative gate networks," *IEEE TC*, Vol. C-28, No. 10, pp. 768-72, Oct. 1979.

[289] S. Nakamura, S. Murai, C. Tanaka, M. Terai, H. Fujiwara, and K. Kinoshita, "LORES: Logic reorganization system," *DAC*, pp. 250-260, June 1978.

[290] T. Nanya and Y. Thoma, "On universal single transition time asynchronous state assignments," *IEEE TC*, Vol. C-27, No. 8, p. 781, Aug. 1978.

[291] T. Nanya and Y. Thoma, "Universal multicode STT state assignments for asynchronous sequential machines," *IEEE TC*, Vol. C-28, No. 11, p. 811, Nov. 1979.

[292] T. Nanya, "Challenges to dependable asynchronous processor design," in Chapter 9 of [355], pp. 191-213.

[293] NEC, *Switching Circuit Theory: History of Research at NEC* (in Japanese), 1989.

[294] R. J. Nelson, "Simplest normal truth functions," *J. Symb. Logic*, pp. 105-108, June 1954.

[295] V. P. Nelson, H. T. Nagle, B. D. Carroll, and J. D. Irwin, *Digital Logic Circuit Analysis and Design*, Prentice-Hall, New Jersey, 1995.

[296] A. R. Newton, "Techniques for logic synthesis," *VLSI'85*, pp. 32-48.

[297] I. Ninomiya, "A theory of coordinate representation of switching functions," *Men. Fac. Eng., Nagoya Univ.*, Vol. 10, No. 2, pp. 175-190, 1959. (Review *IEEE TEC*, Vol. EC-12, No. 2, p. 152, 1963.)

[298] I. Ninomiya, "A study of the structure of Boolean functions and its application to the synthesis of switching circuits," *Men. Fac. Eng., Nagoya Univ.*, Vol. 13, Ph.D. Thesis, Univ. Tokyo, 1961. (Tables reprinted in [157]).

[299] S. Note, F. Chattoor, G. Goossens, and H. De Man, "Combined hardware selection and pipelining in high-level performance data-path design," *IEEE TCAD*, Vol. CAD-11, pp. 413-423, April 1992.

[300] S. M. Nowick and D. L. Dill, "Exact two-level minimization of hazard-free logic with multiple-input changes," *IEEE TCAD*, Vol. 14, No. 8, pp. 986-97, Aug. 1995.

[301] T. Nozaki, *Switching Theory* (in Japanese), Kyoritsu Shuppan, 1972.

[302] H. Ochi, N. Ishiura, and S. Yajima, "Breadth-first manipulation of SBDD of Boolean functions for vector processing," *DAC*, pp. 413-416, 1991.

[303] H. Ochi, K. Yasuoka, and S. Yajima, "Breadth-first manipulation of very large binary-decision diagrams," *ICCAD*, pp. 48-55, 1993.

[304] L. O'Connor, "A new lower bound on the expected size of irredundant forms for Boolean functions," *Inf. Process. Lett.*, Vol. 53, No. 6, pp. 347-354, March 1995.

[305] D. L. Ostapko and S. J. Hong, "Generating test examples for heuristic Boolean minimization," *IBM J. Res. and Develop.*, Vol. 18, pp. 459-464, Sept. 1974.

[306] H. Ozaki and K. Kinoshita, *Digital Algebra* (in Japanese), Kyoritu Shuppan, 1966.

[307] C. A. Papachristou, "Characteristic measure of switching functions," *Information Sciences*, Vol. 13, No. 1, pp. 51-75, 1977.

[308] G. Papakonstantinou, "Minimization of modulo-2 sum of products," *IEEE TC*, C-28, pp. 163-167, 1979.

[309] M. C. Paull and S. H. Unger, "Minimizing the number of states in incompletely specified sequential switching functions," *IRE TEC*, EC-8, No. 3, pp. 356-366, Sept. 1959.

[310] F. J. Pelayo, A. Prieto, A. Lloris, and J. Ortega, "CMOS current-mode multi-valued PLA's," *IEEE TCS*, Vol. 38, No. 4, April 1991.

[311] D. Pellerin and M. Holley, *Practical Design Using Programmable Logic*, Prentice-Hall, New York, 1991.

[312] M. Perkowski and M. Chrzanowska-Jeske, "An exact algorithm to minimize mixed-radix exclusive sums of products for incompletely specified Boolean functions," *ISCAS*, pp. 1652-1655, June 1990.

[313] S. R. Petric, "A direct determination of the irredundant forms of Boolean function from the set of prime implicants," *Tech. Rept. AFCRC-TR-56-110*, Air Force Cambridge Res. Center, Cambridge, MA. April 1956.

[314] H. Piesch, "Begriff der Allgemeinen Schaltungstechnik," *Arch f. F. Elektrotech*, Vol. 33, pp. 672-686, 1939, (Austria).

[315] I. Pomeranz and S. M. Reddy, "On determing symmetries in inputs of logic circuits," *IEEE TCAD*, Vol. 13, No. 11, pp. 1428-1433, Nov. 1994.

[316] E. L. Post, *The Two-Valued Iterative System of Mathematical Logic*, Princeton Univ. Press, 1941.

[317] D. K. Pradhan, "Universal test sets for multiple fault detection in AND-EXOR arrays," *IEEE TC*, Vol. C-27, No. 2, pp. 181-187, Feb. 1978.

[318] D. Pradhan, *Fault-Tolerant Computing*, Vol. 1 & 2, Prentice-Hall, 1986.

[319] F. P. Preparata and R. T. Yeh, *Introduction to Discrete Structure*, Addison-Wesley, 1973.

[320] W. V. Quine, "A way to simplify truth functions," *Amer. Math. Mon.*, Vol. 62, pp. 627-631, Nov. 1955.

[321] A. Ramesh, B. Beckert, R. Hahnle, and N. V. Murray, "Fast subsumption checks using anti-links," *Journal of Automated Reasoning*, 18:47-83, 1997.

[322] S. M. Reddy, "Easily testable realization for logic functions," *IEEE TC*, Vol. C-21, No. 11, pp. 1083-1088, Nov. 1972.

[323] I. S. Reed, "A class of multiple-error-correcting codes and the decoding scheme," *IRE TIT*, PGIT-4, pp. 38-49, 1954.

[324] B. Reusch, "Generation of prime implicants from subfunctions and a unifying approach to the covering problem," *IEEE TC*, Vol. C-24, No. 9, pp. 924-930, Sept. 1975.

[325] D. A. Reynolds and G. Metze, "Fault detection capabilities of alternating logic," *IEEE TC*, Vol. C-27, No. 12, pp. 1093-1098, Dec. 1978.

[326] V. T. Rhyne, P. S. Noe, M. H. Mckinney, and U. W. Pooch, "A new technique for the fast minimization of switching function," *IEEE TC*, Vol. C-26, No. 8, pp. 757-764, Aug. 1977.

[327] D. C. Rine (ed.), *Computer Science and Multiple-Valued Logic: Theory and Applications*, North-Holland, Amsterdam, 1984.

[328] J. Riordan and C. E. Shannon, "The number of two-terminal series-parallel

networks," *Journal of Mathematics and Physics*, Col. 21, No. 2, pp. 83-93, 1942.

[329] J. P. Robinson and Chia-Lung Yeh, "A method for modulo-2 minimization," *IEEE TC*, Vol. C-31, No. 8, pp. 800-801, Aug. 1982.

[330] J. P. Roth, "Algebraic topological methods in synthesis," *Proc. of Int. Symp. on Theory of Switching April 1957*, In Annals of Computational Laboratory of Harvard University, Vol. 29, pp. 57-73, 1959.

[331] J. P. Roth, "Diagnosis of automata failures: A calculus and a method," *IBM J. Res. and Develop.*, Vol. 10, No. 4, pp. 278-291, July 1966.

[332] J. P. Roth, W. G. Bouricius, and P. R. Schneider, "Programmed algorithm to compute tests to detect and distinguish between failures in logic circuits," *IEEE TC*, Vol. EC-16, No. 5, pp. 567-580, Oct. 1967.

[333] R. L. Rudell and A. L. Sangiovanni-Vincentelli, "Multiple-valued minimization for PLA optimization," *IEEE TCAD*, Vol. CAD-6, No. 5, pp. 727-750, Sept. 1987.

[334] R. Rudell, "Dynamic variable ordering for ordered binary decision diagrams," *ICCAD*, pp. 42-47, Nov. 1993.

[335] K. K. Saluja and S. M. Reddy, "Fault detecting test sets for Reed-Muller canonic networks," *IEEE TC*, Vol. C-24, No. 10, pp. 995-998, Oct. 1975.

[336] K. K. Saluja and E. H. Ong, "Minimization of Reed-Muller canonic expansion," *IEEE TC*, Vol. C-28, No. 7, pp. 535-537, July 1979.

[337] T. Sasao and K. Kinoshita, "Cascade realization of 3-input 3-output conservative logic circuits," *IEEE TC*, Vol. C-27, No. 3, pp. 214-221, March 1978.

[338] T. Sasao and K. Kinoshita, "Realization of minimum circuits with two-input conservative logic elements," *IEEE TC*, Vol. C-27, No. 8, pp. 749-752, Aug. 1978.

[339] T. Sasao, "An application of multiple-valued logic to a design of programmable logic arrays," *ISMVL*, pp. 65-72, May 1978.

[340] T. Sasao and K. Kinoshita, "Conservative logic elements and their universality," *IEEE TC*, Vol. C-28, No. 9, pp. 682-685, Sept. 1979.

[341] T. Sasao and K. Kinoshita, "On the number of fanout-free functions and unate cascade functions," *IEEE TC*, Vol. C-28, No. 1, pp. 866-72, Jan. 1979.

[342] T. Sasao, "Multiple-valued decomposition of generalized Boolean functions and the complexity of programmable logic arrays," *IEEE TC*, Vol. C-30, No. 9, pp. 635-643, Sept. 1981.

[343] T. Sasao, "Input variable assignment and output phase optimization of PLA's," *IEEE TC*, Vol. C-33, No. 10, pp. 879-894 Oct. 1984.

[344] T. Sasao, "Tautology checking algorithm for multi-valued input binary functions and their application," *ISMVL*, pp. 242-250, May 1984.

[345] T. Sasao, "HART: A hardware for logic minimization and verification," *ICCD*, pp. 713-718, Oct. 1985.

[346] T. Sasao, "An algorithm to derive the complement of a binary function with multiple-valued input," *IEEE TC*, Vol. C-34, No. 2, pp. 131-140, Feb. 1985.

[347] T. Sasao, *Programmable Logic Array: How to Use and How to Make* (in Japanese), Nikkan Kougyo Shinbun Pub., May 1986.

[348] T. Sasao, "MACDAS: Multi-level AND-OR circuit synthesis using two- variable function generators," *DAC*, Las Vegas, pp. 86-93, June 1986.
[349] T. Sasao, "Multiple-valued logic and optimization of programmable logic arrays," *IEEE Computer*, Vol. 21, No. 4, pp. 71-80, April 1988.
[350] T. Sasao, "On the optimal design of multiple-valued PLA's," *IEEE TC*, Vol. 38, No. 4, pp. 582-592, April 1989.
[351] T. Sasao, "Application of multiple-valued logic to a serial decomposition of PLA's," *ISMVL*, Zangzou, China, pp. 264-271, May 1989.
[352] T. Sasao and Ph. Besslich, "On the complexity of MOD-2 Sum PLA's," *IEEE TC*, Vol. 39, No. 2, pp. 262-266, Feb. 1990.
[353] T. Sasao, "Bounds on the average number of products in the minimum sum-of-products expressions for multiple-valued input two-valued output functions," *IEEE TC*, Vol. 40, No. 5, pp. 645-651, May 1991.
[354] T. Sasao, "A transformation of multiple-valued input two-valued output functions and its application to simplification of exclusive-or sum-of-products expressions," *ISMVL-91*, pp. 270-279, May 1991.
[355] T. Sasao (ed.), *Logic Synthesis and Optimization*, Kluwer Academic Publishers, Boston, 1993.
[356] T. Sasao, "FPGA design by generalized functional decomposition," Chapter 11 in [355].
[357] T. Sasao, "Logic synthesis with EXOR logic gates," Chapter 12 in [355].
[358] T. Sasao, "AND-EXOR expressions and their optimization," Chapter 13 in [355].
[359] T. Sasao, "EXMIN2: A simplification algorithm for exclusive-OR- sum-of-products expressions for multiple-valued input two-valued output functions," *IEEE TCAD*, Vol. 12, No. 5, pp. 621-632, May 1993.
[360] T. Sasao, "Optimization of pseudo-Kronecker expressions using multiple-place decision diagrams," *IEICE Transactions on Information and Systems*, Vol. E76-D, No. 5, pp. 562-570, May 1993.
[361] T. Sasao, "An exact minimization of AND-EXOR expressions using reduced covering functions," *Proc. of the Synthesis and Simulation Meeting and International Interchange (SASIMI'93)*, pp. 46-53, Oct. 1993.
[362] T. Sasao, "A design method for AND-OR-EXOR three-level networks," *ACM/IEEE Int. Workshop on Logic Synthesis*, pp. 8:11-8:20, Tahoe City, California, May 1995.
[363] T. Sasao and M. Fujita (ed.), *Representation of Discrete Functions*, Kluwer Academic Publishers, Boston, 1996.
[364] T. Sasao, "Representation of logic functions using EXOR operators," Chapter 2 in [363].
[365] T. Sasao and F. Izuhara, "Exact minimization of FPRMs using multi-terminal EXOR TDDs," Chapter 8 in [363].
[366] T. Sasao, "Ternary decision diagrams and their applications," Chapter 12 in [363].
[367] T. Sasao, "OR-AND-OR three-level networks," Chapter 13 in [363].
[368] T. Sasao and D. Debnath, "Generalized Reed-Muller expressions: Complex-

ity and an exact minimization algorithm," *IEICE Transactions Fundamentals*, Vol. E79-A, No. 12, pp. 2123-2130, Dec. 1996.

[369] T. Sasao, "Easily testable realizations for generalized Reed-Muller expressions," *IEEE TC*, Vol. 46, No. 6, pp. 709-716, June 1997.

[370] T. Sasao and J. T. Butler, "On bi-decompositions of logic functions," *ACM/IEEE Int. Workshop on Logic Synthesis*, Tahoe City, California, May 1997.

[371] T. Sasao, "Ternary decision diagrams: survey" *(invited paper), ISMVL*, pp. 241-250, Nova Scotia, Canada, May 1997.

[372] T. Sasao and J. T. Butler, "Comparison of the worst and best sum-of-products expressions for multiple-valued functions," *ISMVL*, pp. 55-60, Nova Scotia, Canada, May 1997.

[373] G. Saucier, "Encoding for asynchronous sequential networks," *IEEE TEC*, Vol. EC-16, No. 3, pp. 365-369, June 1967.

[374] G. Saucier, M. Crastes de Paulet and P. Sicards, "ASYL: A rule-based system for controller synthesis," *IEEE TCAD*, Vol. CAD-6, No. 6, pp. 1088-1097, Nov. 1987.

[375] J. M. Saul, "An algorithm for the multi-level minimization of Reed-Muller representation," *ICCD-91*, Oct. 1991.

[376] J. E. Savage, *The Complexity of Computing*, Wiley-Interscience, 1976.

[377] F. F. Sellers, M. Y. Hsiao, and L. W. Bearnson, "Analyzing errors with the Boolean differences," *IEEE TC*, Vol. C-17, No. 7, pp. 678-683, July 1968.

[378] S. C. Seth and K. L. Kodandapani, "Diagnosis of faults in linear tree networks," *IEEE TC*, Vol. C-26, No. 1, pp. 29-33, Jan. 1977.

[379] C. E. Shannon, "A symbolic analysis of relay and switching circuits," *Trans. AIEE*, Vol. 57, pp. 713-723, 1938.

[380] C. E. Shannon, "A mathematical theory of communication," *Bell System Tech. J.*, Vol. 27, pp. 379-423, pp. 623-656, 1948.

[381] C. E. Shannon, "The synthesis of two-terminal switching circuits," *Bell Syst. Tech. J.*, Vol. 28, No. 1, pp. 59-98, Jan. 1949.

[382] I. Schäfer and M. A. Perkowski, "Multiple-valued input generalized Reed-Muller forms," *IEE Proceedings-E*, Vol. I39, No. 6, Nov. 1992.

[383] V. I. Shestakov, "Some mathematical methods for construction and simplification of two-terminal electrical networks of class A" *(in Russian), Dissertation*, Lomonosov State University, Moscow, 1938.

[384] V. Y-S. Shen and A. C. Mckellar, "An algorithm for the disjunctive decomposition of switching functions," *IEEE TC*, Vol. C-19, No. 3, pp. 239-248, March 1970.

[385] V. Y-S. Shen, A. C. Mckellar, and P. Weiner, "A fast algorithm for the disjunctive decomposition of switching functions," *IEEE TC*, Vol. C-20, No. 3, pp. 304-309, March 1971.

[386] J. R. Slagle, C. L. Chang, and R. C. T. Lee, "A new algorithm for generating prime implicants," *IEEE TC*, Vol. C-19, No. 4, pp. 304-310, April 1970.

[387] D. Slepian, "On the number of symmetry types of Boolean functions on n variables," *Can. J. Math.*, Vol. 5, No. 2, pp. 185-193, 1953.

[388] R. A. Smith, "Minimal three-variable NOR and NAND logic circuits," *IEEE TEC*, Feb. 1965.

[389] J. E. Smith, "Detection of faults in programmable logic arrays," *IEEE TC*, Vol. C-28, No. 11, pp. 845-853, Nov. 1979.

[390] K. C. Smith and P. G. Glulak, "Prospect for Multiple-valued integrated circuits," *IEICE Trans. Electron.*, Vol. E76-C, No. 3, March 1993.

[391] N. Song and M. A. Perkowski, "Minimization of exclusive sum of products expressions for multiple-valued input, incompletely specified functions," *IEEE TCAD*, Vol. CAD-15, No. 4, pp. 385-395, April 1996.

[392] R. S. Stanković, T. Sasao, and C. Moraga, "Spectral transforms decision diagrams," Chapter 3 in [363].

[393] C. E. Stroud, R. R. Munoz, and D. A. Pierce, "Behavioral Model synthesis with CONES," *IEEE Design and Test of Computers*, Vol. 5, No. 3, pp. 22-30, June 1988.

[394] Y. H. Su and P. T. Cheung, "Computer minimization of multi-valued switching functions," *IEEE TC*, Vol. C-21, pp. 995-1003, 1972.

[395] S. Y. H. Su and P. Y. Cheung, "Computer simplification of multi-valued switching functions," D. Rine (ed.), *Computer Science and Multiple-Valued Logic North -Holland*, pp. 189-220, 1977.

[396] A. Svoboda and D. E. White, *Advanced Logical Circuit Design Techniques*, Garland Press, New York, 1979.

[397] H. Takahashi (ed.), *Parametron Computers* (in Japanese), Iwanami Shoten, 1968.

[398] N. Takagi, H. Yasuura, and S. Yajima, "High-speed VLSI multiplication algorithm with a redundant binary addition tree," *IEEE TC*, Vol. C-34, No. 9, pp. 789-796, Sept. 1985.

[399] Y. Takamatsu and K. Kinoshita, "CONT: A concurrent test generation system," *IEEE TCAD*, Vol. CAD-8, No. 9, pp. 966-972, Sept. 1989.

[400] M. A. Tapia, "Boolean integral calculus for digital systems," *IEEE TC*, Vol. C-34, No. 1, pp. 78-81, Jan. 1985.

[401] A. Thayse and M. Davio, "Boolean differential calculus and its applications in switching theory," *IEEE TC*, Vol. C-22, No. 4, pp. 409-420, 1971.

[402] A. Thayse, M. Davio, and J.-P. Deschamps, "Optimization of multiple-valued decision diagrams," *ISMVL-79, Rosemont*, IL., pp. 171-177, May 1978.

[403] P. Thomson and J. F. Miller, "Symbolic method for simplifying AND-EXOR representations of Boolean functions using a binary decision technique and a genetic algorithm," *IEE Proc. Comput. Digt. Tech.*, Vol. 143, No. 2, pp. 151-155, March 1996.

[404] M. A. Thornton and V. S. S. Nair, "BDD-based spectral approach for Reed-Muller circuit realization," *IEE Proc. Comput. Digt. Tech.*, Vol. 143, No. 2, pp. 145-150, March 1996.

[405] P. Tison, "Generalization of consensus theory and application to the minimization of Boolean functions," *IEEE TEC*, Vol. EC-16, pp. 446-456, Aug. 1967.

[406] P. L. Tison, "An algebra for logic systems: switching circuits application," *IEEE TC*, Vol. C-20, No. 3, pp. 339-351, March 1971.

[407] I. Toda, "On the number of types of self-dual logical functions," *IRE TEC*, Vol. EC-11, pp. 282-284, April 1963.

[408] Y. Tohma, "Decomposition of logical functions using majority decision elements," *IEEE TEC*, Vol. EC-13, No. 6, pp. 698-705, 1964.

[409] S. Toida and N. S. V. Rao, "On test generation for combinational circuits consisting of AND and EXOR gates," *1992 IEEE VLSI Test Symposium. 10th Anniversary. Design, Test and Application: ASICs and Systems-on-a-Chip*, 113-118, 1992.

[410] N. Tokura, T. Kasami, and A. Hashimoto, "Failsafe logic nets," *IEEE TC*, Vol. C-20, No. 3, pp. 323-30, March 1971.

[411] P P. Tirumalai and Jon T. Butler, "Minimization algorithm for multiple-valued programmable logic arrays," *IEEE TC*, Vol. 40, No. 2, Feb. 1991.

[412] C-C. Tsai and M. Marek-Sadowska, "Generalized Reed-Muller forms as a tool to detect symmetries," *IEEE TC.*, Vol. 45, No. 1, pp. 33-40, Jan. 1996.

[413] S. H. Unger, *Asynchronous Sequential Switching Circuits*, John Wiley, New York, 1969.

[414] D. Varma and E. A. Trachtenberg, "Design automation tools for efficient implementation of logic functions by decomposition," *IEEE TCAD*, Vol. CAD-8, No. 8, 1989.

[415] D. Varma and E. A. Trachtenberg, "Computation of Reed-Muller expansions of incompletely specified Boolean functions from reduced representation," *IEE Proc. Part E*, Vol. 138, No. 2, pp. 85-92, March 1991.

[416] D. Varma and E. A. Trachtenberg, "Efficient spectral techniques for logic synthesis," Chapter 10 in [355], pp. 215-232.

[417] J. Venn, *Symbolic Logic*, 2nd edition, London, Macmillan, 1894. (Reprinted by Chelsea Pub. Co., New York, 1971).

[418] T. Villa and A. Sangiovanni-Vincentelli, "NOVA: State assignment for finite state machines for optimal two-level logic implementation," *IEEE TCAD*, Vol. CAD-9, No. 9, pp. 905-924, Sept. 1990.

[419] Y. Wang and C. McCrosky, "Negation trees: A unified approach to Boolean functions complementation," *IEEE TC*, Vol. 45, No. 5, pp. 626-630, May 1996.

[420] Y. Watanabe and R. Brayton, "Heuristic minimization of multiple-valued relations," *IEEE TCAD*, Vol. CAD-12, No. 10, pp. 1458-1472, Oct. 1993.

[421] I. Wegener, *The Complexity of Boolean Functions*, John Wiley & Sons, Stuttgart, 1987.

[422] I. Wegener, "The size of reduced OBDD's and optimal read-once branching program for almost all Boolean functions," *IEEE TC*, Vol. C-43, No. 11, pp. 1262-1269, Nov. 1994.

[423] A. Weinberger, "High speed programmable logic array adders," *IBM J. Res. and Develop.*, Vol. 23, pp. 163-178, March 1979.

[424] N. H. E. Weste and K. Eshragnian, *Principles of CMOS VLSI Design: A Systems Perspective*, Addison-Wesley Pub., Reading, MA. 1985.

[425] K-N. Wong, "New heuristic for the exact minimization of logic functions," *ISCAS*, pp. 1865-1868, 1988.

[426] R. A. Wood, "High-speed dynamic programmable logic array chip," *IBM J.*

Res. and Develop., Vol. 19, pp. 379-383, July 1975.

[427] X. Wu, X. Chen, and S. L. Hurst, "Mapping of Reed-Muller coefficients and the minimisation of Exclusive-OR switching functions," *Proc. IEE-E*, Vol. 129, pp. 15-20, 1982.

[428] S. V. Yablonskii, "Functional computations in k-valued logic," *Trudy Matem, Inst. Sktek.*, Vol. 51, pp. 5-142, 1958.

[429] T. Yaita, N. Yonezawa, and Y. Hirose, "Equipment to generate prime implicants by searching truth table," *(in Japanese), The Institute of Electronics and Communication Engineers of Japan*, EC-85, March 1986.

[430] S. Yajima and T. Ibaraki, "A lower bound on the number of threshold functions," *IEEE TEC*, Vol. EC-14, No. 6, pp. 926-929, 1965.

[431] S. Yajima and K. Inagaki, "Power minimization problem of logic networks," *IEEE TC*, Vol. C-23, No. 2, pp. 153-165, Feb. 1974.

[432] T. Yamada and T. Nanya, "Stuck-at fault tests in the presence of undetectable bridging faults," *IEEE TC*, Vol. C-33, No. 8, pp. 758-61, Aug. 1984.

[433] S. Yang, "Logic synthesis and optimization benchmark user guide, Version 3.0," *MCNC*, Jan. 1991.

[434] Saeyang Yang and M. J. Ciesielski, "Optimum and suboptimum algorithm for input encoding and its relationship to logic minimization," *IEEE TCAD*, Vol. 10, No. 1, pp. 4-12, Jan. 1991.

[435] M. H. Young and S. Muroga, "Symmetric minimal covering problem and minimal PLA's with symmetric variables," *IEEE TC*, Vol. C-34, No. 6, pp. 523-541, June 1985.

[436] H. Yasuura, N. Takagi, and S. Yajima, "The parallel enumeration sorting scheme for VLSI," *IEEE TC*, Vol. C-31, No. 12, pp. 1192-1201, Dec. 1982.

[437] S. S. Yau and C. K. Tang, "Universal logic modules and their applications," *IEEE TC*, Vol. C-10, pp. 141-149, 1970.

[438] N. Yoshida, *Logical Mathematics*, Vol. II and III (in Japanese), Kyoritsu Shuppan, 1978.

[439] I. I. Zhegalkin, "The technique of calculation of statements in symbolic logic," *(in Russian) Mathe. Sbornik*, Vol. 34, pp. 9-28, 1927.

[440] Z. Zilc and Z. G. Vranesic, "Using decision diagrams to design ULMs for FPGAs," *IEEE TC*, Vol. 47, No. 9, pp. 971-982, Sept. 1998.

INDEX